Statistical Analysis and Stochastic Modelling of Hydrological Extremes

Statistical Analysis and Stochastic Modelling of Hydrological Extremes

Special Issue Editor
Hossein Tabari

MDPI • Basel • Beijing • Wuhan • Barcelona • Belgrade

MDPI

Special Issue Editor
Hossein Tabari
University of Leuven (KU Leuven)
Belgium

Editorial Office
MDPI
St. Alban-Anlage 66
4052 Basel, Switzerland

This is a reprint of articles from the Special Issue published online in the open access journal *Water* (ISSN 2073-4441) from 2018 to 2019 (available at: https://www.mdpi.com/journal/water/special_issues/hydrological_extremes_model).

For citation purposes, cite each article independently as indicated on the article page online and as indicated below:

LastName, A.A.; LastName, B.B.; LastName, C.C. Article Title. *Journal Name* **Year**, *Article Number*, Page Range.

ISBN 978-3-03921-664-2 (Pbk)
ISBN 978-3-03921-665-9 (PDF)

Contents

About the Special Issue Editor

Hossein Tabari ,received his Ph.D. from KU Leuven in 2018 and currently is FWO Postdoctoral Fellow at the same university, where he investigates the impact of climate change on hydrological extremes. He has published over 80 peer-reviewed articles in world class high-impact journals that are also well cited (>2400 citations with a corresponding h-index of 28 based on the Web of Science). He is currently serving as an Editorial Board Member for the MDPI journals of *Geosciences* and *Water*. He has received several awards such as the 2018 Allianz Climate Risk Research Award (2nd prize) for his postdoctoral research and the 2016 Ernest du Bois Prize for his PhD research.

water

MDPI

Editorial

Statistical Analysis and Stochastic Modelling of Hydrological Extremes

Hossein Tabari

Department of Civil Engineering, Hydraulics Section, KU Leuven, 3000 Leuven, Belgium;
hossein.tabari@kuleuven.be

Received: 31 August 2019; Accepted: 4 September 2019; Published: 7 September 2019

Abstract: Analysis of hydrological extremes is challenging due to their rarity and small sample size and the interconnections between different types of extremes and gets further complicated by an untrustworthy representation of meso-scale processes involved in extreme events by coarse spatial and temporal scale models as well as biased or missing observations due to technical difficulties during extreme conditions. The special issue "Statistical Analysis and Stochastic Modelling of Hydrological Extremes"—motivated by the need to apply and develop innovative stochastic and statistical approaches to analyze hydrological extremes under current and future climate conditions —encompass 13 research papers. Case studies presented in the papers exploit a wide range of innovative techniques for hydrological extremes analyses. The papers focus on six topics: Historical changes in hydrological extremes, projected changes in hydrological extremes, downscaling of hydrological extremes, early warning and forecasting systems for drought and flood, interconnections of hydrological extremes and applicability of satellite data for hydrological studies. This Editorial provides an overview of the covered topics and reviews the case studies relevant for each topic.

Keywords: extreme events; innovative methods; downscaling; forecasting; compound events; satellite data

---

1. Introduction

Assessment of hydrological extremes is of paramount importance, as they have the potential to affect society in terms of human health and mortality, and the ecosystem and the economy (e.g., infrastructure and agriculture) [1,2]. In the last decades, millions of people have been affected by hydrological extremes. The risk of these hazards will increase in the future as a result of climate change and as population and infrastructure continue to increase and occupy areas exposed to higher risks [3–6].

Analyzing extremes in a complex interacting hydro-climatology system is challenging on the following grounds. First, extremes are rare events in the tail of distribution, characterized by either very small or very large values, and therefore, have a different statistical behavior. Being rare, extreme events have a small sample size, which adds a large uncertainty to the results of statistical analyses. With regard to observations, extreme events might be biased or missed altogether due to technical difficulties during extreme events [7].

For future hydrological extremes, the projections are mainly derived from global climate models (GCMs) which provide coarse spatial and temporal scale data that cannot be implemented directly in the hydrological impact analysis of climate change. For instance, for design applications in urban hydrology, precipitation data with a temporal resolution of a few minutes and a spatial resolution of $1–10 \text{ km}^2$ are needed especially when simulating flood events for small urban catchments with fast runoff processes and short response times [8]. To meet the needs of the end user for hydrological applications, information provided by climate models needs to be downscaled to much finer spatial and temporal scales [9]. It adds an extra tier of complexity to the hydrological extremes analyses.

The statistical analysis and stochastic modelling of hydrological extremes in developing countries is further hampered by unavailability or scarcity of high quality, fine-scale ground observations. The analysis of hydrological extremes is particularly important for vulnerable developing countries because they will shoulder a far greater burden of loss and damage due to inadequately built infrastructure, weaker preparedness and low levels of capacity to respond to disasters [10]. In the case rain-gauge observations are not available, satellite estimations with a spatially and temporally homogeneous precipitation data at a quasi-global to global scale are excellent alternatives [11]. Before the use of satellite data, their quality must, nevertheless, be validated next to the spatial downscaling of the data to make them suitable for the applications that require fine-resolution data such as in urban hydrology.

The complexity of analyzing hydrological extremes calls for robust statistical methods for the treatment of such events. This Special Issue is motivated by the need to apply and develop innovative stochastic and statistical approaches to analyze hydrological extremes under current and future climate conditions.

2. Overview of the Special Issue Contributions

The Special Issue includes 13 papers exploiting a broad range of innovative statistical methods for hydrological extremes analyses. The papers were published between 31 March 2018 and 14 June 2019 with an average time of 45 days from initial submission to online publication. In only a few months after the online publication, the papers have received 6 (7) citations in the literature indexed in Scopus (Google Scholar) (Table 1), indicating the significance and immediate impacts of the published studies. The abstract and full-text of the papers published in the Special Issue have been viewed on average 33 and 44 times per day, showing the broad reach of the published research.

Table 1. Metrics of the papers published in this Special Issue (average full-text and abstract views per day were calculated until 30 August 2019).

Paper Reference	Publication Date	Full-Text Views	Abstract Views	Citations Google Scholar	Scopus
Mahmud et al. [12]	31 March 2018	2	4	1	1
Rhee and Yang [13]	14 June 2018	2	2	0	0
Khan et al. [14]	27 July 2018	2	2	1	1
Mousavi et al. [15]	16 October 2018	2	3	1	1
Amnatsan et al. [16]	9 November 2018	3	3	0	0
Bafitlhile and Li [17]	6 January 2019	3	3	1	1
Pan et al. [18]	22 January 2019	2	2	0	0
Ávila et al. [19]	22 February 2019	4	5	2	1
Pham et al. [20]	3 March 2019	3	3	1	1
Tung et al. [21]	8 March 2019	2	2	0	0
Dawley et al. [22]	5 April 2019	3	3	0	0
Zhang and Wang [23]	4 June 2019	2	5	0	0
Mehmood et al. [24]	14 June 2019	3	5	0	0

The papers of this Special Issue focus on six topics associated with hydrological extremes which are reviewed in the following sections:

- Section 3: Historical changes in hydrological extremes;
- Section 4: Projected changes in hydrological extremes;
- Section 5: Downscaling of hydrological extremes;
- Section 6: Early warning and forecasting systems for drought and flood;
- Section 7: Interconnections of hydrological extremes;
- Section 8: Applicability of satellite data for hydrological studies.

3. Historical Changes in Hydrological Extremes

3.1. Background

Trend analysis of hydrological extremes provides essential information for regional water resource planning, risk assessment of hydrological hazards and adaptation and mitigation strategies to climate change [25,26]. To this end, various parametric and non-parametric methods have been developed and used over time. Non-parametric tests have gained popularity for the trend analysis of hydrological variables owing to their insensitivity to skewed data (non-normal distribution) which is likely the case for extreme events as well as their more resilience to outliers in time series [27,28]. These methods have, nonetheless, some shortcomings. One of these shortcomings is the sensitivity of trend test results to serial correlation in time series. In fact, a positive serial correlation, mostly the case for hydroclimatological data, increases the possibility of rejecting the null hypothesis of no trend while it is true [28]. Similar to serial correlation, a cross-correlation among the time series of neighboring stations or gridcells increases the false rejection of the null hypothesis of no field significance of trends [29,30]. These limitations need to be thoroughly addressed to avoid biased and misleading trend analysis results, e.g., using the effective sample size (ESS) for the former [31] and a bootstrapping method for the latter [32].

3.2. Case Studies

In a case study for the High Basin of the Cauca River in Southwestern Colombia, Ávila et al. [19] examined temporal trends in eight extreme precipitation indicators by the Mann-Kendall (MK) test and the Sen slope estimator for the period 1970–2013. The results showed a decreasing trend in precipitation intensity indices of annual maximum 1-day precipitation amount (Rx1day) and annual maximum 5-day precipitation amount (Rx5day). An increasing trend was also observed in the September–October–November period for consecutive dry days index and in the December–January–February period for total precipitation and number of wet days. The extreme precipitation indices were found to have a concurrent correlation with sea surface temperature in the equatorial Pacific and a lagged correlation (a lag of 2–3 months) with ENSO.

In another study, Mehmood et al. [24] analyzed temporal trends in annual maximum streamflow using the MK test at 29 stations with varying records between 30 (1987–2016) and 55 (1962–2016) years in the Kabul River Basin, Pakistan. Stationary and non-stationary Bayesian models were then used for flood frequency estimation and their results were compared using the corresponding flood frequency curves (FFCs). The results revealed a mixture of increasing and decreasing trends across different stations, implying a signal of clear non-stationarity in the flood regime. The non-stationarity was also confirmed by the findings of the Bayesian models where reliable results were found from the non-stationary Bayesian model, while the stationary Bayesian model either overestimated or underestimated flood frequencies.

4. Projected Changes in Hydrological Extremes

4.1. Background

Hydrological extremes can change due to natural climate variability and/or anthropogenic climate change. While climate variability, referred to natural processes influencing the atmosphere, is a periodic variation (yearly, decadal or multidecadal) in average or range of weather conditions either above or below a long-term average value without causing the long-term average itself to change, climate change is a long-term continuous change (increase or decrease) in these statistics. As for anthropogenic climate change whose hydrological impact was investigated in three papers in this Special Issue, extreme precipitation is expected to increase due to an increase in the atmospheric water-holding capacity under warmer climates dictated by the Clausius–Clapeyron (CC) relation of $\approx 7\%\,°C^{-1}$ [33,34]. The extreme precipitation intensification has, therefore, been projected under future climate change [35,36]. At the regional scale, the extreme precipitation-temperature rate varies with regional climatic settings (available water vapor and ranges of air temperature variation) [18] and different scaling rates were reported such as sub-CC (~3% °C^{-1}), close-CC (~7% °C^{-1}), super-CC (~14% °C^{-1}), peak-like CC (positive and negative), and negative CC [37–40]. The scaling rate also changes with the intensity and duration of extreme precipitation, being higher for more extreme and shorter duration precipitation [37,39,41].

4.2. Case Studies

Hourly extreme precipitation-temperature scaling rate regarding storm characteristics (types) and event process-based temperature variations in South China was investigated by Pan et al. [18]. They found a different magnitude of air temperature fluctuations prior to and after different storm types, and more reliable scaling rates from the 24-h mean air temperature prior to storms than the naturally daily mean air temperature. A peak-like scaling relation with a break at 28 °C temperature was reported between precipitation extremes and the 24-h mean air temperature. They obtained a positive scaling rate below 28 °C and a negative one for above 28 °C. The former was attributed to a high availability of relative humidity (80–100%), and the latter was triggered by a lack of moisture in the atmosphere instead of by the atmospheric water vapor-holding capacity. Comparing heavy storm producing weather systems in South China (e.g., warm-front storms, cold-front storms, monsoon storms, convective storms, and typhoon storms), a small influence of the storm types on the scaling rates was deduced.

China's extreme precipitation response of the next few decades to emission reductions based on the implement intended nationally determined contributions (INDCs) under the Paris Agreement was investigated by Zhang and Wang [23] using an ensemble of GCMs from the Fifth Coupled Climate Model Intercomparison Project (CMIP5). The maximum consecutive five-day precipitation and number of heavy precipitation days over China are projected to increase by 16% and up to 20%, respectively. The population exposure to heavy precipitation events will also increase in almost all Chinese regions, e.g., by 10% for extreme precipitation of > the 20-year return period.

The climate change impact on hydro-climatology and the potential of hydropower generation in the Dez Dam Basin in Iran was also investigated [15]. The results showed a remarkable reduction of up to 33% in future streamflows. Different climate change impacts on the electricity generation potential were found in two hydropower plants considered: 3% decrease and 33% increase.

5. Downscaling of Hydrological Extremes

5.1. Background

The primary tool for future hydroclimatic projections is GCM. However, the resolution of the current generation of Earth System Models (see CMIP5) is still coarse and unable to capture sub-grid scale processes [6,42–44]. The processes that cannot be resolved in horizontal grid spacing of GCMs are parameterized, which is a source of large bias and uncertainty in the simulations [45–49]. A more trustworthy representation of these processes and features is provided at finer spatial resolutions of regional climate models (RCMs) which are typically between 50 and 12 km, for instance 50 km for RCMs implemented and simulated in the project PRUDENCE [50] and NARCCAP [51], 25 km in ENSEMBLES [52] and 12 km in EURO-CORDEX [53]. Even if the spatial resolution of RCMs is much higher than that of GCMs, the grid size is still too large to adequately represent convective rain which is of primary importance for flood risk analysis [54]. To explicitly represent deep convection, very high resolution climate models (<4 km) termed convection-permitting models (CPMs) are needed. The major issue still plaguing CPMs is the representation of rain-on-snow events which are responsible for flash flooding in urban watersheds.

Another alternative to circumvent the intrinsic deficiency of GCMs and RCMs to represent fine-scale physical processes is statistical downscaling [55,56]. Nevertheless, the results of downscaling methods are often compromised with bias and limitations [57,58] due to assumptions and approximations made within each method. Some of these assumptions cast doubt on the reliability of downscaled projections and may limit the suitability of downscaling methods for some applications [59]. As there is no single best downscaling method, the assumptions that led to the final results for different methods require evaluation. Therefore, end users can select an appropriate method based on their strengths and limitations.

5.2. Case Study

To address the need for a statistical downscaling of extreme precipitation to a finer scale (e.g., catchment), Pham et al. [20] proposed a two-step statistical downscaling approach. Precipitation was first classified into wet and dry day or dry day, non-extreme precipitation day, and extreme precipitation day using the linear discriminant analysis (LDA), random forest (RF), and support vector classification (SVC). Afterwards, the precipitation amount for each precipitation state was predicted using the least square support vector regression (LS-SVR). Predictors of classification and prediction were obtained from the large-scale climate variables of the NCEP reanalysis data during 1964–1999 and 2000–2013 for calibration and validation, respectively. The results showed an outperformance of RF compared to LDA and SVC for precipitation classification. The extreme precipitation downscaling was found to be improved using RF for the classification of three-precipitation-states and using LS-SVR for the prediction of precipitation amount.

6. Early Warning and Forecasting Systems for Drought and Flood

6.1. Background

Hydrological hazards (flood and drought) are manageable by implementing appropriate emergency preparedness and mitigation strategies. One of the effective measures to mitigate the negative impacts of drought and flood is the early warning and forecasting systems. Forecasting of hydrological events is performed using conceptual or data-driven models [60]. Each group of forecasting models has its own cons and pros. The conceptual models usually incorporate simplified forms of physical laws and are generally nonlinear, deterministic, and time-invariant, with parameters that characterize watershed features [61]. Nevertheless, the main limitation of these models is that when they are calibrated to a set of time series, they may not provide an accurate prediction for values beyond the range of calibration or validation values [62]. Data-driven models are numerical models which represent causal relationships or patterns between sets of input and output time series data, independent of the physics of the real-world situation. Although a limited prior knowledge requirement of internal functions of the system being modeled [63] along with a high ability to represent non-linear processes [64] and time–space variability [60] are the main pros of data-driven models, the prediction entirely based on mathematics without explicit physical consideration is the main limitation of these models.

6.2. Case Studies

Rhee and Yang [13] developed a hybrid model for the meteorological drought prediction of the 6-month Standardized Precipitation Index (SPI) for areas with a sparse gauge network using the APEC Climate Center Multi-Model Ensemble seasonal climate forecast and machine learning models of Extra-Trees and Adaboost. To overcome the limitation of the sparse network, dynamically downscaled historical climate data from the Weather Research and Forecasting (WRF) model were used to train machine learning models instead of in-situ data as a reference. In another study, Khan et al. [14] developed two artificial neural network (ANN)-based models and two wavelet-based artificial neural network (W-ANN) models for meteorological and hydrological droughts characterized by Standard Index of Annual Precipitation (SIAP) and Standardized Water Storage Index (SWSI), respectively.

Owing to the importance of reservoir inflow forecasting for appropriate reservoir management, especially in the flood season, the variation analogue method (VAM), the W-ANN, and the weighted mean analogue method (WMAM) were used to forecast reservoir inflows by Amnatsan et al. [16]. In another study, Bafitlhile and Li [17] applied ε-Support Vector Machine (ε-SVM) and ANN for the simulation and forecasting streamflow of three catchments with humid, semi-humid and semi-arid climates. To optimize the ANN and SVM sensitive parameters, the Evolutionary Strategy (ES) optimization method was used.

7. Interconnections of Hydrological Extremes

7.1. Background

Hydrological extremes are often investigated in isolation, while in reality hydrologic processes in the water cycle are interconnected. Or the complex interconnected water systems are oversimplified such as the relation between precipitation and groundwater table fluctuations. Isolated analysis of hydrological extremes or oversimplification of their complex interactions results in an underestimation of the impact associated with extreme conditions. Hydrological extremes must, therefore, be analyzed in a compound manner for a more realistic estimate of the overall impact. This ensures a better decision-making to curb the growing impacts of the extremes and to plan and build more resilient water systems [10].

7.2. Case Studies

The interconnection between hydrological extremes was addressed by Dawley et al. [22] who correlated surface and subsurface hydrological extreme events by investigating the possible effects of extreme storm events of different properties on the fluctuations in surface and subsurface water systems. They applied three probability density functions (PDFs), Gumbel, stable, and stretched Gaussian distributions, to capture the distribution of extremes and the full-time series of storm properties (storm duration, intensity, total precipitation, and inter-storm period), stream discharge, lake stage, and groundwater head values. The potentially non-stationary statistics of hydrological extremes were quantified by computing the time-scale local Hurst exponent (TSLHE) for the time series data recording both the surface and subsurface hydrological processes. The results indicated that groundwater recharge has a strong relationship with storm duration and intensity and a weak one with total precipitation. The surface water and groundwater series were found to be persistent because of their relatively slow evolving nature, while storm properties were anti-persistent because of their rapid temporal evolution. They also showed that a single distribution cannot most effectively capture all of hydrological extremes and different distributions depending on the variable under study should be used.

The difficulty of establishing a joint distribution function for multiple correlated random variables with a mixture of non-normal marginal distributions affecting the design and safety evaluation of hydro-infrastructural systems was tacked by Tung et al. [21]. They presented a framework for a practical normal transform based on the third-order polynomial with an explicit consideration of sampling errors in sample L-moments and correlation coefficients. The modeling framework was then applied to establish an at-site precipitation intensity–duration-frequency (IDF) relationship for 27-year annual maximum precipitation records with seven durations (1–72 h). The results showed that the proposed framework is able to deal with multivariate data having a mixture of non-normal distributions.

8. Applicability of Satellite Data for Hydrological Studies

8.1. Background

The precipitation measurements can be obtained through different sources such as surface networks, weather radars and satellite estimations. Among them, the most common practice is the derivation of precipitation estimates over land areas from surface rain-gauge observations at automated or human-operated sites [65]. Despite the different types of errors included in surface network measurements such as instrument/human errors, change of instrument/observer, change of observing technique and changes in station surroundings like urbanization, they provide the most accurate measurements. Notwithstanding being the most reliable and longer records [66], they are limited in sampling precipitation for continental and global applications [67]. In most regions of the world, rain-gauges do not provide a reliable spatial representation of precipitation [68], especially over oceans, deserts and mountainous areas. In addition, other possible problems of surface rain-gauge observations are the inhomogeneous spatial distribution and the existence of missing data resulting in inadequate temporal and spatial sampling [65] especially in developing countries.

Satellite estimations with a spatially and temporally homogeneous precipitation information at a quasi-global to global scale are excellent alternatives wherever/whenever rain-gauge observations are not available [11]. Nevertheless, their quality has to be validated before any application. The typical sources of errors in satellite precipitation data are sensor-related errors, retrieval errors and spatial and temporal sampling errors [69]. The most common practice for the verification of satellite data is to compare satellite estimations with local station-based observations, by considering the spatial scale mismatch between the point station observations and gridded outputs as the latter represent area averages rather than point values [54]. Specifically, extreme precipitation values obtained from station observations are expected to be more intense compared with the ones from gridded outputs [70], because of the smoothing associated with the spatial averaging of precipitation characteristics over gridcells [71]. Another limitation of satellite data is their coarse spatial resolution, which is not suitable for practical applications in hydrology, calling for the spatial downscaling of the data.

8.2. Case Study

Addressing the need for the spatial downscaling of satellite data, Mahmud et al. [12] developed a spatial downscaling algorithm to produce finer-scale satellite precipitation data in humid tropics. They used the potential of the low precipitation variability in Peninsular Malaysia and monsoon characteristics (period, location, and intensity) at the local scale as a proxy to spatially downscale TRMM (Tropical Rainfall Measuring Mission) satellite precipitation data. To this end, a site-specific coefficient (SSC) was first derived for each individual pixel by comparing the high-resolution areal precipitation (0.05°) from a dense gauge network and re-gridded TRMM satellite precipitation data (from the initial resolution of 0.25° to 0.05°) and then the SSC was validated to produce high-resolution precipitation maps.

9. Conclusions

The research published in this Special Issue applied or developed a broad range of novel methods for the statistical analysis and stochastic modelling of hydrological extremes. The case studies presented in the 13 published papers have touched on six research areas: (1) Historical changes in hydrological extremes; (2) projected changes in hydrological extremes; (3) downscaling of hydrological extremes; (4) early warning and forecasting systems for drought and flood; (5) interconnections of hydrological extremes; and (6) applicability of satellite data for hydrological studies. Contributions to this Special Issue are expected to be greatly beneficial for researchers, policy-makers and risk managers dealing with hydrological hazards. Yet, innovative statistical methods have to be developed to keep up with the accelerating pace of socio-environmental changes. Hence, of particular interests for further research, are the topics concerning future hydrological extremes, for instance:

- Assessment of the decadal natural oscillations of hydrological extremes and their concurrent and lagged relationships with large-sale atmospheric circulation patterns as done in Tabari and Willems [1];

- Attribution analysis of changes in the intensity, duration and frequency of hydrological extremes to anthropogenic influences;
- Dynamical downscaling of hydrological extremes and exploring the added value of CPMs and RCMs;
- Climate-proof hydraulic designs based on projected IDF curves;
- Assessment of uncertainties in hydrological projections and observations;
- Socioeconomic risk analysis of future hydrological hazards;
- Hydrological hazard mitigation and adaptation strategies.

Funding: This research received no external funding.

Acknowledgments: The guest editor would like to thank all the authors who contributed to this special issue, as well as the anonymous reviewers for their insightful comments on the submitted manuscripts which greatly helped the authors improve the scientific quality of their papers. Sincere gratitude goes to the Editorial Board and the Editorial Office of Water, especially to the managing editor Evelyn Ning, for the invaluable help and assistance provided for the design of this special issue, and during the review and publication of the submitted manuscripts.

Conflicts of Interest: The author declares no conflict of interest.

References

1. Tabari, H.; Willems, P. Lagged influence of Atlantic and Pacific climate patterns on European extreme precipitation. *Sci. Rep.* **2018**, *8*, 5748. [CrossRef] [PubMed]
2. Tabari, H.; Willems, P. More prolonged droughts by the end of the century in the Middle East. *Environ. Res. Lett.* **2018**, *13*, 104005. [CrossRef]
3. Zulkafli, Z.; Buytaert, W.; Manz, B.; Rosas, C.V.; Williams, P.; Lavado-Casimiro, W.; Guyot, J.L.; Santini, W. Projected increases in the annual flood pulse of the Western Amazon. *Environ. Res. Lett.* **2016**, *11*, 014013. [CrossRef]
4. Donat, M.G.; Angélil, O.; Ukkola, A.M. Intensification of precipitation extremes in the world's humid and water-limited regions. *Environ. Res. Lett.* **2019**, *14*, 065003. [CrossRef]
5. Tabari, H.; Willems, P. Seasonally varying footprint of climate change on precipitation in the Middle East. *Sci. Rep.* **2018**, *8*, 4435. [CrossRef] [PubMed]
6. Lee, T.; Park, T. Nonparametric temporal downscaling with event-based population generating algorithm for RCM daily precipitation to hourly: Model development and performance evaluation. *J. Hydrol.* **2017**, *547*, 498–516. [CrossRef]
7. Casati, B.; Wilson, L.J.; Stephenson, D.B.; Nurmi, P.; Ghelli, A.; Pocernich, M.; Damrath, U.; Ebert, E.E.; Brown, B.G.; Mason, S. Forecast verification: Current status and future directions. *Meteor. Appl.* **2008**, *15*, 3–18. [CrossRef]
8. Willems, P.; Arnbjerg-Nielsen, K.; Olsson, J.; Beecham, S.; Pathirana, A.; Gregersen, I.B.; Madson, H.; Nguyen, V.-T.-V. *Impacts of Climate Change on Rainfall Extremes and Urban Drainage*; IWA: London, UK, 2012.
9. Maraun, D.; Wetterhall, F.; Ireson, A.M.; Chandler, R.E.; Kendon, E.J.; Widmann, M.; Brienen, S.; Rust, H.W.; Sauter, T.; Themebl, M. Precipitation downscaling under climate change: Recent developments to bridge the gap between dynamical models and the end user. *Rev. Geophys.* **2010**, *48*. [CrossRef]
10. Tabari, H. The dry facts. In *Proceedings of the A Compendium of Essays for the Allianz Climate Risk Research Award 2018—Weathering the Storms: Fostering Our Understanding of Climate-Related Risks and Our Capacity to Respond*; Allianz Climate Solutions: Munich, Germany, 2018.
11. Tabari, H.; AghaKouchak, A.; Willems, P. A perturbation approach for assessing trends in precipitation extremes across Iran. *J. Hydrol.* **2014**, *519*, 1420–1427. [CrossRef]
12. Mahmud, M.; Hashim, M.; Matsuyama, H.; Numata, S.; Hosaka, T. Spatial downscaling of satellite precipitation data in humid tropics using a site-specific seasonal coefficient. *Water* **2018**, *10*, 409. [CrossRef]
13. Rhee, J.; Yang, H. Drought prediction for areas with sparse monitoring networks: A case study for Fiji. *Water* **2018**, *10*, 788. [CrossRef]
14. Khan, M.; Muhammad, N.; El-Shafie, A. Wavelet-ANN versus ANN-based model for hydrometeorological drought forecasting. *Water* **2018**, *10*, 998. [CrossRef]

15. Mousavi, R.; Ahmadizadeh, M.; Marofi, S. A multi-GCM assessment of the climate change impact on the hydrology and hydropower potential of a semi-arid basin (A Case Study of the Dez Dam Basin, Iran). *Water* **2018**, *10*, 1458. [CrossRef]

16. Amnatsan, S.; Yoshikawa, S.; Kanae, S. Improved forecasting of extreme monthly reservoir inflow using an analogue-based forecasting method: A case study of the Sirikit Dam in Thailand. *Water* **2018**, *10*, 1614. [CrossRef]

17. Bafitlhile, T.M.; Li, Z. Applicability of ε-support vector machine and artificial neural network for flood forecasting in humid, semi-humid and semi-arid basins in China. *Water* **2019**, *11*, 85. [CrossRef]

18. Pan, C.; Wang, X.; Liu, L.; Wang, D.; Huang, H. Characteristics of heavy storms and the scaling relation with air temperature by event process-based analysis in South China. *Water* **2019**, *11*, 185. [CrossRef]

19. Ávila, Á.; Guerrero, F.C.; Escobar, Y.C.; Justino, F. Recent precipitation trends and floods in the Colombian Andes. *Water* **2019**, *11*, 379. [CrossRef]

20. Pham, Q.B.; Yang, T.C.; Kuo, C.M.; Tseng, H.W.; Yu, P.S. Combing random forest and least square support vector regression for improving extreme rainfall downscaling. *Water* **2019**, *11*, 451. [CrossRef]

21. Tung, Y.K.; You, L.; Yoo, C. Third-order polynomial normal transform applied to multivariate hydrologic extremes. *Water* **2019**, *11*, 490. [CrossRef]

22. Dawley, S.; Zhang, Y.; Liu, X.; Jiang, P.; Tick, R.G.; Sun, H.G.; Zheng, C.; Chen, L. Statistical analysis of extreme events in precipitation, stream discharge, and groundwater head fluctuation: Distribution, memory, and correlation. *Water* **2019**, *11*, 707. [CrossRef]

23. Zhang, J.; Wang, F. Extreme precipitation in China in response to emission reductions under the Paris Agreement. *Water* **2019**, *11*, 1167. [CrossRef]

24. Mehmood, A.; Jia, S.; Mahmood, R.; Yan, J.; Ahsan, M. Non-stationary Bayesian modeling of annual maximum floods in a changing environment and implications for flood management in the Kabul River Basin, Pakistan. *Water* **2019**, *11*, 1246. [CrossRef]

25. Pandey, B.K.; Khare, D. Identification of trend in long term precipitation and reference evapotranspiration over Narmada river basin (India). *Glob. Planet Chang.* **2018**, *161*, 172–182. [CrossRef]

26. Nashwan, M.S.; Shahid, S.; Wang, X. Uncertainty in estimated trends using gridded rainfall data: A case study of Bangladesh. *Water* **2019**, *11*, 349. [CrossRef]

27. Yu, Y.-S.; Zou, S.; Whittemore, D. Non-parametric trend analysis of water quality data of rivers in Kansas. *J. Hydrol.* **1993**, *150*, 61–80. [CrossRef]

28. Tabari, H.; Hosseinzadeh Talaee, P. Temporal variability of precipitation over Iran: 1966–2005. *J. Hydrol.* **2011**, *396*, 313–320. [CrossRef]

29. Yue, S.; Pilon, P.; Phinney, B. Canadian streamflow trend detection: Impacts of serial and cross-correlation. *Hydrol. Sci. J.* **2003**, *48*, 51–64. [CrossRef]

30. El Kenawy, A.; Lopez-Moreno, J.I.; Vicente-Serrano, S.M. Recent trends in daily temperature extremes over northeastern Spain (1960–2006). *Nat. Hazards Earth Syst. Sci.* **2011**, *11*, 2583–2603. [CrossRef]

31. Hamed, K.H.; Rao, A.R. A modified Mann–Kendall trend test for autocorrelated data. *J. Hydrol.* **1998**, *204*, 182–196. [CrossRef]

32. Douglas, E.M.; Vogel, R.M.; Kroll, C.N. Trends in floods and low flows in the United States: Impact of spatial correlation. *J. Hydrol.* **2000**, *240*, 90–105. [CrossRef]

33. Berg, P.; Moseley, C.; Haerter, J.O. Strong increase in convective precipitation in response to higher temperatures. *Nat. Geosci.* **2013**, *6*, 181–185. [CrossRef]

34. Lenderink, G.; Barbero, R.; Loriaux, J.M.; Fowler, H.J. Super-Clausius–Clapeyron scaling of extreme hourly convective precipitation and its relation to large-scale atmospheric conditions. *J. Clim.* **2017**, *30*, 6037–6052. [CrossRef]

35. Prein, A.F.; Rasmussen, R.M.; Ikeda, K.; Liu, C.; Clark, M.P.; Holland, G.J. The future intensification of hourly precipitation extremes. *Nat. Clim. Chang.* **2017**, *7*, 48–52. [CrossRef]

36. Ge, F.; Zhu, S.; Peng, T.; Zhao, Y.; Sielmann, F.; Fraedrich, K.; Zhi, X.; Liu, X.; Tang, W.; Ji, L. Risks of precipitation extremes over Southeast Asia: Does 1.5 or 2 degrees global warming make a difference? *Environ. Res. Lett.* **2019**, *14*, 044015. [CrossRef]

37. Utsumi, N.; Seto, S.; Kanae, S.; Maeda, E.E.; Oki, T. Does higher surface temperature intensify extreme precipitation? *Geophys. Res. Lett.* **2011**, *38*, 239–255. [CrossRef]

38. Maeda, E.E.; Utsumi, N.; Oki, T. Decreasing precipitation extremes at higher temperatures in tropical regions. *Nat. Hazards* **2012**, *64*, 935–941. [CrossRef]

39. Panthou, G.; Mailhot, A.; Laurence, E.; Talbot, G. Relationship between surface temperature and extreme rainfalls: A multi-time-scale and event-based analysis. *J. Hydrol.* **2014**, *15*, 1999–2011. [CrossRef]

40. Wang, G.; Wang, D.; Trenberth, K.E.; Erfanian, A.; Yu, M.; Bosilovich, M.G.; Parr, D.T. The peak structure and future changes of the relationships between extreme precipitation and temperature. *Nat. Clim. Chang.* **2017**, *7*, 268–274. [CrossRef]

41. Drobinski, P.; Alonzo, B.; Bastin, S.; Silva, N.D.; Muller, C. Scaling of precipitation extremes with temperature in the French Mediterranean region: What explains the hook shape? *J. Geophys. Res. Atmos.* **2016**, *121*, 3100–3119. [CrossRef]

42. Wilby, R.L.; Wigley, T.M.L. Precipitation predictors for downscaling: Observed and general circulation model relationships. *Int. J. Climatol.* **2000**, *20*, 641–661. [CrossRef]

43. Frei, C.; Schöll, R.; Schmidli, J.; Fukutome, S.; Vidale, P.L. Future change of precipitation extremes in Europe: An intercomparison of scenarios from regional climate models. *J. Geophys. Res.* **2006**, *111*. [CrossRef]

44. Hassanzadeh, E.; Hassanzadeh, A.; Elshorbagy, A. Quantile-based downscaling of precipitation using genetic programming: Application to IDF curves in Saskatoon. *J. Hydrol. Eng.* **2014**, *19*, 943–955. [CrossRef]

45. Prein, A.F.; Gobiet, A.; Suklitsch, M.; Truhetz, H.; Awan, N.K.; Keuler, K.; Georgievski, G. Added value of convection permitting seasonal simulations. *Clim. Dyn.* **2013**, *41*, 2655–2677. [CrossRef]

46. Kendon, E.J.; Roberts, N.M.; Fowler, H.J.; Roberts, M.J.; Chan, S.C.; Senior, C.A. Heavier summer downpours with climate change revealed by weather forecast resolution model. *Nat. Clim. Chang.* **2014**, *4*, 570–576. [CrossRef]

47. Olsson, J.; Berg, P.; Kawamura, A. Impact of RCM spatial resolution on the reproduction of local, subdaily precipitation. *J. Hydrometeorol.* **2015**, *16*, 534–547. [CrossRef]

48. Mendoza, P.A.; Mizukami, N.; Ikeda, K.; Clark, M.P.; Gutmann, E.D.; Arnold, J.R.; Brekke, L.D.; Rajagopalan, B. Effects of different regional climate model resolution and forcing scales on projected hydrologic changes. *J. Hydrol.* **2016**, *541*, 1003–1019. [CrossRef]

49. Hosseinzadehtalaei, P.; Tabari, H.; Willems, P. Precipitation intensity–duration–frequency curves for central Belgium with an ensemble of EURO-CORDEX simulations, and associated uncertainties. *Atmos. Res.* **2018**, *200*, 1–12. [CrossRef]

50. Christensen, J.H.; Christensen, O.B. A summary of the PRUDENCE model projections of changes in European climate by the end of this century. *Clim. Chang.* **2007**, *81*, 7–30. [CrossRef]

51. Mearns, L.O.; Gutowski, W.J.; Jones, R.; Leung, L.-Y.; McGinnis, S.; Nunes, A.M.B.; Qian, Y. A regional climate change assessment program for North America. *EOS* **2009**, *90*, 311–312. [CrossRef]

52. van der Linden, P.; Mitchell, J.F.B. *ENSEMBLES: Climate Change and Its Impacts: Summary of Research and Results from the ENSEMBLES Project*; Technical Report; Met Office Hadley Centre: Exeter, UK, 2009; p. 160.

53. Giorgi, F.; Jones, C.; Asrar, G. Addressing climate information needs at the regional level: The CORDEX framework. *WMO Bull.* **2009**, *58*, 175–183.

54. Tabari, H.; De Troch, R.; Giot, O.; Hamdi, R.; Termonia, P.; Saeed, S.; Brisson, E.; van Lipzig, N.; Willems, P. Local impact analysis of climate change on precipitation extremes: Are high-resolution climate models needed for realistic simulations? *Hydrol. Earth Syst. Sci.* **2016**, *20*, 3843–3857. [CrossRef]

55. Bi, E.G.; Gachon, P.; Vrac, M.; Monette, F. Which downscaled rainfall data for climate change impact studies in urban areas? Review of current approaches and trends. *Theor. Appl. Climatol.* **2017**, *127*, 685–699.

56. Keller, D.E.; Fischer, A.M.; Liniger, M.A.; Appenzeller, C.; Knutti, R. Testing a weather generator for downscaling climate change projections over Switzerland. *Int. J. Climatol.* **2017**, *37*, 928–942. [CrossRef]

57. Trzaska, S.; Schnarr, E. *A Review of Downscaling Methods for Climate Change Projections*; Technical Report for United States Agency for International Development by Tetra Tech ARD; United States Agency for International Development: Pasadena, CA, USA, 2014; pp. 1–42.

58. Maraun, D.; Widman, M.; Gutierrez, J.M.; Kotlarski, S.; Chandler, R.E.; Hertig, E.; Wibig, J.; Huth, R.; Wilcke, R.A.I. VALUE: A framework to validate downscaling approaches for climate change studies. *Earth's Future* **2015**, *3*, 1–14. [CrossRef]

59. Hall, A. Projecting regional change. *Science* **2014**, *346*, 1461–1462. [CrossRef] [PubMed]

60. Nourani, V.; Komasi, M.; Mano, A. A multivariate ANN-Wavelet approach for rainfall–runoff modeling. *Water Resour. Manag.* **2009**, *23*, 2877. [CrossRef]

61. Panagoulia, D. Artificial neural networks and high and low flows in various climate regimes. *Hydrol. Sci. J.* **2006**, *51*, 563–587. [CrossRef]
62. Panagoulia, D. Hydrological modeling of a medium-sized mountainous catchment from incomplete meteoro-logical data. *J. Hydrol.* **1992**, *137*, 279–310. [CrossRef]
63. Rezaeianzadeh, M.; Tabari, H.; Yazdi, A.A.; Isik, S.; Kalin, L. Flood flow forecasting using ANN, ANFIS and regression models. *Neural Comp. Appl.* **2014**, *25*, 25–37. [CrossRef]
64. Tabari, H.; Kisi, O.; Ezani, A.; Hosseinzadeh Talaee, P. SVM, ANFIS, regression and climate based models for reference evapotranspiration modeling using limited climatic data in a semi-arid highland environment. *J. Hydrol.* **2012**, *444*, 78–89. [CrossRef]
65. Lockhoff, M.; Zolina, O.; Simmer, C.; Schulz, J. Evaluation of satellite-retrieved extreme precipitation over Europe using gauge observations. *J. Clim.* **2013**, *27*, 607–623. [CrossRef]
66. Yatagai, A.; Arakawa, O.; Kamiguchi, K.; Kawamoto, H.; Nodzu, M.I.; Hamada, A. A 44-year daily gridded precipitation dataset for Asia based on a dense network of rain gauges. *SOLA* **2009**, *5*, 137–140. [CrossRef]
67. Nastos, P.T.; Kapsomenakis, J.; Douvis, K.C. Analysis of precipitation extremes based on satellite and high-resolution gridded data set over Mediterranean basin. *Atmos. Res.* **2013**, *131*, 46–59. [CrossRef]
68. Gruber, A.; Levizzani, V. *Assessment of Global Precipitation Products*; WCRP Report No. 128, WMO/TD No. 1430 (50p); Technical Report for a Project of the World Climate Research Programme Global Energy and Water Cycle Experiment (GEWEX) Radiation Panel: Geneva, Switzerland, 2008.
69. AghaKouchak, A.; Mehran, A.; Norouzi, H.; Behrangi, A. Systematic and random error components in satellite precipitation data sets. *Geophys. Res. Lett.* **2012**, *39*. [CrossRef]
70. Chen, C.-T.; Knutson, T. On the verification and comparison of extreme rainfall indices from climate models. *J. Clim.* **2008**, *21*, 1605–1621. [CrossRef]
71. Sivapalan, M.; Blöschl, G. Transformation of point rainfall to areal rainfall: Intensity-duration-frequency curves. *J. Hydrol.* **1998**, *204*, 150–167. [CrossRef]

water

MDPI

Article

Recent Precipitation Trends and Floods in the Colombian Andes

Álvaro Ávila [1,2,*], Faisury Cardona Guerrero [1], Yesid Carvajal Escobar [1,*] and Flávio Justino [2]

[1]   Research group in Water Resources Engineering and Soil (IREHISA), Universidad del Valle, Cali,
      Valle del Cauca 76001, Colombia; faisury.cardona@correounivalle.edu.co
[2]   Department of Agricultural Engineering, Universidade Federal de Viçosa, Viçosa, MG 36570-900, Brasil;
      fjustino@ufv.br
*   Correspondence: alvaro.diaz@ufv.br (Á.Á.); yesid.carvajal@correounivalle.edu.co (Y.C.E.);
      Tel.: (+57)-3164485361 (Y.C.E.)

Received: 13 December 2018; Accepted: 5 February 2019; Published: 22 February 2019

Abstract: This study aims to identify spatial and temporal precipitation trends by analyzing eight extreme climate indices of rainfall in the High Basin of the Cauca River in Southwestern Colombia from 1970 to 2013. The relation between historical floods and El Niño Southern Oscillation (ENSO) is also analyzed. Results indicate that in general, the reduction of precipitation, especially in the center of the basin with negative annual and seasonal trends in intensity indices, namely, the annual maximum 1-day precipitation amount (RX1day) and annual maximum 5-day precipitation amount (RX5day). Sixty-four percentage of the stations exhibit an increasing trend in September–October–November in the consecutive dry days. In December–January–February interval, positive trends in most of the stations is noted for total precipitation and for the number of wet days with rainfall greater than or equal to 1 mm. The findings also show that sea surface temperature (SST) in the equatorial Pacific is statistically correlated (r) with indices of extreme precipitation (r $\geq$ −0.40). However, the effect of ENSO is evident with a time lag of 2–3 months. These results are relevant for forecasting floods on a regional scale, since changes in SST of the equatorial Pacific may take place 2–3 months ahead of the basin inundation. Our results contribute to the understanding of extreme rainfall events, hydrological hazard forecasts and climate variability in the Colombian Andes.

Keywords: climate change; the Cauca River; climate variability; ENSO; extreme rainfall; trends

---

1. Introduction

Local intense rainfall events, as well as environmental alterations (deforestation and/or urbanization), may often trigger the incidence of hydrological hazards, such as floods, flash floods and landslides, especially in the tropical areas [1]. In Colombia, hydrological hazards mostly occur as a result of local heavy precipitation during the cold phase of the El Niño Southern Oscillation (ENSO), known as La Niña [2–6]. Hydrological basins located in the Andes are prone to hazards due to complex geographical terrain combined with spatial and temporal climate variability [5]. These events have a strong impact on a large portion of the local economy, since agriculture accounts for a significant amount of income generation [7,8].

Between 1970 and 2013, floods affected about 16 million Colombians and claimed the lives of more than 3000 people, according to the Emergency Events Database (EM-DAT). In most cases, this is related to the "La Niña" that can lead to water-related hydrological hazards with catastrophic consequences for the livelihood of people and water resources. Indeed, the 2010–2011 La Niña affected 5.2 million people and caused 683 deaths and losses of more than US$7.8 billion [9,10]. It is crucial to understand the causes of precipitation changes on global, continental, and regional scales and identify the link between these hydrological events and weather phenomena and/or climate variability at inter-decadal and inter-annual timescales [11–15].

The study of hydrometeorological threats, such as floods, flash floods, and landslides, requires a multidisciplinary approach to understanding the impact of climate variability and climate change on water pathways, water storage, and related hazards. Floods are the most frequent hydrological threats in the Colombia River Basins that is present along the Andes [4,6,9]. The increased frequency and intensity of extreme hydrometeorological events has become the main socio-environmental issue in the 21st century [13,16–19]. Therefore, efforts are needed to detect temporal changes in extreme rainfall and the main climate-associated mechanisms to mitigate the damage to society [20–22].

This study focuses on the High Basin of the Cauca River in Southwestern Colombia located in the Andes region. This region is susceptible to extreme precipitation events that cause flooding [6,23], affecting agricultural production sectors, thereby hampering regional and national economic development [4,9,10]. In lowlands, the natural land cover has been replaced by intensive agriculture, namely maize, sugar cane, and yucca, which are also cultivated for biofuel production [24]. More importantly, agriculture, livestock and the human population in the Colombian Andes have also increased exponentially over the last two centuries, with major peaks between the 1970s and 2000s [25]. Consequently, combined climate–topography–socio-economic factors create a highly vulnerable scenario for catastrophic events.

Currently, only a few studies have addressed long-term trends in precipitation extremes in the Colombian Andes [11,23,26,27]. In this study, 44 years (1970–2013) of daily precipitation data are used to analyze spatiotemporal trends of rainfall at an inter-annual time scale. Moreover, eight extreme precipitation indices in the High Basin of the Cauca River are analyzed. Specifically, the study explores the link between climate rainfall indices, historical floods, and ENSO, which can be used in association with climate indicators for flooding forecasting systems.

2. Methodology

2.1. Study Area

The High Basin of the Cauca River (75°42′–76°58′ W, 2°06′–05°2′ N) rises in the Colombian Massif close to the Ecuadorian border and meanders along the Western and Central Andean Cordilleras. The basin extends over 18.111 km^2 (Figure 1a). The mountains are higher (4635 m) in the east but lower (891 m) in the center and north, including the plains and hills.

This basin is a matter of concern in Colombia due to its important economic and ecological role as a natural resource of water for the La Salvajina Dam (hydropower plant), domestic water supply, and industrial and irrigation systems [28]. The population is approximately 4.5 million (9.8% of the Colombian population in 2010), according to the National Administrative Department of Statistics. The High Basin of the Cauca River has been identified as the strongest potential leader in agriculture in Colombia, since it accounts for more than 90% of the sugarcane-planting area and sugar production [29], and trade and consumption of ethanol [30].

Flood events in the High Basin of the Cauca River occur due to extreme rainfall, drastic reductions in areas with natural forest [24,31], and the increased level of the population. For example, the city of Santiago de Cali (the most populous city in Southwestern Colombia) has regularly been vulnerable to river-based flooding. According to the Government of Colombia's National Administrative Department of Statistics (DANE, for its acronym in Spanish), the population of the city presented a strong growth of 31.538 people per year between 1985 and 2010, from 1.42 million to 2.24 million. This still raises concerns of land use changes due to increased agricultural activities and livestock, rural/urban migration and the expansion of urban edges, which imply that local geographic characteristics, anthropogenic factors and climate risks may lead to the onset of catastrophic floods.

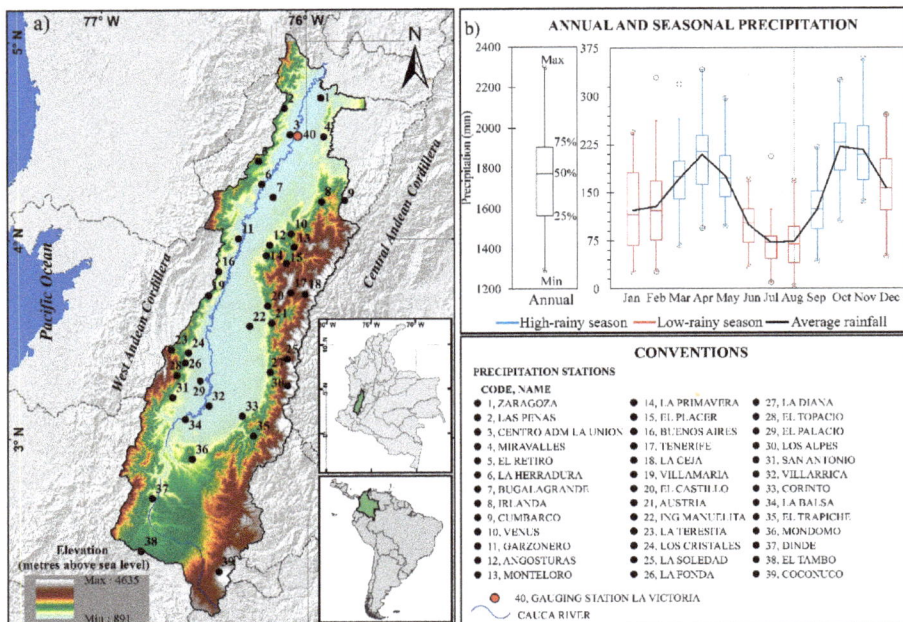

Figure 1. (**a**) Study area. Location of the High Basin of the Cauca River showing the pluviometric stations (black points) and location of the river gauging station (red point). (**b**) Annual and monthly variability of precipitation over the High Basin of the Cauca River over the 1970–2013 period. Source of cartography: Corporación Autónoma Regional del Valle del Cauca (CVC).

The hydroclimatology of the Colombian Andes is dominated by climatic mechanisms such as the latitudinal migration of the Intertropical Convergence Zone (ITCZ), associated with the trans-equatorial dynamics of the moisture induced by the eastern trade winds. The Chocó jet activity and the behavior of meso-scale convective systems also play a role as well [32,33]. Other hydroclimatological features in the area are discussed by [3,34–36]. The precipitation is affected by a double-ITCZ migration that flows from north to south and then back to north and goes across the geographic valley of the Cauca River twice a year (Figure 1b); this is a consequence of the semi-annual cycle of the march of the sun, and the circulation of the trade winds. For these reasons, there are two rainy seasons: March–April–May (MAM), and September–October–November (SON), which alternate with two reduced-rain seasons—December–January–February (DJF) and June–July–August (JJA). The total annual precipitation based on observational data over the 1970–2013 period is between 1294 and 2299 mm (Figure 1b).

2.2. Data Quality Control and Homogeneity

The data used in this study are provided by the Corporación Autónoma Regional del Valle del Cauca (CVC), based on historical daily precipitation from 108 rainfall stations. From these, only the stations with at least 90% of daily information between 1970 and 2013 were used. The dataset homogeneity and identification of possible biased records were investigated by performing RhtestV3 software developed and maintained by Wang and Feng [37] at the Climate Research Branch of Meteorological Service of Canada, a software running on R which are freely available online at http://etccdi.pacificclimate.org/software.shtml. After homogeneity testing, only thirty-nine (39) rainfall stations met the established criteria to compute the extreme precipitation indices.

2.3. Extreme Precipitation Indices

We used the RClimDex package developed by Zhang and Yang (2004) [38,39] at Climate Research Branch Environment of Canada. This package is a statistical tool for the R platform, proposed by the Expert Team on Detection and Climate Change Indices (ETCCDI; http://etccdi.pacificclimate.org/). In this study, eight precipitation-related indices were selected (Table 1). These indices are calculated at annual and seasonal scales (DJF, MAM, JJA, and SON). The seasonal analysis is justified by the marked inter-annual variability of precipitation, as shown above (Figure 1b). These extreme precipitation indices are also used to detect possible relationships between hydrological hazards and areas potentially vulnerable to catastrophic events [13,15,22,40–42].

Table 1. Definition of indices selected for analysis of extreme precipitation recommended by the Expert Team on Climate Change Detection and Indices (ETCCDI).

Index	Indicator	Definition	Unit
PRCPTOT	Annual total wet-day precipitation	Total wet-day precipitation (RR * $\geq$ 1 mm)	mm
RX1day	Maximum 1-day precipitation amount	Highest 1-day precipitation amount	mm
RX5day	Maximum 5-day precipitation amount	Highest 5-day precipitation amount in consecutive days	mm
R95p	Very wet days	Precipitation due to very wet days (>95th percentile)	mm
NW	Number of wet days	Number of days for precipitation $\geq$ 1 mm	days
R30mm	Number of very heavy precipitation days	Number of days for precipitation $\geq$ 30 mm	days
CWD	Consecutive wet days	Maximum length of wet spell (RR $\geq$ 1 mm)	days
CDD	Consecutive dry days	Maximum length of dry spell (RR < 1 mm)	days

* RR is the daily rainfall ($\geq$1 mm) amount on a wet day. Note: Indices are calculated using daily data precipitation, computed at the annual and seasonal scales in the 1970–2013 period.

The extreme rainfall indices can be divided in three categories (Table 1): (1) the intensity indices describe the amount of maxima (or maximum) precipitation in one day (RX1day) and maximum accumulated precipitation in 5 consecutive days (RX5day), respectively. The very wet days (R95p) represents the daily amount of precipitation that surpasses the 95th percentile value. The RX1day, RX5day and R95p were used to describe floods and flash flood risks (e.g., [13,41]); (2) frequency indices: the R30mm index represents the number of heaviest precipitation days and indicates the seasonal/annual count of days when the daily rainfall is greater than or equal to 30 mm. The number of wet days (NW) counts the number of days with rainfall of $\geq$1 mm; and (3) duration indices: the maximum number of consecutive wet days (CWD) and consecutive dry days (CDD). The PRCPTOT index is the seasonal/annual total wet-day precipitation with daily rainfall greater than or equal to 1 mm. PRCPTOT does not necessarily have a direct relationship with the precipitation extremes but provides relevant information on the climatological aspects and in wet or dry periods [42].

Temporal trends of precipitation indices are examined by the Mann-Kendall (MK) trend test and calculated by the Sen slope estimator. The MK test [43,44] is a wildly applied nonparametric method to characterize trends of extreme precipitation indices [11,14,18,45], whereas the nonparametric statistical test developed by Sen (1968) [46] is used to estimate the trend magnitude. These methods are less sensitive to outliers than parametric statistics [11,45]. A more detailed description of these methods can be found in Yue et al. [47]. Trends are considered to be statistically significant at the 5% significance level.

2.4. Regional Anomalies of Extreme Precipitation, ENSO and Flood Events

Despite the small number of stations in the south (2°06′–3°0′ N) and a larger coverage in the center (3°0′–4°0′ N) and north (4°0′–05°2′ N), the stations are found to be reasonably well distributed over the basin. The study area was divided into three regions according to the latitude: northern, central and southern. It is important to note that 12%, 72 % and 15% of the basin's population are distributed in the north, central and south regions, respectively. We calculated the aggregate time series to evaluate the possible existence of large subregional asymmetries as well as exploring spatial coherence of natural variability in the precipitation extremes, ENSO, and floods events. This method has been widely used in the analysis of climate precipitation extremes [13,17,26,48]. The regional averaged anomaly series for each index was calculated using the following equation:

$$x_{r,t} = \sum_{i=1}^{n_t} (x_{i,t} - \overline{x}_i) / n_t \qquad (1)$$

where $x_{r,t}$ is the regional averaged index at year t; $x_{i,t}$ is the index i at year t; $\overline{x}_i$ is the 1970–2013 index mean at series and n_t is the number of stations with data in year t. The regional series are expressed in millimeters (PRCPTOT, RX1day, RX5day, and R95p) and days (R30mm, NW, CDD, and DWD).

The impacts of ENSO on precipitation-related indices are also analyzed. The ENSO regimes are defined by the Niño 3.4, in line with previous studies by Poveda et al., Ávila et al., Morán-Tejeda et al., Maldonado et al. and Vicente-Serrano et al. [49–53]. The Niño 3.4 (0–10S, 90W–80W) characterizes Eastern Central Tropical Pacific Sea Surface Temperature (SST) anomalies. The series of the ENSO index over the 1970–2013 period are extracted from the National Oceanic and Atmospheric Administration website (NOAA; https://www.esrl.noaa.gov/psd/data/climateindices/list/). To characterize the effect of SST of the Pacific Ocean on precipitation, Pearson's correlation test was performed between El Niño 3.4 index and the regional time series of the precipitation indices.

According to the U.S. Geological Survey, a flood is "an overflow of water onto lands that are used or usable by man and not normally covered by water. Floods have essential features: The inundation of land is temporary; the land is adjacent to and inundated by overflow from a river".

We explored the data from thirteen catastrophic floods, recorded since 1970, provided by the CVC and available in the "Hydrological Analysis of the Historical Flooding of the Cauca River" report [54], to analyze which is the pattern how extreme precipitation indices characterize the behavior of a flooding. The report is available at https://www.cvc.gov.co/. The CVC's report shows inundation maps (flooded area), dates and the maximum flow registered at the Victoria gauging station (Station ID 40; see the red point in Figure 1; further details in Appendix A). The flooding events occurring in the Cauca River after the onset of heavy rains are concentrated in flooded areas between 50 km^2–700 km^2 (Table A1).

3. Results and Discussion

3.1. Annual and Seasonal Extreme Precipitation Trends

Trends and percentages of stations have been calculated for eight precipitation indices on annual (Figure 2a–i) and seasonal scales (Figures A1–A5), over the study area during the 1970–2013 period.

The total annual precipitation in wet days (PRCPTOT) has experienced positive trends in the north (Figure 2a) and negative trends over the center (3°00′–04°0′ N) and south (2°06′–3°00′ N) of the basin. The seasonal analyses also demonstrate negative trends for the PRCPTOT index (Figure 3b–d). In MAM, 59% of stations show negative trends, with 59% in JJA and 69% in SON (Figures A2–A4). However, for DJF (Figure A1), positive trends predominated in 74% of the stations. In general, the PRCPTOT has decreased over the past four decades, in particular in the central region. These results suggest that precipitation is decreasing in particular for both rainy seasons (MAM and SON) and in the dry seasons (JJA). These results are in accordance with those obtained by Puertas et al. [27], who analyzed the rainfall trends at annual and seasonal timescales in the 1975–2006 period. Differences between the current study and Puertas et al. [27] appear in the seasonal analyses. Puertas et al. [27] found statistically significant positive trends for DJF and SON, whereas in our study, this is true only for DJF where the positive trends predominated in 74% of the stations. Differences between these studies may arise from the longer time series and time intervals used in our analyses. We have investigated 12 years more than Puertas et al. [27] which indeed modifies the characteristics of the trends.

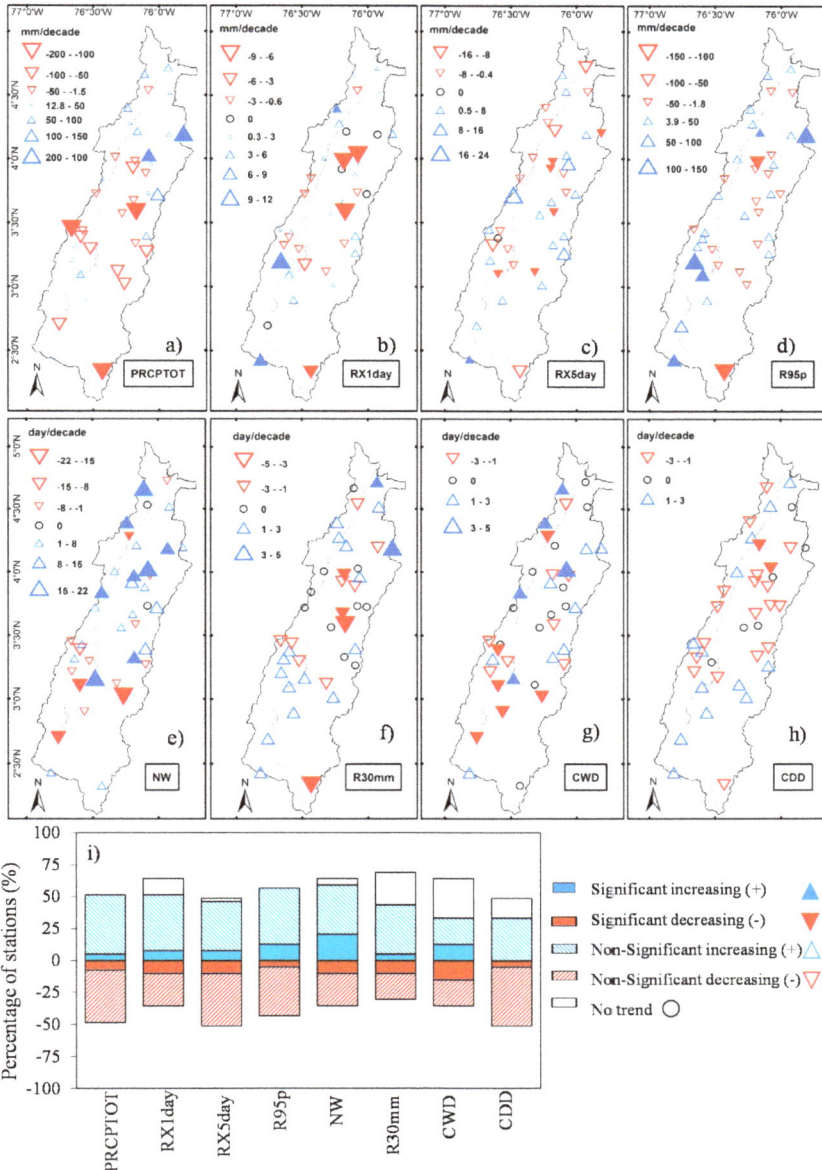

Figure 2. Spatial distribution of linear trends for annual precipitation indices: (**a**) PRCPTOT, (**b**) RX1day, (**c**) RX5day, (**d**) R95p, (**e**) NW, (**f**) R30mm, (**g**) CWD (**h**) CDD and (**i**) Percentage of stations with positive, negative and no change trends, out of the total stations examined over the Cauca River High Basin over the 1970–2013 period. Upward (downward) triangles refer to positive (negative) trends. Saturated triangles indicate trends significant at a 5% level and circles indicate no change trends.

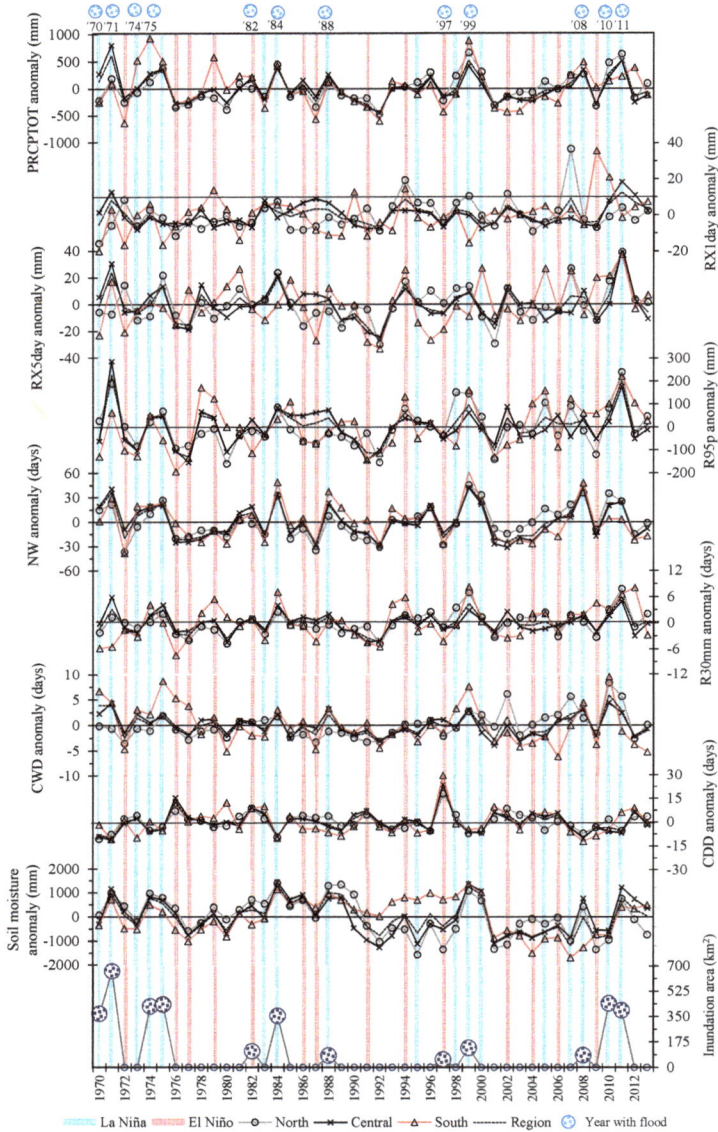

Figure 3. Temporal evolution of the anomalies for annual precipitation indices over north, central and south regions in terms of PRCPTOT, RX1day, RX5day and R95p, NW, R30mm, CDW and CDD. La Niña and El Niño occurrences are displayed by the vertical blue and red bars, respectively.

Celleri et al. [55] detected significant positive trends in rainfall over the Equatorial Andean for DJF and MAM, and negative tendency for JJA, from 1963 to 1993. Casimiro et al. [56] found similar seasonal trends over the Peruvian Andes for the 1965–2007 interval. In general, previous studies corroborate the generalized trend of increased precipitation for DJF (Figure A1), and decreased precipitation for JJA and SON results (Figures A3 and A4, respectively), which support our findings with the exception of MAM (Figure A2).

On the annual scale, the negative trend of PRCPTOT is consistent with the results of Skansi et al. [11], who determined local trends of climate extremes indices on the annual scale for South America in the period 1969–2009. Likewise, this result agrees with those obtained by Aguilar et al. [26], who analyzed extreme precipitation indices for Central America and Northern South America over the 1961–2003. The PRCPTOT delivered negative trends over stations located in the lower and middle zones of our study area (Figure 2a).

The annual maximum 1-day precipitation amount (RX1day), annual maximum 5-day precipitation amount (RX5day), the number of days for precipitation of $\geq$30 mm (R30mm), and the maximum number of consecutive days (CWD) do not show similar trend characteristics (negative or positive) on the annual scale (Figure 2). Nevertheless, RX1day and RX5day were characterized by significant negative trends over the center of the basin, and the same happens at the seasonal level, except in DJF (Figures A1–A4). It is important to note that R95p and NW also show 56% and 59% of stations with positive trends, respectively, especially with significant upward trends in the center and north of the basin. The trends of R95p in the central region indicate that although PRCPTOT has been decreasing, the maximum events of rain have become more intense, especially if these tendencies correspond to the same stations (19, 22, 24, 26 and 30). The NW delivers positive trends in DJF and MAM which increased by 69% and 54%, respectively. It is important to note that the R30mm has shown no change of rainfall in most stations during all seasons. Similarly, CWD demonstrates in DJF and JJA that 54% and 62% of the stations do not experience trends, respectively, with the exception of SON, with a general negative trend.

Additionally, RX1day, RX5day and R95p show positive trends, specifically in DJF and MAM (Figures A1 and A2). This is important because intense rainfall has been the main factor in the generation of historical flood of the Cauca River (Enciso, Carvajal, and Saldoval, 2016). Additionally, in these seasons, the maximum peaks in the flow of the Cauca River during floods have been presented (Table A1). In addition, these indices have statistically significant (p value < 0.05) positive trends in the northern region, except for R95p that presents this type of trends in the southwest region, particularly in DJF. The positive trends of RX1day, RX5day and R95p in the north and south of the basin, at annual and seasonal levels, are important because it is where the sub-basins are located that contributed to the maximum flow of the historical floods of the Cauca River.

Skansi et al. [11] and Aguilar et al. [26] show positive trends on the annual scale for PRCPTOT, RX1day, RX5day and R95p, which is in line with our results in the northern region (Figure 2). However, this is not true in all cases for the center and south, where for example the PRCPTOT decreases. Our results provide an update of the extreme precipitation indices for the region, which were previously determined by Cardona et al. [23] during 1982–2011 on the annual scale in two Andean sub-basins of Cauca River Basin, located in the southwest of the region. The results coincide for PRCPTOT and CWD with negative trends and for RX1day with a positive trend. Additionally, Cuartas et al. [57] shows indices of climatic extremes at a monthly level between 1998 and 2013 in the flat area of the Cauca River Basin, where PRCPTOT and RX5day had positive trends. Our results show decreasing trends for these indices in the same area. This can be explained by the difference in the evaluation period and the analysis scale.

3.2. Regional Anomalies of Extreme Precipitation, ENSO and Flood Events

In order to verify whether the extreme precipitation is related with ENSO flood events from 1970 to 2013, we evaluated the temporal evolution of annual precipitation indices and soil moisture, which are shown in Figure 3. The stronger events tend to occur during La Niña years (e.g., 1970–1971, 1999, and 2011). We have found that during La Niña years, the anomalies of the PRCPTOT, RX5day, R95p, R30mm and NW are positive and higher than those related to El Niño years (Figure 3; vertical blue and red bars). The disaster areas of the 1988, 1999 and 2008 were years dictated by precipitation anomalies related to La Niña, and show however the lowest recorded (less than 134 km²). This has occurred, because a year before the flood, the contribution of soil moisture was low and the non-homogenous behavior of total precipitation and extreme events in comparison with the floods of greater disaster area. For example, in 1987 and 1998, there were negative anomalies in all indices in the southern region and in soil moisture in the northern region. Equally, in 2007, soil moisture anomalies were negative in all regions.

It is interesting to note that the CDD play an important role in the magnitude of the floods due to its implication on soil moisture before the flood occurrence. At the annual scale, the CDD index shows negative trends in the central and northern regions (Figure 3). The CDD present a positive trend in SON in 64% of the stations (Appendix A). These results can be associated to the minimum occurrence of rainfall due to a decrease in the PRCPTOT, RX1day and NW in this season. It should be emphasized that, currently, this is the High Basin of the Cauca River rainy season (SON). Furthermore, the decreasing in precipitation (PRCPTOT) in SON can be changes in DJF patterns rainfall, while the unbalanced spatial-temporal distribution of rainfall can play an important role in the formation of flood disasters, mainly in DJF and MAM that present the maximum peaks in the flow (Table A1).

The CDD reflects positive anomalies between extreme precipitation indices and years with floods (Figure 3). Furthermore, the magnitude of anomalies of the longest dry periods are associated with the El Niño events. For instance, El Niño conditions are related to reduced precipitation in the High Basin of the Cauca River. In 1982 and 1997 (Figure 3), the excess precipitation was not sufficient to cause major floods, because the soil could not store most of the rain since it was particularly dry. Two of the strongest El Niños in the last century occurred in these years [58]. In the 12 months preceding floods (1982, and 1997), positive anomalies were identified for two extreme precipitation indices (PRCPTOT and NW). This may have influenced a gradual increase in soil moisture, reducing the infiltration basin capacity and thus leading to flooding.

Avila et al. [13] investigated the daily extreme precipitation events and their link to the number of flash floods in Southeastern Brazil and found statistically significant positive correlation coefficients between flash flood and extreme precipitation for RX1day ($r > 0.49$) and RX5day ($r > 0.39$). Using similar indices, Wu and Huang [41] demonstrated that maximum RX1day and RX5day values coincide with the years when floods occur. However, in the current investigation, the RX1day was the least significant index for indicating floods in the High Basin of the Cauca River, as shown in Figure 3.

Aiming to further explore the climatological evolution of floods on the study area, we examined the large-scale forcing of the ENSO index (Niño 3.4), and its effects on extreme rainfall indices on a monthly scale. Considering the delayed precipitation in response to the ENSO index, we used different time lags (1 lag = 1 month). The maximum coefficients of the Niño 3.4 (Figure 4) and precipitation-related index have been found for the time lag of 2–3 months. The PRCPTOT, RX1day, RX5day, R95p, NW, R30mm and CWD demonstrated significant negative correlations for all three areas. Central and south regions reflects the highest values for 2-lag in the PRCPTOT, RX5day, NW and CWD ($r \leq -0.45$), whereas the lowest values were generally found in RX1day, R95p, and R30 mm (≥ -0.40) over the north region. On the other hand, the CDD reflect maximum significant positive correlations in central and south areas for the 3-lag with values between 0.41 and 0.44.

The Niño 3.4 index delivered negative correlations with seven precipitation indices, which indicates that they are out of phase in this region and have an indirect relationship with extreme precipitation events. Indeed, negative values of the Equatorial Pacific SST, on the other hand, increase rainfall events over the basin. According to Morán-Tejada et al. and Vicente-Serrano et al. [51,52], the index of 3.4 explains most of the rainfall in the equatorial mountainous region (the Andes). This is supported by the results discussed here, since the correlation coefficients are statistically significant for all extreme precipitation indices related to dry as well as very humid periods.

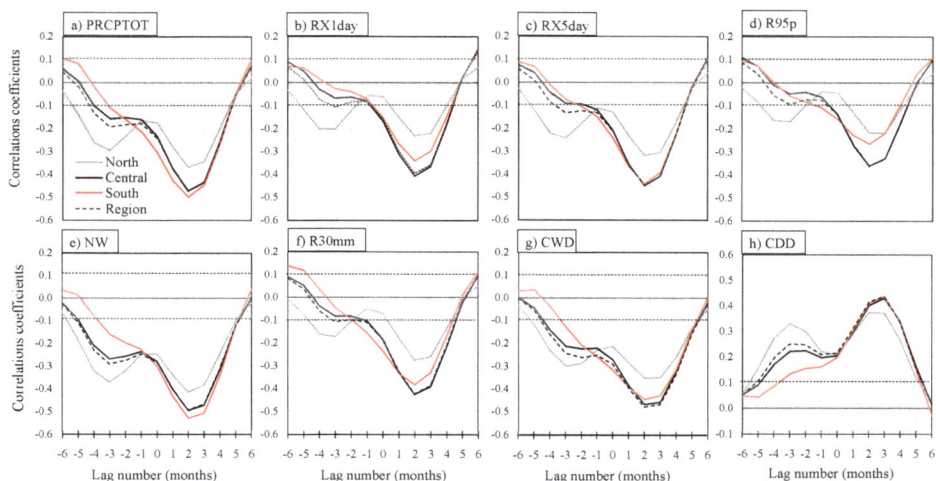

Figure 4. Pearson's correlation coefficients between Niño 3.4 and extreme monthly precipitation indices: (**a**) PRCPTOT, (**b**) RX1day, (**c**) RX5day, (**d**) R95p, (**e**) NW, (**f**) R30mm, (**g**) CWD and (**h**) CDD. The values in the gray zone indicate statistically insignificant correlations (at *p*-value = 0.05). The Y-axis refers to the correlation coefficient values and the X-axis refers to the lag number (1 lag = 1 month).

In general terms, the north, central and south regions have similar temporal evolutions of extreme precipitation and highly correlated behavior within the index of 3.4 (Figure 4). In particular, central, southern and regional (region area: all stations) time series have analogues values of the anomalies and correlations coefficients (Figures 3 and 4). These results may have important implication in terms of preventing floods and being used in forecasting systems in key areas of the basin. In addition, the CVC and Enciso et al. [6,54] found that the highest frequency of flooding is concentrated in the center of the High Basin of the Cauca River (3°0′–4°0′ N).

We explored the data from eleven catastrophic floods between 1970 and 2013 to investigate the ENSO influence (see Appendix A). Figure 5 only presents results for the central region because the majority of population is placed in this region. Further details of index anomalies over northern and southern regions are found in Appendix A (Figure A5). The most important factor to cause floods are the heavy precipitations associated with La Niña conditions, which generates the probability of elevated discharges of the Cauca River's and its tributaries rivers [6]. For this reason, the CDD index, which represents the maximum length of dry spell (dry events), is not included in this analysis.

Figure 5 shows the monthly evolution of seven precipitation extremes and the Niño 3.4 index before and after the flood event. Based on the 95% confidence interval (hatched area), it was observed that these indices are positive or close to zero for 3–11 months (phase 1) prior to the month with the peak flood discharge. At 0–2 months prior to the peak discharge (phase 2), during this phase, all extreme precipitation indices exhibit the maximum positive anomalies. This suggests that heavy precipitation (RX1day, R95p, and R30mm) and rainstorms (RX5day, NW, and CWD) contribute to the evolution and onset of the flooding.

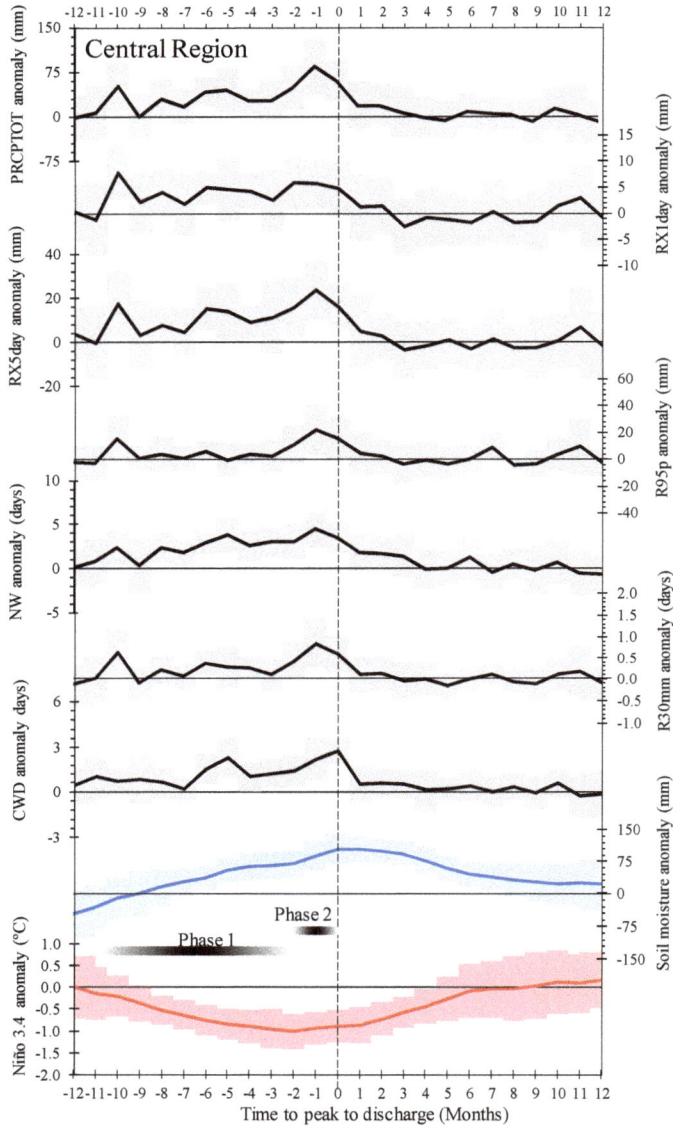

Figure 5. Index anomalies over the central region for PRCPTOT, RX1day, RX5day, R95, NW, R30mm, CDW, and soil moisture Niño 3.4 for floods that occurred under La Niña conditions (n = 11) in relation to the time of the month with a peak discharge. The mean (line) and the 95% confidence interval (shaded bands) are shown for each index. The vertical dotted line refers to the peak of discharge; horizontal lines denote a zero anomaly for each index calculated.

The influence of climate features on extreme hydro-meteorological processes (such as floods) depends on the capability of understanding the role of abnormal pattern of atmospheric pressure, SST and soil moisture patterns [59,60]. An attempt is done here by evaluation of the synoptic hydro-meteorological conditions by constructing composite anomalies maps during the 1970–2013 period (Figure 6), according to Welhouse et al. and Munoz and Dee [59,61]. With it being important to identify whether climate regimes, occurrence in far-off regions of interest may lead to floods in the Cauca River.

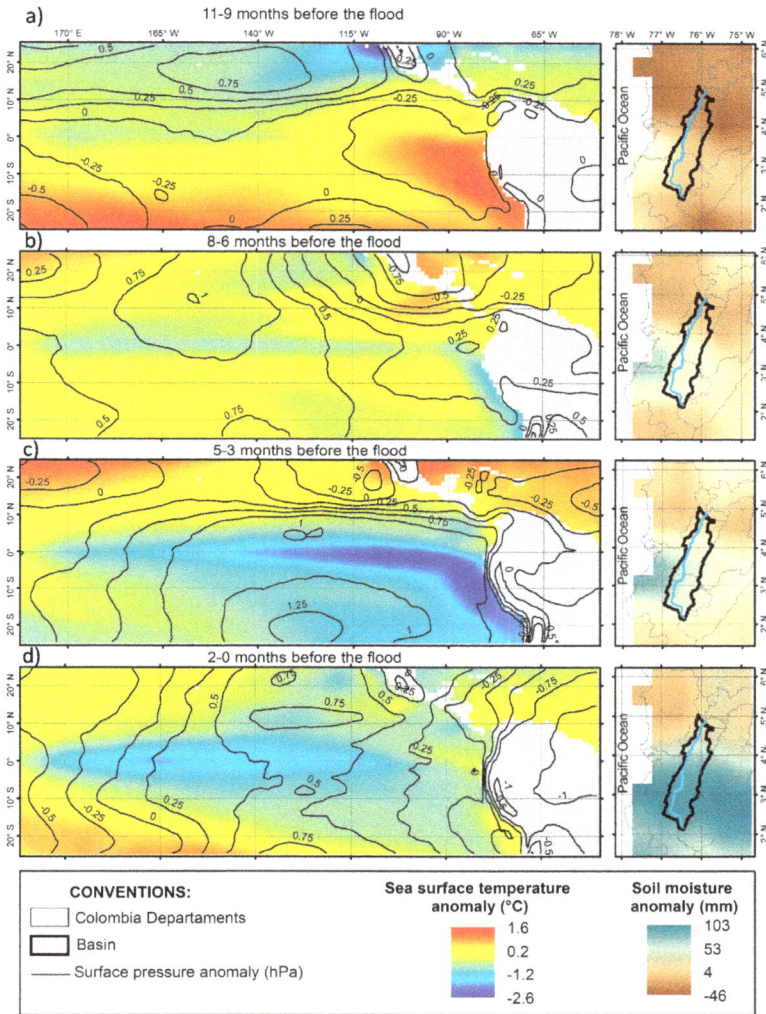

Figure 6. Composited anomalies of surface temperature over the Pacific sea surface, sea level pressure (contours), and soil moisture over the High Basin of the Cauca River: (**a**) 11–9 months, (**b**) 8–6 months, (**c**) 5–3 months, and (**d**) 2–0 months prior to floods events (n = 11).

Henceforth, the monthly pressure data [62] at $1° \times 1°$ spatial resolution from the Princeton University Terrestrial Hydrology Research Group (http://hydrology.princeton.edu/data.pgf.php) is used. The Centennial In Situ Observation-Based Estimates (COBE SST2) provides reliable monthly gridded datasets of global SST [63]. The COBE-SST2 ($1° \times 1°$) dataset was downloaded from NOAA's official website (https://www.esrl.noaa.gov/psd/data/gridded/data.cobe2.html). The Climate Prediction Center (CPC; https://www.esrl.noaa.gov/psd/data/gridded/data.cpcsoil.html) provides a monthly soil moisture dataset with a spatial resolution of $0.5° \times 0.5°$ [64].

Figure 6 displays the composited anomalies of surface temperature, sea level pressure, and soil moisture over the over the tropical Pacific Ocean and Western South America. We calculated monthly subregional anomalies in the 12 months prior to the floods and calculated the mean and 95% confidence intervals for eight extreme precipitation climate indices.

Based on Figure 6, it is clear that by about 11–9 months prior to the floods, El Niño-like conditions dominate the equatorial Pacific, with much higher SST anomalies nearby the South America west coast. This pattern induces changes in surface pressure and negative SM, further leading to dry conditions in the Cauca basin (Figure 6a). The 8–6 month anomalies show a different pattern with lower SST anomalies and increased surface pressure along the South America margin. This pattern is intensified between 5 and 3 months before the flood, indicating the development of La Niña conditions. During this interval, remarkable changes are initiated, in particular in the central-south part of the Cauca River Basin which experiences a drop in surface pressure, and increased precipitation, which result in positive values of soil moisture (Figure 6b,c). Two–Zero months before the flood event, negative anomalies of surface pressure dominate the north part of South America, in phase with substantial changes in soil water availability.

Evaluation of extreme precipitation indices delivered two phases of climatological evolution. The first phase is characterized by positive and high-magnitude extreme precipitation indices before the flooding. The second phase is characterized by the increase of soil moisture anomalies over the basin and saturated water storage over the period before the flood events.

These results imply that ENSO plays an important role in the flood area and magnitude, due to changes in precipitation extremes in a given year. Moreover, the results indicated that the highest negative values of the NIÑO 3.4 index occurs 3 months before the peak discharge (Figures 5 and 6). These results are fundamental in addressing the challenge of forecasting floods, since the changes in the Pacific SST are seen 3 months before the basin inundation. Similar results were found by Munoz and Dee [59] in the Lower Mississippi River floods (USA).

4. Conclusions

This study is the result of substantial data collection efforts, and it significantly builds upon previous studies [23,27,61] by providing updated spatial and temporal coverage in the years leading up to 2013. The results in the study area show that the annual total precipitation on wet days (PRCPTOT) has decreased over the central and southern regions. Likewise, more than 59% of the stations showed decreasing trends in MAM, JJA and SON. However, an increase in total wet-days and the intensity of the extreme events (RX1day and RX5day) can be observed, as shown by a general increase in the events with days with rainfall of $\geq$1 mm (NW), for the DJF season, which is the dry interval.

In SON, decreased intensity, frequency and accumulated rainfall were observed (RX5day, NW and PRCPTOT), and this has generated the increase of CDD. It should be emphasized that, currently, this is the Cauca River rainy season.

Equally, JJA, which is a dry season, shows a decrease of accumulated and intensity rain and a decreased frequency of the days with rainfall, especially in the central region. Although it has not been our focus in the present study, the region has experienced a systematic increase in the number of vegetation fires, particularly in JJA [24,65], with substantial economic losses. The decreasing trends in precipitation indices pose a potential threat for the development of more erratic fires, due to the vulnerability of the region.

We found that the region El Niño 3 + 4 shows significant correlation with the extreme precipitation indices. The SST field, principally at the time lag of 2–3 months, are highlighted, evidencing the strong relationship that ENSO has over the hydro-meteorology of Colombia. Consistent with these results, Córdoba-Machado et al [66] displayed through a Principal Component Analysis that a pattern based on global precipitation, mean temperature over land and SST of the ENSO regions (e.g. Niño 3, and Niño 4) is strongly linked to the hydro-meteorology of country. Nonetheless, the relationship between extreme precipitation indices and the Pacific SST is not considered in those studies, as has been evaluated in the present study. The understanding of precipitation extremes is necessary to undertake coordinated projects in order to alleviate the response of different productive sectors and reduce the threat to local populations. The occurrence of historical floods causes significant economic losses and strong stress on hydric resources. Changes in land use in the region have affected the natural capacity of the basin during extreme climate events.

In general terms, our results will benefit current and future studies on climate resilience for hydrological hazard forecasts on a regional scale, since the findings indicate that flooding events may be predicted by using a series of precipitation extreme indices, soil moisture and ENSO.

Author Contributions: Conceptualization, A.A. Y.C.E. and F.J.; Data curation, A.A. and F.C.G.; Formal analysis, A.A. and F.C.G.; Methodology, A.A. F.C.G. and F.J.; Project administration, F.C.G. and Y.C.E.; Writing – review & editing, A.A., F.C.G., Y.C.E. and F.J.

Funding: This research received no external funding

Acknowledgments: We are thankful to the research groups IREHISA and Gesp-Group of Epidemiology and Population Health from the Universidad del Valle-Colombia. This work was supported by the Universidade Federal de Viçosa (Brazil), the Minas Gerais Research Foundation (FAPEMIG) and to the Coordination for the Improvement of Higher Education Personnel (CAPES). We also appreciate the Corporación Autónoma Regional del Valle del Cauca (CVC) for sponsoring the project: "Constructing a Conceptual Model for Recovering the Conservation Corridor and the Sustainable Use of Cauca River System in Its High Valley". We show gratitude to the Administrative Department of Science, Technology, and Innovation (COLCIENCIAS) in the framework of the program for Young Researchers and Innovators open call 2014 and the project "Regional analysis of the droughts related to climate variability for the implementation of adaptation strategies in agricultural production systems in Valle del Cauca".

Conflicts of Interest: The authors declare no conflicts of interest.

Appendix A

Table A1. Floods recorded at the gauging station at Victoria (Station ID 40; see the red point in Figure 1) in relation to ENSO events and flooded area in High Basin of the Cauca River.

Peak Stage Date (Day/Month/Year)			Flow (m^3/s)	Flooded Area (km^2)	Corresponding La Niña [1]			
					Start Date		End Date	
18	November	1970	1117	367.7	June	1970	December	1971
5	April	1971	1222	663.8	June	1970	December	1971
24	March	1974	1219	419.1	May	1973	June	1974
30	December	1975	1317	431.2	September	1974	February	1975
18	April	1982 [2]	972	110.0		–		
8	November	1984	1214	353.9	September	1984	May	1985
8	December	1988	1148	128.8	April	1988	April	1989
30	January	1997 [2]	993	54.0		–		
2	March	1999	1166	133.7	June	1998	February	2001
30	November	2008	1055	82.9	July	2007	May	2008
4	December	2010	1202	440.2	July	2010	March	2011
28	April	2011 [3]	1188	393.2	July	2010	March	2011
16	December		1205		July	2011	January	2012

[1] Refers to a La Niña event that ended within a year prior to the major flood stage. [2] Refers to an El Niño event, according to the Oceanic Niño Index (ONI) and its definition in Jan 2017. [3] Information available only for the annual total area affected by the floods.

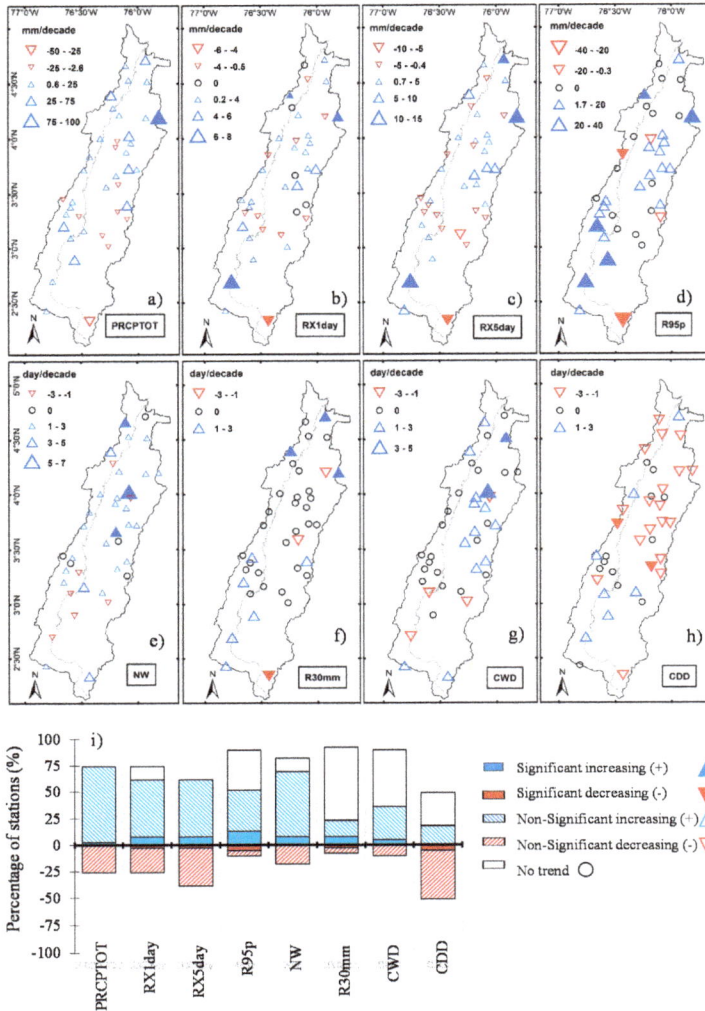

Figure A1. Spatial distribution of decadal trends for December–January–February (DJF). (**a**) PRCPTOT, (**b**) RX1day, (**c**) RX5day, (**d**) R95p, (**e**) NW, (**f**) R30mm, (**g**) CWD (**h**) CDD and (**i**) Percentage of stations with positive, negative and no change trends, out of the total stations examined over the Cauca River High Basin over the 1970–2013 period. Upward (downward) triangles refer to positive (negative) trends. Saturated triangles indicate trends significant at the 5% level and circles indicate no change trends.

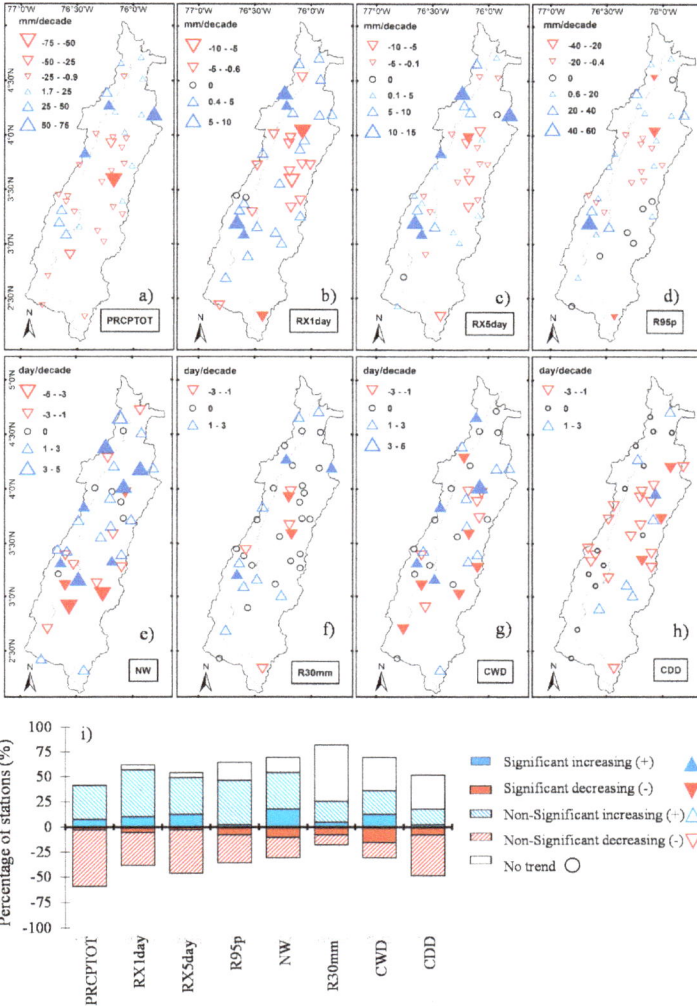

Figure A2. Spatial distribution of decadal trends for March–April–May (MAM). (**a**) PRCPTOT, (**b**) RX1day, (**c**) RX5day, (**d**) R95p, (**e**) NW, (**f**) R30mm, (**g**) CWD (**h**) CDD and (**i**) Percentage of stations with positive, negative and no change trends, out of the total stations examined over the Cauca River High Basin over the 1970–2013 period. Upward (downward) triangles refer to positive (negative) trends. Saturated triangles indicate trends significant at the 5% level and circles indicate no change trends.

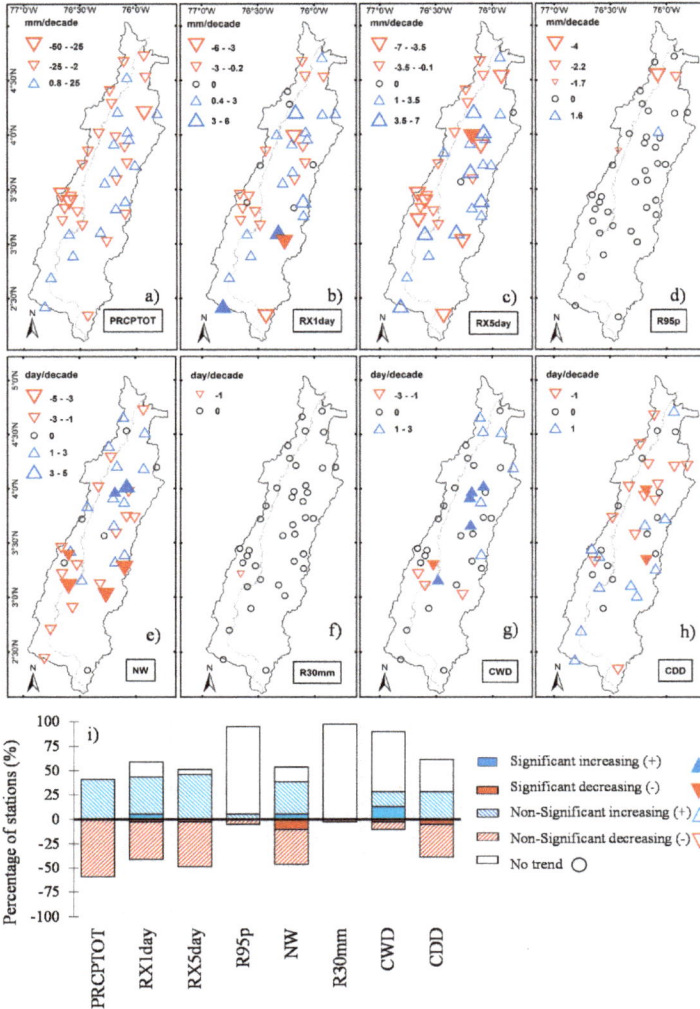

Figure A3. Spatial distribution of decadal trends for June–July–August (JJA). (**a**) PRCPTOT, (**b**) RX1day, (**c**) RX5day, (**d**) R95p, (**e**) NW, (**f**) R30mm, (**g**) CWD (**h**), CDD and (**i**) Percentage of stations with positive, negative and no change trends, out of the total stations examined over the Cauca River High Basin over the 1970–2013 period. Upward (downward) triangles refer to positive (negative) trends. Saturated triangles indicate trends significant at the 5% level and circles indicate no change trends.

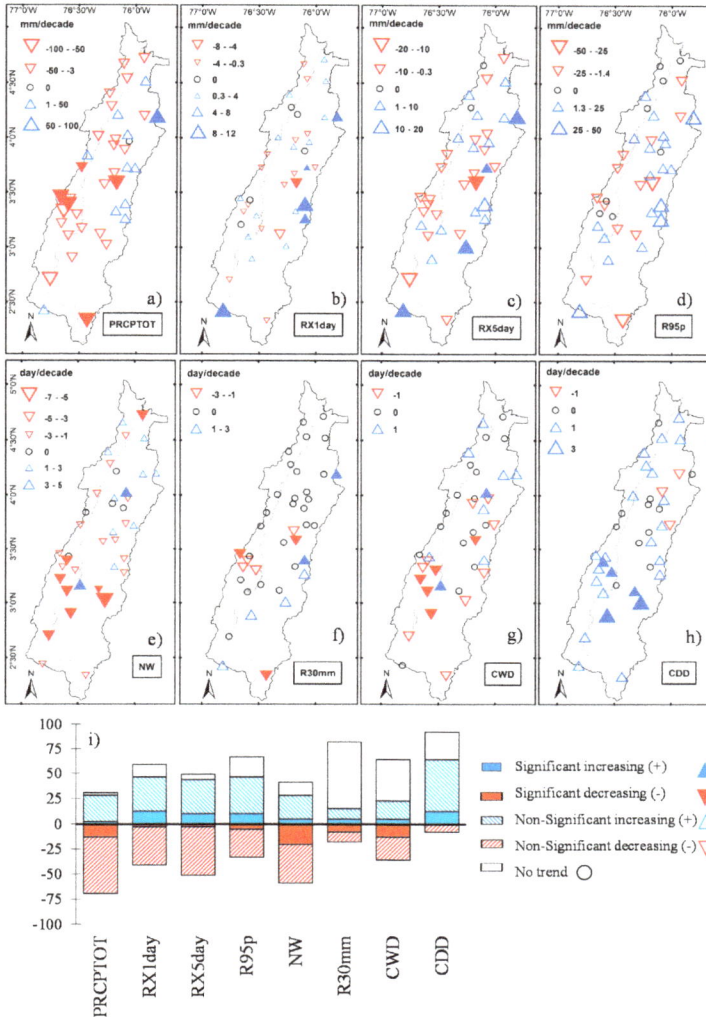

Figure A4. Spatial distribution of decadal trends for September–October–November (SON). (**a**) PRCPTOT, (**b**) RX1day, (**c**) RX5day, (**d**) R95p, (**e**) NW, (**f**) R30mm, (**g**) CWD, (**h**) CDD and (**i**) Percentage of stations with positive, negative and no change trends, out of the total stations examined over the Cauca River High Basin over the 1970–2013 period. Upward (downward) triangles represent positive (negative) trends. Saturated triangles indicate trends significant at the 5% level and circles indicate no change trends.

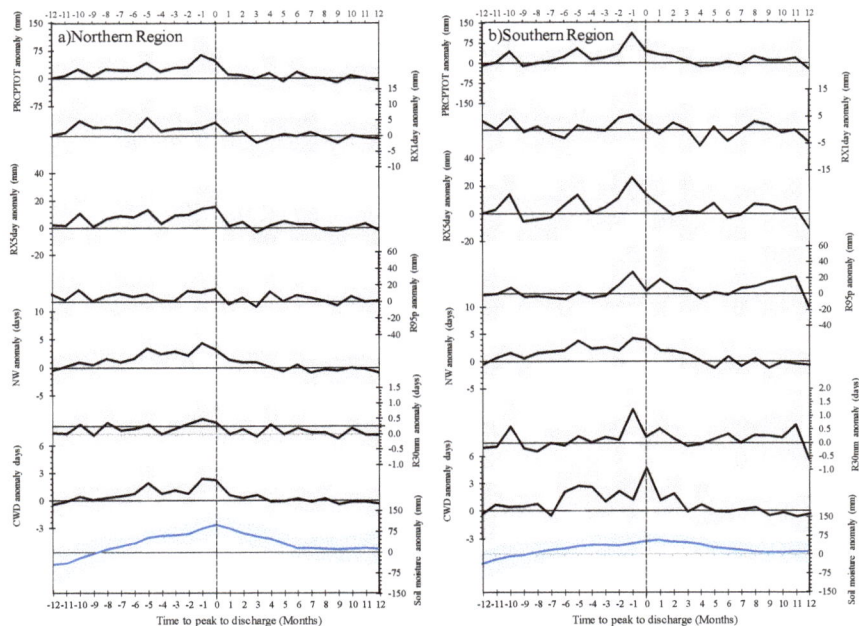

Figure A5. Index anomalies over northern (**a**) and southern (**b**) regions for PRCPTOT, RX1day, RX5day, R95, NW, R30mm, CDW, and soil moisture Niño 3.4 for floods that occurred under La Niña conditions (n = 11) in relation to the time of the month with a peak discharge. The mean (line) and 95% confidence interval (shaded bands) are shown for each index. The vertical dotted lines refer to the peak of discharge; horizontal lines denote a zero anomaly for each index calculated.

References

1. Smyth, C.G.; Royle, S.A. Urban landslide hazards: Incidence and causative factors in Niteroi, Rio de Janeiro state, Brazil. *Appl. Geogr.* **2000**, *20*, 95–117. [CrossRef]
2. Poveda, G. La Hidroclimatología de Colombia: Una Síntesis Desde la Escala Inter-Decadal Hasta la Escala Diurna. *Rev. la Acad. Colomb. Ciencias* **2004**, *XXVIII*, 201–222.
3. Poveda, G.; Álvarez, D.M.; Rueda, Ó.A. Hydro-climatic variability over the Andes of Colombia associated with ENSO: A review of climatic processes and their impact on one of the Earth's most important biodiversity hotspots. *Clim. Dyn.* **2011**, *36*, 2233–2249. [CrossRef]
4. World Bank. El cambio climático y los fenómenos de El Niño y La Niña. In *Análisis de la gestión del riesgo de desastres en Colombia: Un aporte para la construcción de políticas públicas*; Campos, G., Holm-Nielsen, N., Díaz, C., Rubiano, D., Costa, C., Ramírez, F., Dickson, E., Eds.; World Bank: Bogotá, Colombia, 2012; Volume 2, pp. 50–51.
5. Hoyos, I.; Baquero-Bernal, A.; Jacob, D.; Rodríguez, B.A. Variability of extreme events in the Colombian Pacific and Caribbean catchment basins. *Clim. Dyn.* **2013**, *40*, 1985–2003. [CrossRef]
6. Enciso, A.; Carvajal, Y.; Sandoval, M. Hydrological analysis of historical floods in the upper valley of Cauca river (in Spanish). *Ing. Y Compet.* **2016**, *57*, 46–57.
7. Ragettli, S.; Immerzeel, W.W.; Pellicciotti, F. Contrasting climate change impact on river flows from high-altitude catchments in the Himalayan and Andes Mountains. *Proc. Natl. Acad. Sci. USA* **2016**, *113*, 9222–9227. [CrossRef] [PubMed]
8. Campozano, L.; Vázquez-Patiño, A.; Tenelanda, D.; Feyen, J.; Samaniego, E.; Sánchez, E. Evaluating extreme climate indices from CMIP3&5 global climate models and reanalysis data sets: a case study for present climate in the Andes of Ecuador. *Int. J. Climatol.* **2017**, *37*, 363–379.

9. Sedano, K.; Carvajal, Y.; Ávila, Á. Analysis of the aspects which increase the risk of floods in Colombia (in Spanish). *Luna Azul* **2013**, *37*, 219–238.

10. Hoyos, N.; Escobar, J.; Restrepo, J.; Arango, A.; Ortiz, J. Impact of the 2010-2011 La Niña phenomenon in Colombia, South America: The human toll of an extreme weather event. *Appl. Geogr.* **2013**, *39*, 16–25. [CrossRef]

11. Skansi, M.; Brunet, M.; Sigró, J.; Aguilar, E.; Arevalo, G.; Bentancur, O.; Castellón, G.; Correa, A.; Jácome, H.; Malheiros, R.; et al. Warming and wetting signals emerging from analysis of changes in climate extreme indices over South America. *Glob. Planet. Change* **2013**, *100*, 295–307. [CrossRef]

12. Mallakpour, I.; Villarini, G. The changing nature of flooding across the central United States. *Nat. Clim. Chang.* **2015**, *5*, 250–254. [CrossRef]

13. Ávila, A.; Justino, F.; Wilson, A.; Bromwich, D.; Amorim, M. Recent precipitation trends, flash floods and landslides in southern Brazil. *Environ. Res. Lett.* **2016**, *11*, 1–13. [CrossRef]

14. Donat, M.G.; Lowry, A.L.; Alexander, L.V.; O'Gorman, P.A.; Maher, N. More extreme precipitation in the world's dry and wet regions. *Nat. Clim. Chang.* **2016**, *6*, 508–513. [CrossRef]

15. Debortoli, N.; Camarinha, P.; Marengo, J.; Rodrigues, R. An index of Brazil's vulnerability to expected increases in natural flash flooding and landslide disasters in the context of climate change. *Nat. Hazards* **2017**, *86*, 557–582. [CrossRef]

16. Marengo, J.; Valverde, M.; Obregon, G. Observed and projected changes in rainfall extremes in the Metropolitan Area of São Paulo. *Clim. Res.* **2013**, *57*, 61–72. [CrossRef]

17. Duan, W.; He, B.; Takara, K.; Luo, P.; Hu, M.; Alias, N.E.; Nover, D. Changes of precipitation amounts and extremes over Japan between 1901 and 2012 and their connection to climate indices. *Clim. Dyn.* **2015**, *45*, 2273–2292. [CrossRef]

18. Wang, R.; Li, C. Spatiotemporal analysis of precipitation trends during 1961–2010 in Hubei province, central China. *Theor. Appl. Climatol.* **2015**, *124*, 385–389. [CrossRef]

19. Tanoue, M.; Hirabayashi, Y.; Ikeuchi, H. Global-scale river flood vulnerability in the last 50 years. *Sci. Rep.* **2016**, *6*, 36021. [CrossRef] [PubMed]

20. Chadwick, R.; Good, P.; Martin, G.; Rowell, D. Large rainfall changes consistently projected over substantial areas of tropical land. *Nat. Clim. Chang.* **2016**, *6*, 177–181. [CrossRef]

21. Fischer, E.M.; Knutti, R. Observed heavy precipitation increase confirms theory and early models. *Nat. Clim. Chang.* **2016**, *6*, 986–991. [CrossRef]

22. Zhou, X.; Bai, Z.; Yang, Y. Linking trends in urban extreme rainfall to urban flooding in China. *Int. J. Climatol.* **2017**, *37*, 4586–4593. [CrossRef]

23. Cardona, F.; Ávila, Á.; Carvajal, Y.; Jiménez, H. Trends into rainfall time series of two Andes basins of Valle del Cauca (in Spanish). *Tecnológicas* **2014**, *17*, 85–95.

24. Armenteras, D.; Rodríguez, N.; Retana, J.; Morales, M. Understanding deforestation in montane and lowland forests of the Colombian Andes. *Reg. Environ. Chang.* **2011**, *11*, 693–705. [CrossRef]

25. Restrepo, J.; Kettner, A.; Syvitski, J. Recent deforestation causes rapid increase in river sediment load in the Colombian Andes. *Antrhropocene* **2015**, *10*, 13–28. [CrossRef]

26. Aguilar, E.; Peterson, T.C.; Obando, P.R.; Frutos, R.; Retana, J.A.; Solera, M.; Soley, J.; García, I.G.; Araujo, R.M.; Santos, A.R.; et al. Changes in precipitation and temperature extremes in Central America and northern South America, 1961–2003. *J. Geophys. Res.* **2005**, *110*, 1–15. [CrossRef]

27. Puertas, O.; Carvajal, Y.; Quintero, M. Study of monthly rainfall trends in the upper and middle Cauca River Basin, Colombia. *Dyna* **2011**, *169*, 112–120.

28. *El Río Cauca en su valle alto*; CVC: Santiago de Cali, Colombia, 2007; ISBN 9789588332109.

29. Pérez, M.A.; Peña, M.R.; Alvarez, P. Agro Industria Cañera y uso del agua: análisis crítico en el contexto de la política de agrocombustibles en Colombia. *Ambient. Soc.* **2011**, *XIV*, 153–178.

30. Ávila, Á.; Carvajal, Y. Agrofuels and Food Sovereignty in Colombia (in Spanish). *Cuad. Geogr.* **2015**, *24*, 43–60.

31. Berrío, J.C.; Hooghiemstra, H.; Marchant, R.; Rangel, O. Late-glacial and Holocene history of the dry forest area in the south. *J. Quat. Sci.* **2002**, *17*, 667–682. [CrossRef]

32. Poveda, G.; Waylen, P.R.; Pulwarty, R.S. Annual and inter-annual variability of the present climate in northern South America and southern Mesoamerica. *Palaeogeogr. Palaeoclimatol. Palaeoecol.* **2006**, *234*, 3–27. [CrossRef]

33. Jaramillo, L.; Poveda, G.; Mejía, J.F. Mesoscale Convective Systems and other Precipitation Features over the Tropical Americas and Surrounding Seas as seen by TRMM. *Int. J. Climatol.* **2000**, *37*, 380–397. [CrossRef]

34. Poveda, G.; Mesa, O.J. Feedbacks between hydrological processes in tropical South America and large-scale ocean-atmospheric phenomena. *J. Clim.* **1997**, *10*, 2690–2702. [CrossRef]

35. Poveda, G.; Salazar, L.F. Annual and interannual (ENSO) variability of spatial scaling properties of a vegetation index (NDVI) in Amazonia. *Remote Sens. Environ.* **2004**, *93*, 391–401. [CrossRef]

36. Poveda, G.; Mesa, O.J.; Salazar, L.F.; Arias, P.A.; Moreno, H.A.; Vieira, S.C.; Agudelo, P.A.; Toro, V.G.; Alvarez, J.F. The Diurnal Cycle of Precipitation in the Tropical Andes of Colombia. *Mon. Weather Rev.* **2005**, *133*, 228–241. [CrossRef]

37. Wang, X.; Feng, Y. *RHtestV3 User Manual, Climate Research Division, Atmospheric Science and Technology Directorate*; Science and Technology Branch, Environment Canada: Downsvies, ON, Canada, 2009; p. 29.

38. Zhang, X.; Yang, F. *RClimDex_ User Manual*; Climate Research Branch Environment Canada: Toronto, ON, Canada, 2004; p. 22.

39. Zhang, X.; Alexander, L.; Hegerl, G.; Jones, P.; Tank, A.; Peterson, T.; Trewin, B.; Zwiers, F. Indices for monitoring changes in extremes based on daily temperature and precipitation data. *Wiley Interdiscip. Rev. Clim. Chang.* **2011**, *2*, 851–870. [CrossRef]

40. Santos, M.; Fragoso, M.; Santos, J. Regionalization and susceptibility assessment to daily precipitation extremes in mainland Portugal. *Appl. Geogr.* **2017**, *86*, 128–138. [CrossRef]

41. Wu, C.; Huang, G. Changes in heavy precipitation and floods in the upstream of the Beijiang River basin, South China. *Int. J. Climatol.* **2015**, *35*, 2978–2992. [CrossRef]

42. Sillmann, J.; Kharin, V.; Zwiers, W.; Zhang, X.; Bronaugh, D. Climate extremes indices in the CMIP5 multimodel ensemble: Part 2. Future climate projections. *J. Geophys. Res. Atmos.* **2013**, *118*, 2473–2493. [CrossRef]

43. Kendall, M. *Rank Correlation Methods*, 4th ed.; Charles Griffin: London, UK, 1990; p. 160. ISBN 0852641990.

44. Mann, H.B. Nonparametric tests against trend. *Econometrica* **1945**, *13*, 245–259. [CrossRef]

45. Liu, M.; Xu, X.; Sun, A.; Wang, K.; Liu, W.; Zhang, X. Is southwestern China experiencing more frequent precipitation extremes? *Environ. Res. Lett.* **2014**, *9*, 1–15. [CrossRef]

46. Sen, P. Estimates of the Regression Coefficient Based on Kendall's Tau. *J. Am. Stat. Assoc.* **1968**, *63*, 1379–1389. [CrossRef]

47. Yue, S.; Pilon, P.; Cavadias, G. Power of the Mann–Kendall and Spearman's rho tests for detecting monotonic trends in hydrological series. *J. Hydrol.* **2002**, *259*, 254–271. [CrossRef]

48. New, M.; Hewitson, B.; Stephenson, D.; Tsiga, A.; Kruger, A.; Manhique, A.; Gomez, B.; Coelho, C.; Masisi, D.; Kululanga, E.; et al. Evidence of trends in daily climate extremes over southern and west Africa. *J. Geophys. Res. Atmos.* **2006**, *111*, 1–11. [CrossRef]

49. Poveda, G.; Velez, J.; Mesa, O.; Hoyos, C.; Mejía, F.; Barco, O.; Correa, P. Influencia de fenómenos macroclimáticos sobre el ciclo anual de la hidrología colombiana: cuantificación lineal, no lineal y percentiles probabilísticos. *Meteorol. Colomb.* **2002**, *6*, 121–130.

50. Ávila, Á.; Carvajal, Y.; Gutiérrez, S. El Niño and La Niña analysis influence in the monthly water supply at Cali River basin (in Spanish). *Tecnura* **2014**, *18*, 120–133.

51. Morán-Tejeda, E.; Bazo, J.; López-Moreno, J.I.; Aguilar, E.; Azorín-Molina, C.; Sanchez-Lorenzo, A.; Martínez, R.; Nieto, J.J.; Mejía, R.; Martín-Hernández, N.; et al. Climate trends and variability in Ecuador (1966-2011). *Int. J. Climatol.* **2015**, *11*, 3839–3855. [CrossRef]

52. Vicente-Serrano, S.; Aguilar, E.; Martínez, R.; Martín-Hernández, N.; Azorin-Molina, C.; Sanchez-Lorenzo, A.; El Kenawy, A.; Tomás-Burguera, M.; Moran-Tejeda, E.; López-Moreno, J.I.; et al. The complex influence of ENSO on droughts in Ecuador. *Clim. Dyn.* **2016**, *48*, 405–427. [CrossRef]

53. Maldonado, T.; Rutgersson, A.; Alfaro, E.; Amador, J.; Claremar, B. Interannual variability of the midsummer drought in Central America and the connection with sea surface temperatures. *Adv. Geosci.* **2016**, *42*, 35–50. [CrossRef]

54. Grupo de Hi; dráulica Fluvial y Marítima. Descripción y análisis de las inundaciones históricas. In *Análisis hidráulico de las crecientes históricas del Río Cauca*; Ramírez, C., García, J., Bocanegra, R., Ayala, C., Ojeda, Arias., Hurtado, E., Potes, Y., Eds.; Universidad del Valle: Cali, Colombia, 2013; Volume 1, pp. 17–96.

55. Celleri, R.; Willems, P.; Buytaert, W.; Feyen, J. Space–time rainfall variability in the Paute Basin, Ecuadorian Andes. *Hydrol. Process.* **2007**, *3327*, 3316–3327. [CrossRef]

56. Casimiro, W.; Labat, D.; Ronchail, J.; Espinoza, J.; Guyot, J. Trends in rainfall and temperature in the Peruvian Amazon—Andes basin over the last 40 years (1965–2007). *Hydrol. Process.* **2012**, *27*, 2944–2957. [CrossRef]
57. Cuartas, D.; Caicedo, D.; Ortega, D.; Cardona, F.; Carvajal, Y.; Mendez, F. Spatial and temporal trends of extreme climate events in geographical valley of Cauca River (in Spanish). *Actual. Divulg. Científica* **2017**, *20*, 267–278.
58. Setiawan, A.; Lee, W.; Rhee, J. Spatio-temporal characteristics of Indonesian drought related to El Niño events and its predictability using the multi-model ensemble. *Int. J. Climatol.* **2017**, *37*, 4700–4719. [CrossRef]
59. Munoz, S.; Dee, S. El Niño increases the risk of lower Mississippi River flooding. *Sci. Rep.* **2017**, *7*, 1772. [CrossRef] [PubMed]
60. Ndehedehe, C.E.; Awange, J.L.; Agutu, N.O.; Okwuashi, O. Changes in hydro-meteorological conditions over tropical West Africa (1980–2015) and links to global climate. *Glob. Planet. Change* **2018**, *162*, 321–341. [CrossRef]
61. Welhouse, L.J.; Lazzara, M.A.; Keller, L.M.; Tripoli, G.J.; Hitchman, M.H. Composite analysis of the effects of ENSO events on Antarctica. *J. Clim.* **2016**, *29*, 1797–1808. [CrossRef]
62. Sheffield, J.; Goteti, G.; Wood, E.; New, M.; Hulme, M.; Jones, P.; Sheffield, J.; Goteti, G.; Wood, E.; Chaney, N.; et al. Development of a 50-Year High-Resolution Global Dataset of Meteorological Forcings for Land Surface Modeling. *J. Clim.* **2006**, *19*, 3088–3111. [CrossRef]
63. Hirahara, S.; Ishii, M.; Fukuda, Y. Centennial-Scale Sea Surface Temperature Analysis and Its Uncertainty. *J. Clim.* **2014**, *28*, 57–75. [CrossRef]
64. Fan, Y.; van den Dool, H. Climate Prediction Center global monthly soil moisture data set at 0.5° resolution for 1948 to present. *J. Geophys. Res. D Atmos.* **2004**, *109*, 1–8. [CrossRef]
65. Armenteras, D.; Retana, J.; Molowny, R.; Roman, R.; Gonzalez, F.; Morales, M. Characterising fire spatial pattern interactions with climate and vegetation in Colombia. *Agric. For. Meteorol.* **2011**, *151*, 279–289. [CrossRef]
66. Córdoba-Machado, S.; Palomino-Lemus, R.; Gámiz-Fortis, S.R.; Castro-Díez, Y.; Esteban-Parra, M.J. Seasonal streamflow prediction in Colombia using atmospheric and oceanic patterns. *J. Hydrol.* **2016**, *538*, 1–12. [CrossRef]

water

MDPI

Article

Non-Stationary Bayesian Modeling of Annual Maximum Floods in a Changing Environment and Implications for Flood Management in the Kabul River Basin, Pakistan

Asif Mehmood [1,2], Shaofeng Jia [1,2,]*, Rashid Mahmood [1,2], Jiabao Yan [1,2] and Moien Ahsan [3]

[1]   Key Laboratory of Water Cycle and Related Land Surface Processes/Institute of Geographic Sciences and Natural Resources Research (IGSNRR), Chinese Academy of Sciences (CAS), Beijing 100101, China; engrasifmehmood733@gmail.com (A.M.); rashi1254@gmail.com (R.M.); jiabao.yan@foxmail.com (J.Y.)
[2]   University of Chinese Academy of Sciences, Beijing 100049, China
[3]   Center of Excellence in Water Resources Engineering (CEWRE), UET, Lahore 54890, Pakistan; Moien_ag_2232@yahoo.com
*   Correspondence: jiasf@igsnrr.ac.cn; Tel.: +86-10-6485-6539

Received: 20 May 2019; Accepted: 3 June 2019; Published: 14 June 2019

Abstract: Recent evidence of regional climate change associated with the intensification of human activities has led hydrologists to study a flood regime in a non-stationarity context. This study utilized a Bayesian framework with informed priors on shape parameter for a generalized extreme value (GEV) model for the estimation of design flood quantiles for "at site analysis" in a changing environment, and discussed its implications for flood management in the Kabul River basin (KRB), Pakistan. Initially, 29 study sites in the KRB were used to evaluate the annual maximum flood regime by applying the Mann–Kendall test. Stationary (without trend) and a non-stationary (with trend) Bayesian models for flood frequency estimation were used, and their results were compared using the corresponding flood frequency curves (FFCs), along with their uncertainty bounds. The results of trend analysis revealed significant positive trends for 27.6% of the gauges, and 10% showed significant negative trends at the significance level of 0.05. In addition to these, 6.9% of the gauges also represented significant positive trends at the significance level of 0.1, while the remaining stations displayed insignificant trends. The non-stationary Bayesian model was found to be reliable for study sites possessing a statistically significant trend at the significance level of 0.05, while the stationary Bayesian model overestimated or underestimated the flood hazard for these sites. Therefore, it is vital to consider the presence of non-stationarity for sustainable flood management under a changing environment in the KRB, which has a rich history of flooding. Furthermore, this study also states a regional shape parameter value of 0.26 for the KRB, which can be further used as an informed prior on shape parameter if the study site under consideration possesses the flood type "flash". The synchronized appearance of a significant increase and decrease of trends within very close gauge stations is worth paying attention to. The present study, which considers non-stationarity in the flood regime, will provide a reference for hydrologists, water resource managers, planners, and decision makers.

Keywords: non-stationary; extreme value theory; uncertainty; flood regime; flood management; Kabul river basin; Pakistan

---

1. Introduction

The comprehensive understanding of flood regimes is an important challenge in hydrology. Hydrologists and engineers customarily use flood frequency analysis (FFA) as a tool to understand flood regimes throughout the world. FFA estimates the flood peak for a given return period, but the currently used methods of FFA assume that the flood time series are independent and identically distributed [1–3] or, in other words, have no trends and unanticipated variations [4]. Indeed, the concept of stationarity was and is being adopted to design water resources infrastructure and flood protection works all around the globe. In recent decades, the climate system has been under stress due to natural variations in the global climate, and human activity also has a potential influence on regional climate that is ultimately intensifying the hydrologic cycle [5]. The hypothesis of stationarity has become widely questionable due to this regional and global change. Keeping this point of view, several studies have tried to explore the validity of this hypothesis in flood regimes in many regions around the world, considering the effect of natural climate variability [6–12] or land use changes [13–15]. The results of these studies have shown clear violations of the assumption of stationarity, which is consistent with studies that indicate an intensification of the hydrologic cycle [16,17].

Particularly, the KRB in the Hindu Kush Himalayan Range (HKH) is exposed to disturbances from the South Asian monsoon originating from the Bay of Bengal. Several recent studies represented a paucity of stationarity and indicated the intensification in some elements of the hydrologic cycle at the regional scale. The results of these studies investigated the change in the rainfall regime of the KRB. For instance, the number of consecutive wet days has been increasing significantly in the Peshawar valley, with a total change of 2.16 at a 95% confidence level. Consecutive wet days have also increased at Saidu Sharif in the Swat valley and Chitral [18]. Ahmad et al. [19] investigated trends in rainfall over the entire Swat River basin, a sub-basin of the KRB. They observed the highest positive trend (7.48 mm year^{-1}) at the Saidu Sharif in Swat valley. For annual precipitation time series, statistically insignificant trends were revealed for the whole Swat River basin. However, significant positive increasing trends of precipitation (2.18 mm year^{-1}) were observed in the Lower Swat basin. Saidu Sharif, Mardan, and Charsada stations showed significant positive trends (increased precipitation over time) at the 5% significance level in the annual precipitation time series [20]. The results of these studies revealed the presence of trends in precipitation, and their conclusions suggest an important link between the changes exhibited in hydro-climatic variables [21].

Furthermore, other factors that may affect the magnitude and frequency of floods in the KRB are associated with human-induced alterations, such as changes in land use, deforestation, and dam construction. In the KRB, the human activities that can considerably influence flood frequency are land use changes linked with population increase. For instance, a recent study regarding land use cover change (LUCC) dynamics in the KRB in Afghanistan highlighted that substantial LUCCs have occurred during the time interval 2000–2010; among several land cover classes, forest, cultivated land, and grassland showed dynamical change. During the study period, one-fourth of the forest area was lost, while cultivated land and grassland showed an increase of 13% and 11%, respectively. The forest area was mainly transformed into grassland and barren land. Unused land was changed into built-up areas, up to 2%, and water areas increased by 4%. A total loss of 43% was observed in forest area [22]. Similarly, LUCCs in the Swat valley have also occurred. Deforestation occurring due to agriculture expansion was 11.4% at a rate of 0.29%, 77.6% at a rate of 1.98%, and 129.9% at a rate of 3.3%, annually in Kalam, Malam Jaba, and the Swat district areas, respectively. The rangeland has increased due to the conversion of forest land from 1968–1990, by about 158.7%, 38.18%, and 22.2% in Kalam, Malam Jaba, and Swat regions, respectively, while a 13.22% increase has occurred from 1990 to 2007 due to the conversion of agriculture land to rangeland [23]. Dir Kohistan areas of the Hindu Kush Mountains, the northern regions of Pakistan, also showed a 6.4% decrease in forest cover, 22.1% increase in rangeland, and 2.9% increase in agriculture land [24,25]. Similarly, Ahmad and Nizami [26] reported a 7.64% decrease in total area under rangeland in Kumrat valley, Hindu Kush regions. The Mardan city–Kalpani River basin showed an increase in built-up area by 30–60% during 1990–2010. An increase in built-up area has doubled the impervious surface in Mardan and the agriculture land has shrunk from 42% to 35% [27]. Similar results were presented for Peshawar, the capital city of Khyber Pakhtunkhwa province, Pakistan, indicating a 26.59% increase in built-up area during 1999–2016 [28]. The Peshawar valley, with a rich history of flooding, provides the junctions for the Kabul River and its various right and left tributaries.

The above studies clearly show the presence of trends in rainfall regime as well as land use change in different sub-basins of the KRB. These climate and human interventions may induce non-stationarity in the flood regime. However, no studies have been reported to examine the presence or absence of stationarity in the flood regime of the basin. Therefore, it is imperative to study floods with a non-stationary point of view for the KRB.

Recently, Milly et al. [29] stated that the hypothesis of stationarity must be relinquished and that "stationarity is dead" and "should not be revived". The methods used for estimation of hydrologic indicators should be based on an innovative approach that would be reliable and useful for water management under a changing environment.

In the literature, various approaches have been reported using probabilistic modeling of flood frequency in a non-stationary context. Khaliq et al. [2] presented a comprehensive review, including the incorporation of trends in the parameters of the distributions, the incorporation of trends in statistical moments, the quantile regression method, and the local likelihood method. The studies of FFA under non-stationary conditions have mostly assumed trends in time [30–37]. The present study outlined a Bayesian framework for "at site flood frequency modeling" in stationary and non-stationary conditions. The fundamental concept is based on the generalized extreme value (GEV) distribution, combined with Bayesian inference for uncertainty assessment. For this study, a model with trend (non-stationary) and without trend (stationary) was used.

Previous studies in the KRB were limited to inundation mapping of flood-prone areas with a very little flow gauge station data, using a traditional frequentist approach [38–42].

The main objectives of the study were: (1) to analyze temporal and spatial trends in the annual maximum flood regime for the KRB, Pakistan, because no study has yet been reported in the literature to study the trends in annual extreme data of flood in detail, and (2) to address the non-stationary modeling of the flood regime in the KRB and its implications for flood management in a changing environment. We explored the differences between stationary and non-stationary flood quantile estimates for a given return period using flood frequency curves (FFCs), along with their uncertainty bounds for risk assessment, to analyze the importance of non-stationary models for improving flood management in the study area.

2. Study Area and Data Description

2.1. Study Area

The Kabul River basin (KRB), in Pakistan, stretches from 71°1′55″–72°56′0″ E to 33°20′9″–36°50′0″ N, as shown in Figure 1, which covers an area of 33,709 km². The Kabul River starts at the base of Unai pass from the Hindu Kush Mountains in Afghanistan and flows eastward, covering a distance of 700 km to drain into the Indus River, Pakistan [43]. The entire basin covers an area of 87,499 km². The elevation in the basin varies substantially from 249 m.a.s.l to 7603 m.a.s.l. High elevation mountains are mainly located in the north. The average temperature and average precipitation vary significantly across the River basin. The average temperature is about 13 °C. Most of the precipitation occurs in the northern mountain and highlands, reported up to 1600 mm. [44].

This study explores the part of the KRB that contributes to flooding. The flood problem arises mainly as the Kabul River enters Pakistan. The Logar River basin, Alingar River basin, and Panjshir River basin lie in Afghanistan. Three dams—Naghlu, Surobi, and Darunta—are located in Afghanistan on the Kabul River and Warsak dam is also located on the Kabul River in Pakistan. The study area is further divided into eight sub-basins: Kabul River basin, Chitral River basin, Main Swat River basin, Panjkora River basin, Lower Swat River basin, Kalpani River basin, Jindi River basin, and Bara River basin. The SRTM-DEM (Shuttle Radar Topography Mission–digital elevation model) of 30 m resolution and the geographical location of the sub-basins are also illustrated in Figure 1.

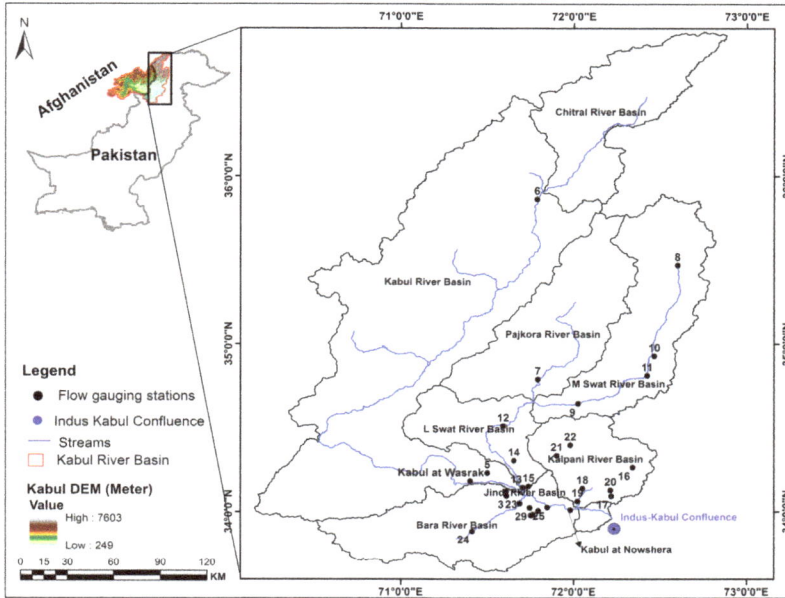

Figure 1. Description of Location, SRTM-DEM (Shuttle Radar Topography Mission–digital elevation model, meters) and flow gauge stations for the Kabul River basin (KRB), Pakistan.

2.2. Flood Data

Twenty-nine flow gauge stations were selected to study the flood regime of the KRB. The annual maximum daily peak flow data for the seven flow gauge stations at main rivers sites were obtained from Surface Water Hydrology Project of Water and Power Development Authority (SWHP–WAPDA). The streamflow data of the remaining study sites were obtained from the Hydrology Section of the Irrigation Department of Khyber Pakhtunkhwa Province, Pakistan. The study sites that had at least 30 years of records were selected. The main characteristics of the sub-basins and the respective flow gauge stations in each sub-basin are presented in Table 1 and Figure 1 describes the geographical locations of the flow gauge stations in each sub-basin.

Table 1. Basic information of flow gauges and sub-basins in the KRB, Pakistan.

Site#	Sub Basin and Flow Gauge Stations	Basin Area (km^2)	Coefficient of Variation (Cv)	Number of Years of Record
	Kabul River Basin	87,499		
1	Kabul River at Warsak		0.292	52 (1965–2016)
2	Kabul River at Nowshera		0.433	55 (1962–2016)
3	Shahalam River		0.724	30 (1987–2016)
4	Naguman River		0.829	30 (1987–2016)
5	Adezai River		0.739	30 (1987–2016)
	Chitral River Basin	11,396		
6	Chitral River		0.2	50 (1964–2013)
	Panjkora River Basin	5917		
7	Panjkora River		0.859	33 (1984–2016)

Table 1. *Cont.*

Site#	Sub Basin and Flow Gauge Stations	Basin Area (km²)	Coefficient of Variation (Cv)	Number of Years of Record
	Main Swat River Basin	6066		
8	Swat River at Kalam		0.2	59 (1961–2009)
9	Swat River at Chakdara		0.336	49 (1961–2009)
10	Swat River at Khawazakela		0.84	34 (1983–2016)
11	Swat River at Ningolai		1.425	31(1986–2016)
	Lower Swat River Basin	2685		
12	Swat River at Munda Head Works		0.744	55 (1962–2016)
13	Khiyali River at Charsada Road		0.815	48 (1969–2016)
14	Jundi Nullah at Tangi		3.06	37 (1974–2011)
	Jindi River Basin	13		
15	Jindi River		0.684	48 (1969–2016)
	Kalpani River Basin	2830		
16	Naranji Nullah		0.975	49 (1968–2016)
17	Badri Nullah		0.893	45 (1966–2010)
18	Kalpani River at Mardan		1.476	33 (1984–2016)
19	Kalpani River at Risalpur		0.752	33 (1984–2016)
20	Dagi Nullah		1.01	33 (1984–2016)
21	Bagiari Nullah		0.917	30 (1987–2016)
22	Lund Khawar West		1.13	30 (1987–2016)
	Bara River Basin	3388		
23	Budni Nullah		1.28	43 (1974–2016)
24	Bara River at Kohat Bridge		1.69	34 (1983–2016)
25	Khuderzai Nullah		1.65	32 (1980–2011)
26	Chillah Nullah at Pabi		1.15	32 (1980–2011)
27	Hakim Garhi Nullah		0.6	31 (1980–2010)
28	Wazir Garhi Nullah		1.69	30 (1981–2010)
29	Muqam Nullah		0.781	30 (1981–2010)

2.3. Flood Generating Mechanism in KRB

The hydrology of floods is linked to weather and climate as well as to physiographical features [45]. The basin has large altitudinal variations from 249 m.a.s.l. to 7603 m.a.s.l. Glacier-melt contribution from the upper part of the basin combined with rainfall in the lower part is the most likely cause of flooding in the region [38]. In the KRB, floods are mostly generated by monsoon rainfall but snow or glacial melt floods have also been observed in some parts of the basin. Snowmelt floods are not common. According to the data used in this study, all of the flood peaks were observed during the monsoon season, from July to August, in almost all the tributaries of the KRB. The historical floods occurred in July 2010, August 1995, and July 1992; all were observed during the monsoon. Anjum et al. [46] provided the details regarding rainfall magnitude, intensity, and spatial extent for the 2010 event. The South Asian monsoon originating from the Bay of Bengal is the dominant weather system for flood generation in the KRB.

However, the flood of 2005 in the Kabul and Indus Rivers was due to snowmelt as well as rainfall in the pre-monsoon period [47]. The flooding behavior of the different tributaries differs according to their catchment characteristics. The riverine floods in the Kabul River usually start below the Warsak dam, and this phenomenon propagates until its confluence with the Indus River at Khairabad near Attock. Riverine floods also occur in the Swat River in the Lower Swat catchment. In the rest of the KRB, flash flooding is a common disaster, along with landslides and torrential rains [45].

3. Methods

3.1. Preliminary Analysis

3.1.1. Trend Analysis

The non-parametric rank-based Mann–Kendall (MK) [48,49] test was used to detect trends in annual maximum flood series. The trend analysis was performed to show a clear understanding of the whole study area, while an objective criterion (means if statistically significant trend exists) was adopted for the non-stationary modeling of flood regime. The Mann–Kendall test was applied at different significance levels. The autocorrelation function (ACF) was also computed, before applying the Mann–Kendall (MK) test to check the presence of serial correlations in the annual maximum flood series.

3.1.2. Selection of Extreme Value Distribution

The Bayesian method using the GEV distribution is getting attention for analyzing hydrological extremes. The current study also utilized the GEV distribution, which is the integration of Gumbel, Fréchet, and Weibull distributions, and developed on the limit theorems for block maxima or annual maxima [50]. Mathematically, the cumulative distribution of the GEV can be written as [51]:

$$\psi(x) = exp\left\{-\left(1 + \xi\left(\frac{x-\mu}{\sigma}\right)\right)^{\frac{-1}{\xi}}\right\}, \left(1 + \xi\left(\frac{x-\mu}{\sigma}\right)\right) > 0, \tag{1}$$

where $\psi(x)$ is expressed as $\left(1 + \xi\left(\frac{x-\mu}{\sigma}\right)\right) > 0$; somewhere else, $\psi(x)$ is either 0 or 1 [52].

The location parameter (μ), describes the center of the GEV distribution, the scale parameter (σ) describes the deviation around (μ), and the shape parameter (ξ) describes the tail behavior of the distribution. When $\xi \to 0$, $\xi < 0$, and $\xi > 0$, GEV approaches the Gumbel, Weibull, and Fréchet distributions, respectively.

3.1.3. Goodness of Fit Statistics to GEV Distribution

The goodness of fit analysis of annual maximum peak flow data to the GEV distribution was performed in order to investigate whether the historical data belonged to the said GEV distribution. The Anderson–Darling (AD) [53] and Kolmogorov–Smirnov (K-S) [54] tests were performed for this purpose, using an EasyFit software (version 5.6, MathWave Technologies) [55]. EasyFit estimated the parameters of the GEV distribution based on maximum likelihood (ML) estimation, using equal probability sampling. The parameters estimated using the EasyFit software were used to assess the goodness of fit by AD and K-S statistics.

The K-S statistic (D) is based on the largest vertical difference between the theoretical and the empirical cumulative distribution function as shown below:

$$D = m_{1 \leq i \leq n}(\psi(x_i) - \frac{i-1}{n}, \frac{i}{n} - \psi(x_i)). \tag{2}$$

The Anderson–Darling procedure compares the fit of an observed cumulative distribution function to an expected cumulative distribution function. This test gives more weight to tails than the K-S test.

$$A^2 = -n - \frac{1}{n}\sum_{i=1}^{n}(2i-1) \times [\ln\psi(x_i) + \ln(1-\psi(x_{n-i+1}))]. \tag{3}$$

H_0: The data follow the specified distribution;
H_a: The data do not follow the specified distribution.

The hypothesis regarding the distributional form was rejected at the significance level of 0.05 (alpha) if the test statistic, A^2 or D, was greater than the critical value of 2.5018 and 0.18482, respectively.

Moreover, the outlier's detection in the annual extreme data of flood series was also performed using the Chauvenet's Criterion [56].

3.2. Model Design

The extreme value theory of stationary random process is based on that the statistical properties of extremes—here, the distribution parameters $\theta = (\mu, \sigma, \xi)$ are free from time dependency [57], while in a non-stationary random process, the parameters of the said distribution function rely on time-dependency, and the properties of the distribution also vary with time [58]. For this study, two cases were considered.

(1) Stationary Case: all the model parameters were considered constant.
(2) Non-stationary Case: the location parameter (μ) was considered a function of time, as shown in Equation (4), while scale and shape parameters were kept constant:

$$\mu(t) = \mu_1 t + \mu_0, \tag{4}$$

where t is time, $\theta = (\mu_1, \mu_0)$ are the regression parameters [50,57–60]. The location parameter was calculated for each study site in the stationary case and non-stationary case.

3.2.1. Bayes Theorem for GEV Distribution

Let θ be the parameter of given distribution and let $Y = \{y_1, y_2, \dots, y_n\}$ be the set of n observations. According to the Bayes theorem, the probability of θ given Y (posterior) is proportional to the product of the probability of θ (prior) and the probability of Y given θ (likelihood function). Assuming the independence between the observations, Y:

$$P(\theta|Y) \propto \prod_{i=1}^{n} P(\theta) \times P(y_i|\theta). \tag{5}$$

Here, the likelihood function is the GEV distribution and θ is the vector containing the parameters of GEV distribution to be estimated. In the stationary case, $\theta = (\mu, \sigma, \xi)$. By assuming independent GEV parameters:

$$P(\mu, \sigma, \xi|Y) \propto \prod_{i=1}^{n} P(\mu) \times P(\sigma) \times P(\xi) \times P\left(y_i|\mu, \sigma, \xi\right). \tag{6}$$

In the case of non-stationary analysis, θ contains an additional parameter, which is time-dependent here, i.e., $\mu(t)$, hence, the Bayes theorem for estimation of GEV parameters under the non-stationary case can be expressed as [57,60]:

$$P(\theta|Y, t) \propto \prod_{i=1}^{n} P(\theta) \times P(y_i|\theta, t). \tag{7}$$

The resulting posterior distributions $P(\theta|Y, t)$ provide information on the distribution parameters $(\mu_1, \mu_0, \sigma, \xi)$.

3.2.2. Prior Distribution

A Bayesian model utilizes a prior belief to calculate the posterior belief. For the current study, we utilized NEVA (non-stationary extreme value analysis, Matlab Package) [60–62] for our analysis. In NEVA, the priors are non-informative normal distributions, for location and scale parameters, while the priors for the shape parameter are a normal distribution, with a standard deviation of 0.3, as suggested by [57,60,62]. Initially, the shape parameter was considered a non-informative prior:

$$\xi \sim N(-5, 5), \tag{8}$$

if the value of the shape parameter in the posterior distribution exceeded beyond the plausible limit (−5, 5), as suggested by Martins and Stedinger [63]. Then, we modified the priors for shape parameter, considering partial pooling of information across sites that had similar flood types, for improving the flood quantiles estimates for "at site modeling" using the regional information. The shape parameter was considered an informative prior and the range of priors for shape parameter was:

$$\xi \sim N\,(0, \text{Ksi}). \tag{9}$$

where, the Ksi stands for the shape parameter value of the site of interest from where it was exchanged. However, the location and scale parameter across sites were not shared.

3.2.3. Parameters Estimation and Convergence Criterion

To estimate the parameters inferred by Bayes, the Differential Evolution Markov Chain (DE-MC) is integrated to generate a large number of realizations from the parameters' posterior distributions [64,65]. The DE-MC attributes to the genetic algorithm Differential Evolution (DE) for global optimization over real parameter space with the Markov Chain Monte Carlo (MCMC) approach [64,65]. Here, the target posterior distributions were sampled through five Markov Chains constructed in parallel. These chains were allowed to learn from each other by generating candidate draws based on two random parent Markov Chains, rather than running independently. Therefore, it had the advantages of simplicity, speed of calculation, and convergence over the conventional MCMC. The initial numbers of burned samples were 6000 and numbers of evaluations were 10,000 for each study site. The R-hat criterion, suggested by Gelman and Shirley [66], was used to assess convergence, where R-hat should remain below 1.1.

Uncertainty estimates for FFCs are crucial for risk assessment and decision making. By combining DE-MC with Bayesian inference, the posterior probability intervals or credible intervals and uncertainty bounds of estimated return levels based on the sampled parameters could be obtained simultaneously for FFCs. For example, for a time series of annual maximum peak flow, the time-variant parameter ($\mu(t)$) was derived by computing the 95th percentile of DE-MC sampled $\mu(t)$, (i.e., the 95th percentile of $\mu(t = 1), \dots, \mu(t = 100)$). These model parameters were then used to develop the stationary and non-stationary FFCs.

FFCs could also be drawn at 50% Bayesian credible intervals or at any other desired intervals.

3.2.4. Model Evaluation

In order to evaluate the suitability of the stationary versus non-stationary models, a Bayes factor K was calculated based on the posterior distributions of sampled parameters of both models. The stationary model was considered a null model M1, while the non-stationary model M2 was considered an alternative.

A value of Bayes factor > 1 denotes the stationary model is favored, while a value < 1 argues in the favor of the non-stationary model. Similarly, a value approaching +infinity favors the stationary model, and −infinity favors non-stationary models. Equation (10) represents the computation of Bayes factor, as follows:

$$K = \frac{Pr(DA|M1)}{Pr(DA|M2)} = \frac{\int Pr(\theta_1|M1)Pr(DA|\theta_1 M1)d\theta_1}{\int Pr(\theta_2|M2)Pr(DA|\theta_2 M2)d\theta_2}. \tag{10}$$

The term DA denotes input data, and θ stands for model parameters. The term $Pr\,(DA|M)$ can be expressed using Monte Carlo integration estimation as follows:

$$Pr(DA|M) = \left\{ \frac{1}{m} \sum_{m_i=1} Pr\,(DA|\theta^{(i)}, M)^{-1} \right\}^{-1}. \tag{11}$$

For more details see [67].

4. Results and Discussion

4.1. Temporal and Spatial Trends in Flood Regime

The trend magnitude for each station is presented in Table 2. Trend analysis results demonstrated the significant trend by 37.93% of the flow gauge stations in the entire basin; among them, 27.6% showed a significant increasing trend at the 0.05 significance level and 10.34% of the stations showed a significant decreasing trend at the significance level of 0.05. Moreover, 6.9% of the flow gauge stations also revealed a significant increasing trend at the significance level of 0.1. Non-significant trends were also exhibited by 31% of the flow gauge stations. The Chitral River at Chitral, the Kalpani River at Risalpur, the Kalpani River at Mardan, the Swat River at Chakdara, the Swat River at Ningolai, the Adezai River, the Naranji Nullah, the Bagiari Nullah, the Lund Khawar West, and the Bara River at Kohat Bridge displayed a significant increasing trend (Site #5, 6, 9, 11, 16, 18, 19, 21, 22, and 24), while the Swat River at Khawazakhela, the Naguman River, and the Badri Nullah showed a significant decreasing trend. The Khiyali River, the Panjkora River, and the Jundi Nullah at Tangi represented a moderate increasing trend, while the Kabul River at Warsak, the Swat River at Kalam, the Swat River at Munda Head Works, the Budni Nullah, and the Khuderzai Nullah displayed a moderate decreasing trend. Moreover, the main flow gauge station—the Kabul River at Nowshera—did not show any significant trend.

Table 2. Description of trends in the annual maximum flood regime across the KRB, Pakistan.

Site #	Mann–Kendall (Test-Z)	Site #	Mann–Kendall (Test-Z)	Site #	Mann–Kendall (Test-Z)
1	−1.54	11	4.78 ***	21	3.28 **
2	−0.35	12	−0.89	22	2.83 **
3	0.41	13	1.18	23	−1.28
4	−2.02 *	14	0.86	24	2.28 *
5	2.61 **	15	−0.37	25	−1.19
6	2.80 **	16	1.79 +	26	−0.67
7	0.93	17	−3.07 **	27	0.34
8	−1.36	18	3.24 **	28	−0.54
9	1.73 +	19	2.13 *	29	−0.83
10	−2.36 *	20	0.16		

*** Trend is significant at $\alpha = 0.001$, ** Trend is significant at $\alpha = 0.01$, * Trend is significant at $\alpha = 0.05$, + Trend significant at $\alpha = 0.1$.

Figure 2 represents the basin-wide spatial distribution of trends in the flood regime, which showed that the flood regime of the Chitral River, the Kalpani River, and the Main Swat River basins exhibited significant increasing trends. However, the Swat River at Khawazakhela and the Badri Nullah represented a significant decreasing trend.

The Lower Swat River, the Kabul sub-basin, and the Jindi River basins showed non-significant trends.

Especially for the southwestern part of the KRB, the Bara River at Kohat Bridge showed non-stationarity in the flood regime due to a significant increasing trend, while all other flow gauge stations in the Bara River basin showed insignificant trends.

The change in flood regime was found to be more evident for the northern and northeastern part of the KRB as compared to the central and southwestern parts of the KRB. Consequently, the overall basin showed large spatial variations. However, the basin did not represent a regular spatial pattern. These spatial variations may be due to climatology, topography, and complex orography of the KRB in the HKH region. However, the temporal changes in the flood regime might be attributable to regional environmental change. The results of trend analysis were consistent with previous studies [68–70], but these studies used only two to three flow gauges.

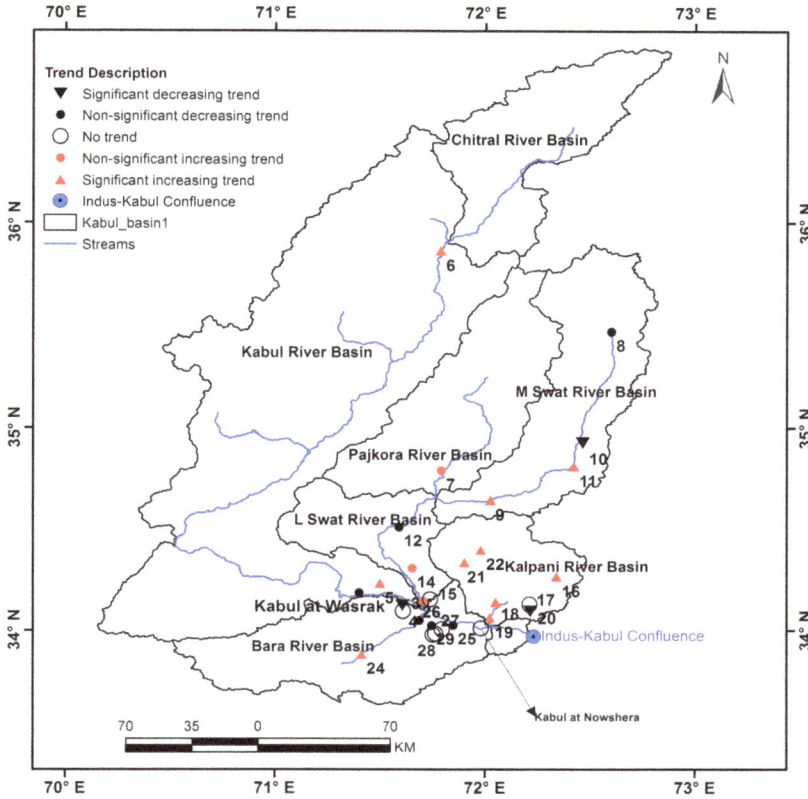

Figure 2. Spatial distribution of trends in the annual maximum flood regime of the KRB, Pakistan.

4.2. Evaluation of Goodness of Fit for Annual Extreme Data of Flood

An objective criterion as suggested by Rosner et al. [71] was adopted to evaluate the goodness of fit of the annual maximum flood series, with GEV and non-stationary temporal trend modeling of the flood regime, i.e., only the sites showing significant trends in their flood regime were selected. The study sites under consideration showed proper fitting using the AD test and the value of test statistics for all the sites using the AD test was less than the critical value of 2.5018. While using the K-S test, all the sites showed proper fitting except site 21, but it was included in the analysis because it satisfied the AD test. Table 3 provides the results of the test statistics and estimated GEV parameters using ML. The *p*-value belongs to K-S test only.

Outliers were also detected in the data of the annual maximum flood series for the KRB. Table 4 demonstrates the outliers present in the data of selected study sites. Sites 5, 6, 16, 18, 19, 21, and 24 displayed the extreme flood of 2010 as an outlier in the data. Site 6 also revealed the flood of 2005 (1603 m^3 s^{-1}) as an outlier, as per evaluation criterion. Similar to site 6, site 9 also represented two outliers in its flood series. The outlier 1918 m^3 s^{-1} represents the flood of 1992, and the corresponding value of 1602 m^3 s^{-1} represents the flood of 1987 for site 9.

For Site 11, the 1475 m^3 s^{-1} value corresponds to the flood of 2016 in the Swat River basin, the sub-basin of the KRB. The flood discharge of 37 m^3 s^{-1} at Lund Khawar West (Site 22) in 1997 was also recorded as an outlier. Despite the existence of outliers, the data series belongs to the GEV distribution as per AD and K-S test statistics.

Table 3. The goodness of fit statistics of annual maximum daily peak flow to generalized extreme value (GEV) distribution.

Site #	Gauge Stations	GEV Parameters	Anderson–Darling Test	Kolmogorov–Smirnov Test	
			A-D Statistics	K-S Statistics	*p*-Value
5	Adezai River	$\xi = 0.07899$ $\sigma = 454.66$ $\mu = 521.18$	0.6903	0.15394	0.43251
6	Chitral River	$\xi = 0.00307$ $\sigma = 143.37$ $\mu = 1026.5$	0.22503	0.06435	0.97732
9	Swat River at Chakdara	$\xi = 0.13247$ $\sigma = 152.1$ $\mu = 646.8$	0.66053	0.10305	0.59055
11	Swat River at Ningolai	$\xi = 0.52162$ $\sigma = 103.82$ $\mu = 83.499$	0.60066	0.1453	0.48501
16	Naranji Nullah	$\xi = 0.25789$ $\sigma = 77.168$ $\mu = 81.939$	0.19219	0.06263	0.98424
18	Kalpani River at Mardan	$\xi = 0.55205$ $\sigma = 106.9$ $\mu = 77.204$	1.2218	0.17818	0.21796
19	Kalpani River at Risalpur	$\xi = 0.20781$ $\sigma = 441.0$ $\mu = 604.88$	0.42201	0.10944	0.7987
21	Bagiari Nullah	$\xi = 0.06073$ $\sigma = 112.05$ $\mu = 94.608$	1.838	0.22761	0.08399
22	Lund Khawar West	$\xi = 0.37899$ $\sigma = 3.7993$ $\mu = 3.164$	0.48511	0.12903	0.7523
24	Bara River at Kohat Bridge	$\xi = 0.57308$ $\sigma = 16.871$ $\mu = 9.4788$	1.2595	0.15782	0.33006

Table 4. Representation of detected outliers, as per Chauvenet's criterion.

Site #	Station Name	Historical Extreme (Outliers)	Observed Value	Critical Value
5	Adezai River	2285	2.449	2.394
6	Chitral River	1633/1603	2.941/2.76	2.576
9	Swat River at Chakdara	1918/1602	4.6/3.35	2.576
11	Swat River at Ningolai	1475	3.447	2.406
16	Naranji Nullah	850	4.748	2.576
18	Kalpani River at Mardan	1499	3.182	2.429
19	Kalpani River at Risalpur	3358	3.316	2.418
21	Bagiari Nullah	473	2.102	2.394
22	Lund Khawar West	37	3.235	2.394
24	Bara River at Kohat Bridge	331	4.234	2.44

4.3. Regionalization of Shape Parameter for Flash Floods Across the KRB

The shape parameter of the GEV distribution is important for the estimation of flood quantiles. Initially, non-informative priors for shape parameter were used for the Bayesian analysis of annual maximum flood regime for the selected study sites that had non-stationarity, due to the existence of temporal trends. About 30% of the study sites yielded a value of shape parameter in posterior distribution that exceeded beyond the plausible limit ($-5, 5$) as suggested by Martins and Stedinger [63], causing the degeneration of the GEV distribution. In order to avoid this, homogeneous sites were identified. Halbert et al., Kuczera, Kyselý et al., Sun et al., and Viglione et al. [72–76] state that the use of regional information will improve the flood frequency estimation and reduce the uncertainty for sites having short records. Table 5 illustrates the correlation matrix for the selected study sites, which demonstrates that hierarchical clustering is possible based on the correlation between the annual maximum flood series of the selected study sites.

Table 5. Correlation matrix for the selected study sites.

Site #	5	6	9	11	16	18	19	21	22	24
5	1	0.24	−0.04	0.63	0.35	0.61	0.14	0.25	0.39	0.48
6	0.24	1	0.29	0.11	0.42	0.33	0.42	0.37	0.38	0.41
9	−0.04	0.29	1	0.11	0.11	−0.05	−0.22	−0.02	−0.17	0.04
11	0.63	0.11	0.11	1	0.2	0.59	0.04	0.49	0.6	0.21
16	0.35	0.42	0.12	0.2	1	0.41	0.47	0.29	0.32	0.63
18	0.61	0.33	−0.05	0.59	0.41	1	0.63	0.53	0.52	0.54
19	0.14	0.42	−0.22	0.04	0.47	0.63	1	0.65	0.64	0.42
21	0.25	0.37	−0.02	0.49	0.29	0.53	0.65	1	0.41	0.2
22	0.39	0.38	−0.17	0.6	0.32	0.52	0.64	0.41	1	0.47
24	0.48	0.41	0.04	0.21	0.63	0.54	0.42	0.2	0.47	1

A positive correlation is present between the sites. Although site 9 possesses the least positive correlation with site 6, site 6 has more positive correlations with other sites, hence why all the sites are considered homogenous. All the sites also possess "flash" flood type. Moreover, all the sites could also be considered homogenous because of the existence of trends in their flood regime. Sun et al. [77] also highlighted the clustering of temporal trends and exchange of shape parameter for the Bayesian analysis of annual maximum floods across Germany.

Furthermore, the utilization of ML for the estimation of the shape parameter for GEV distribution was found reliable for large records—at least 50 year [78]. After considering all the study sites as homogenous based on the correlations between sites, similar flood type, and existence of trends, Naranji Nullah (site 16), with a sufficiently long record, was considered the benchmark site. The shape parameter estimated by ML was approximately 0.26 for site #16. This value of shape parameter (0.26) was further recognized for all the study sites as an informative prior in the Bayesian model. This is like partial pooling of information across homogenous sites, which ultimately improved the flood quantiles estimates using the regional information as compared to non-informative priors on shape parameter. Lima et al. [79] used the basin's average shape parameter value in local and regional hierarchical Bayesian models to solve the issue of sites where the shape parameter value exceeds beyond (−5–5). Lima et al. [79] used the prior for shape parameter as non-informative, but this study considers the priors on shape parameter to be informative priors.

4.4. Comparison between Stationary and Non-Stationary Bayesian Models

Non-stationary FFCs were constructed for the flow gauges with significant increasing trends in their flood series and compared with their stationary FFCs, considering the entire data series without the elimination of outliers. Table 6 demonstrates the results at the 95% Bayesian credible interval for all the selected study sites in the KRB. The stationary model showed overestimation as compared to the non-stationary model for a 100-year flood (The flood having a probability of exceedance of 0.01) by 1494 m^3 s^{-1} (+34.9%) for the Adezai River at Adezai Bridge (site 5). The value of the Bayes factor was 0.0058, which was less than 1, so the non-stationary model was favored. The maximum peak flood observed at the Adezai River was 2285 m^3 s^{-1}, during the historic flood of 2010 in the KRB. The non-stationary model reasonably described the historical peak flood (Figure 3). Similar behavior was observed for the other study sites, 11, 18, 21, 22, and 24 (Table 6 and Figures 4–8).

Table 6. Comparison between 100-year flood estimates using the stationary and non-stationary Bayesian models for "at site modeling" for the KRB, Pakistan.

Site #	Station Name	Historical Extreme m³ s⁻¹	Stationary m³ s⁻¹	Non-Stationary m³ s⁻¹	Difference b/w Stationary & Non-Stationary m³ s⁻¹	Percent Difference (%)	Bayes Factor	% Difference between Preferred Model and Historical Extreme
5	Adezai River	2285	4276	2782	1494	34.9	0.0058	17.86
6	Chitral River	1633	1895	1918	−23	−1.19	0.068	14.85
9	Swat River at Chakdara	1918	1991	2686	−695	−25.8	7.06	3.8
11	Swat River at Ningolai	1475	2891	2528	363	12.5	0.0065	41.65
16	Naranji Nullah	850	1127	1222	−95	−7.7	9.55	24.6
18	Kalpani River at Mardan	1499	3881	2887	1054	27.15	−Infinity	48.14
19	Kalpani River at Risalpur	3358	4918	5140	−222	−4.31	0.4348	34.66
21	Bagiari Nullah	473	1666	819	847	50.8	0.0321	42.24
22	Lund Khawar West	37	76	51	25	32.89	0.11	27.45
24	Bara River at Kohat Bridge	331	686.7	357.5	330.9	48.18	−Infinity	7.2

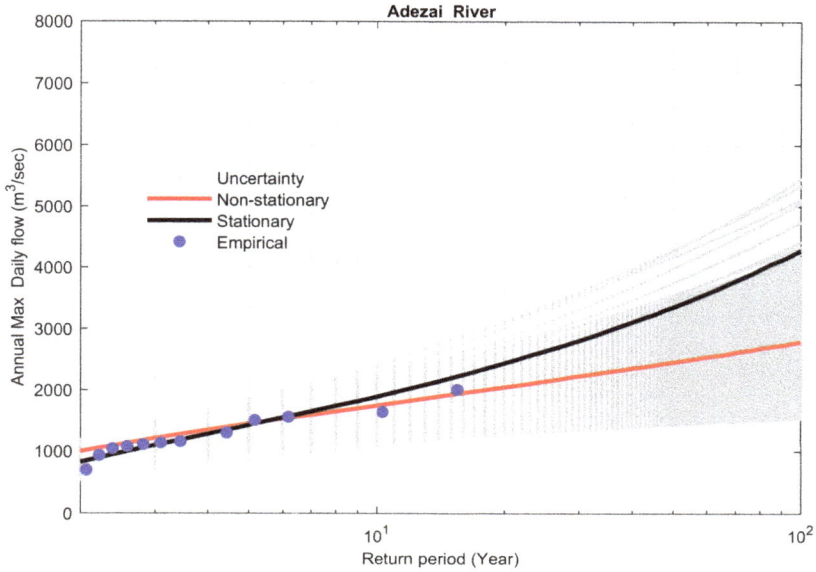

Figure 3. Comparison between non-stationary and stationary flood frequency curves (FFCs) at site 5, along with non-stationary uncertainty bounds for the year 2016.

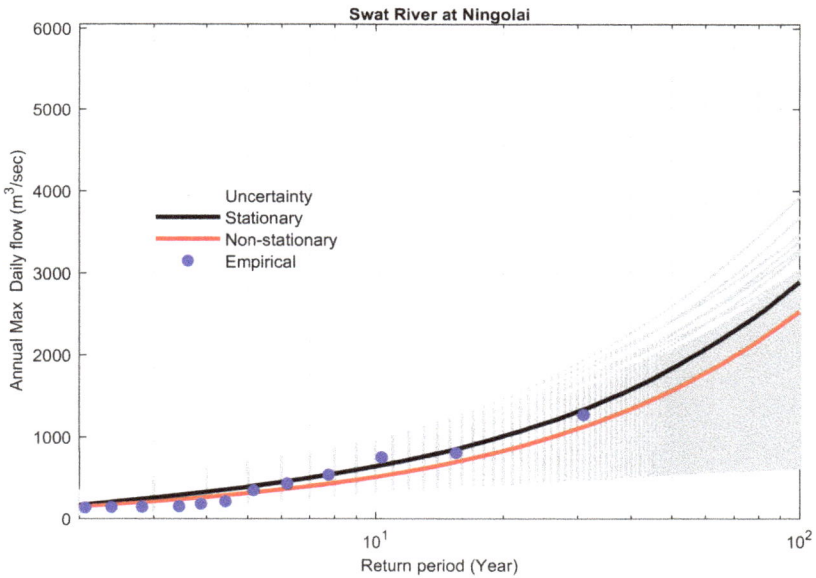

Figure 4. Comparison between non-stationary and stationary FFCs at site 11, along with non-stationary uncertainty bounds for the year 2016.

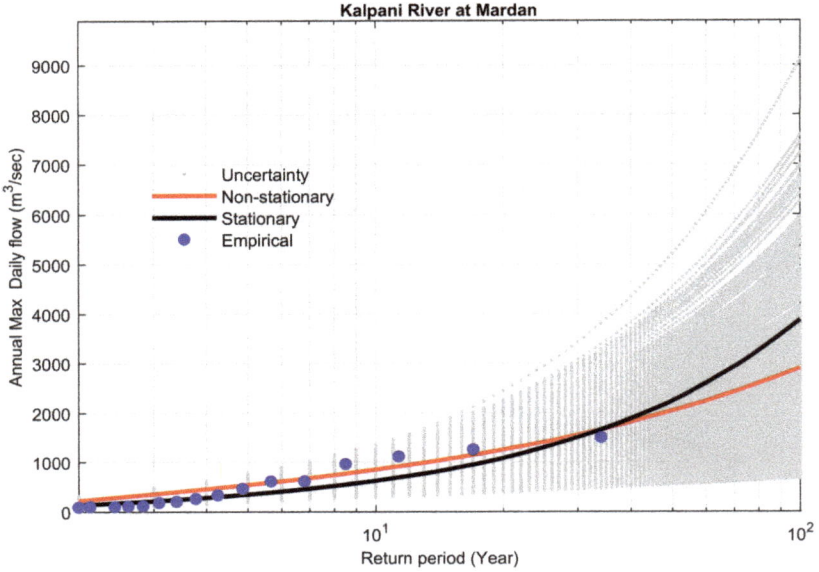

Figure 5. Comparison between non-stationary and stationary FFCs at site 18, along with non-stationary uncertainty bounds for the year 2016.

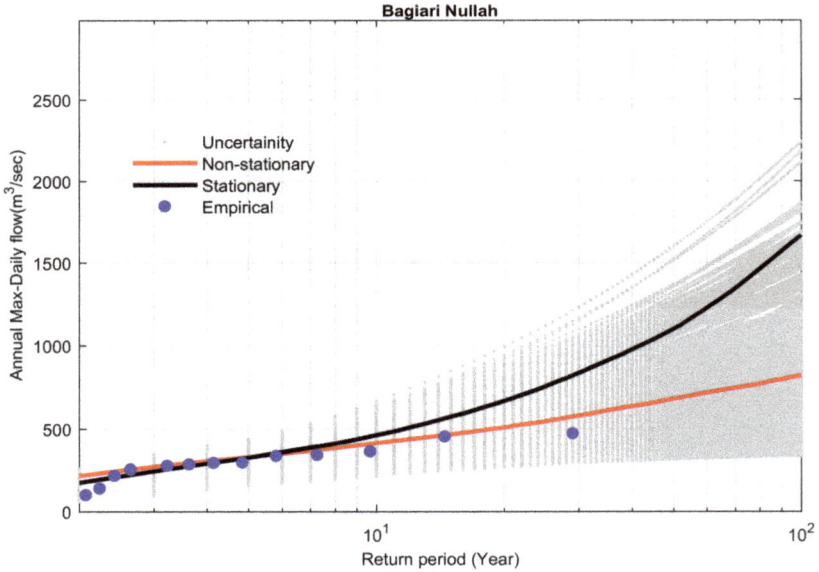

Figure 6. Comparison between non-stationary and stationary FFCs at site 21, along with non-stationary uncertainty bounds for the year 2016.

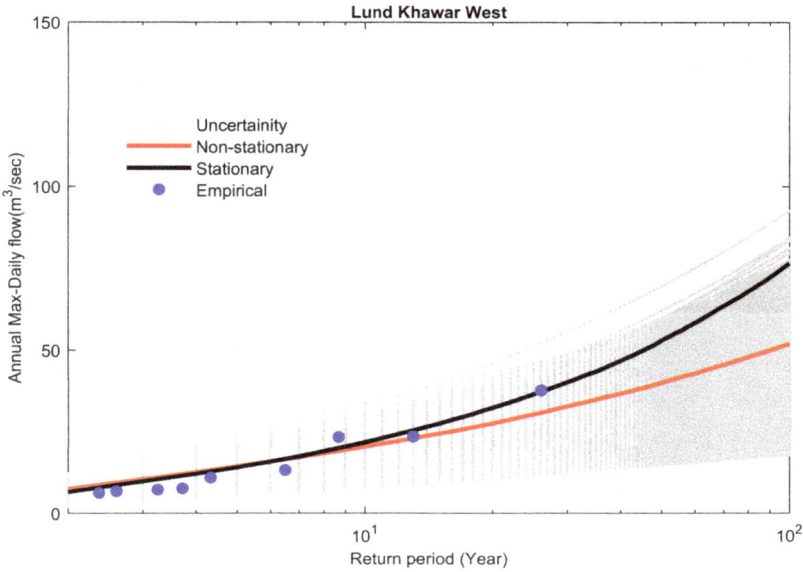

Figure 7. Comparison between non-stationary and stationary FFCs at site 22, along with non-stationary uncertainty bounds for the year 2016.

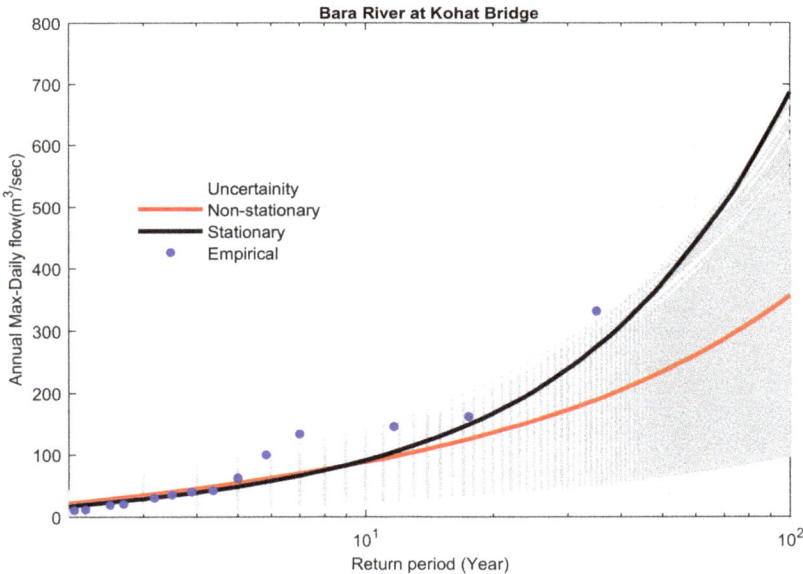

Figure 8. Comparison between non-stationary and stationary FFCs at site 24, along with non-stationary uncertainty bounds for the year 2016.

On the other hand, the stationary model also underestimated the 100-year flood as compared to the non-stationary model. For example, the stationary model underestimated the 100-year flood by 23 m^3 s^{-1} (−1.19%) as compared to the non-stationary model (Figure 9, the Chitral River at Chitral).

The value of the Bayes factor was 0.068, which was less than 1, so the non-stationary model was favored. This behavior was obvious for study sites 6 and 19 (Table 6, Figures 9 and 10).

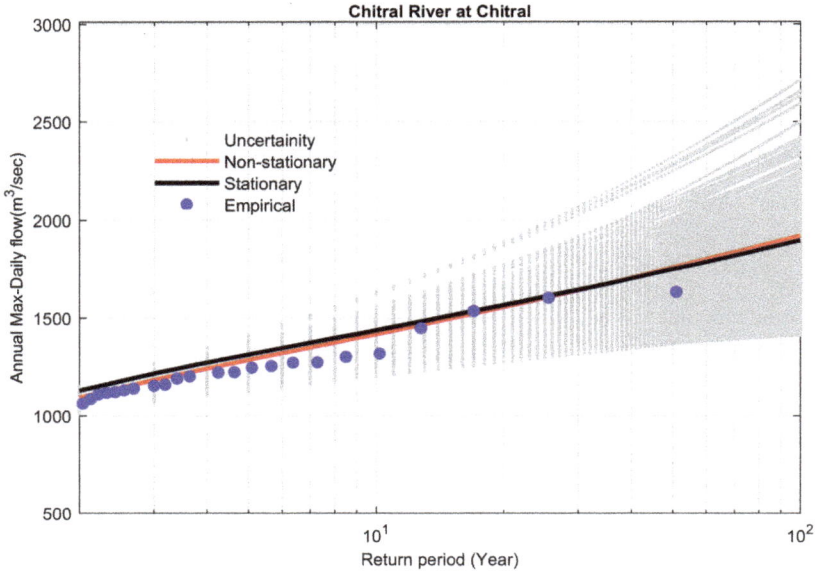

Figure 9. Comparison between non-stationary and stationary FFCs at site 6, along with non-stationary uncertainty bounds for the year 2013.

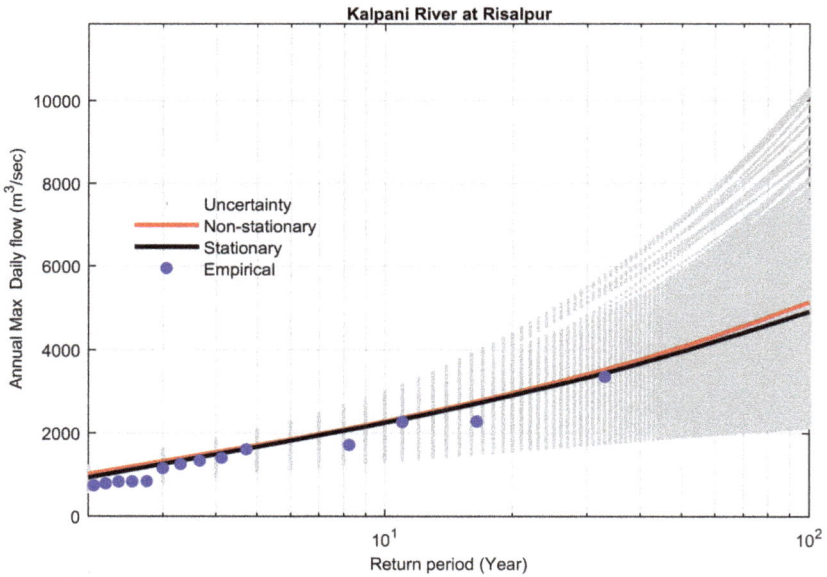

Figure 10. Comparison between non-stationary and stationary FFCs at site 19, along with non-stationary uncertainty bounds for the year 2016.

Furthermore, for study sites possessing trends at the significance level of 0.1 (although the trends were modeled at 0.1% for these sites), the non-stationary model overestimated the 100-year flood as compared to the stationary model, and the corresponding value of the Bayes factor was greater than 1. This ultimately favors the stationary model for sites 9 and 16 (Table 6, Figures 11 and 12).

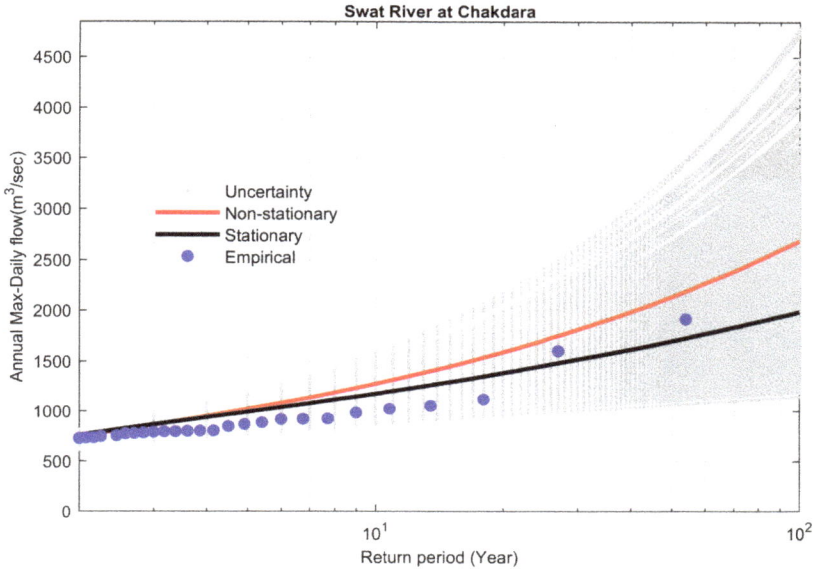

Figure 11. Comparison between non-stationary and stationary FFCs at site 9, along with non-stationary uncertainty bounds for the year 2009.

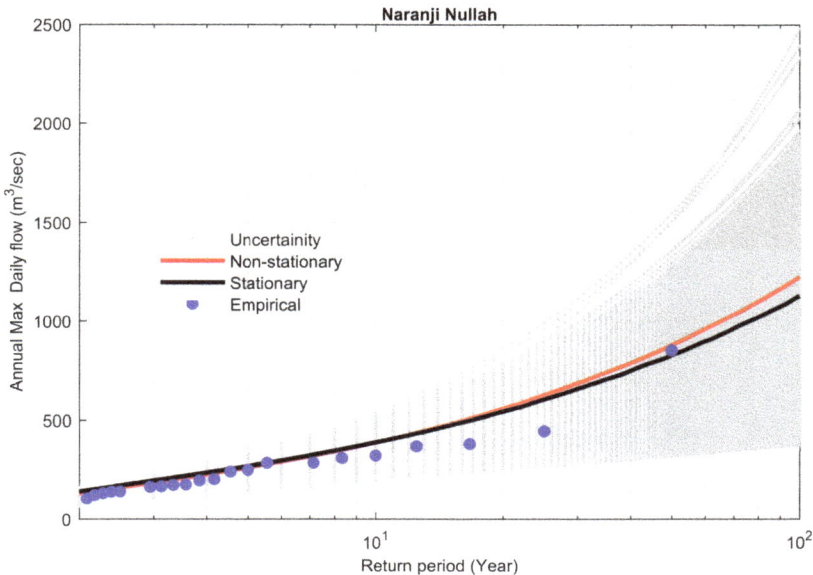

Figure 12. Comparison between non-stationary and stationary FFCs at site 16, along with non-stationary uncertainty bounds for the year 2016.

The non-stationary model was found to be reliable for the sites to study the annual maximum flood regime, which exhibits significant trends at $\alpha = 0.05$. The uncertainty bounds of most of the study sites were high because of the higher value of coefficient of variation, except site 6, where the value of coefficient of variation was 0.2 (see Table 1 and Figure 9). The uncertainty bound could also be higher due to the existence of outliers in the data series. The removal of outliers in the data series can reduce uncertainty.

4.5. Performance of Bayesian Models to Predict the Extreme Floods

The Bayesian models were again re-run for all the study sites before the commencement year of the extreme flood event that we labeled an outlier for all the study sites. Similarly, the objective criterion was adopted again (if a trend exists then modeling was performed using the non-stationary Bayesian model, otherwise only the stationary Bayesian model was used). Table 7 describes the results of the corresponding stationary and non-stationary Bayesian models.

The Bayesian models performed well, predicting the extreme floods satisfactorily for all the study sites except sites 9 and 22 (see Table 7 and Figures 13–22). The stationary Bayesian model was found reliable for sites 5, 6, 11, 16, 21, and 24. However, the non-stationary model was favored as compared to the stationary model as per the Bayes factor criterion (see Table 7) for sites 6, 18, 19, and 21. Despite the existence of a significant trend at site 11, the stationary model was favored as per the Bayes factor criterion, as well as predicting the extreme flood of 2016.

The outlier's removal improved the fitting of the FFCs as compared to considering the entire data series and also reduced the uncertainty (see Figures 13–22).

Statistical modeling of extremes in hydrology is exciting and challenging, and opens the door for further studies. For example, for study site 9 (the Swat River at Chakdara), it is better to consider the entire data series, and for better fitting, the incorporation of monsoon rainfall as a covariate might be fruitful. The modeling could also be performed by considering other distributions in the Bayesian framework or using the traditional frequentist framework.

Finally, from a regional perspective, the region is heterogeneous due to large altitudinal variations. Due to the regional heterogeneity associated with elevation, it seems to be quite difficult to develop a regional Bayesian model for the whole KRB, but efforts should be made to develop a regional Bayesian model at sub-basin or catchment scales in future studies by further pooling of information for other parameters, like location and scale, across sites. Moreover, the studies are also required to deeply understand the impact of climate or dominating weather patterns, such as the South Asian monsoon, low climate variability El-Niño Southern Oscillation and Indian Ocean Dipole (ENSO, IOD, etc.) and human factors, such as land use cover change (LUCC), population increase, and reservoir construction, on the flood regime of the KRB [1,13,80–83].

Table 7. Performance of Bayesian models for predicting the extreme floods in the KRB, Pakistan.

Site #	Time Series Length	Extreme Event (Year)	Mann–Kendall (Test-Z)	Stationary $m^3\,s^{-1}$	Non-Stationary $m^3\,s^{-1}$	Difference between Stationary and Non-Stationary $m^3\,s^{-1}$	Percent Difference (%)	Bayes Factor
5	1987–2009	2010	−0.05	3300	N/A	N/A	N/A	N/A
6	1964–2004	2005/2010	2.88 **	1701	1978	277	14	0.0054
9	1961–1991	1992	1.42	1746	N/A	N/A	N/A	N/A
11	1986–2015	2016	4.49 ***	2211	1295	916	41.43	15.167
16	1968–2009	2010	1.31	850.8	N/A	N/A	N/A	N/A
18	1984–2009	2010	2.29 *	1472	1085	387	26.3	0.1208
19	1984–2009	2010	2.76 **	4580	3595	990	21.6	0.008
21	1987–2009	2010	2.2 *	1469	704	765	52	0.016
22	1987–1996	1997	−0.28	30.38	N/A	N/A	N/A	N/A
24	1983–2009	2010	1.15	339.6	355	15.4	4.33	+Infinity

*** Trend is significant at $\alpha = 0.001$, ** Trend is significant at $\alpha = 0.01$, * Trend is significant at $\alpha = 0.05$, N/A, Non-stationary Bayesian modeling not applicable because of insignificant or no trend.

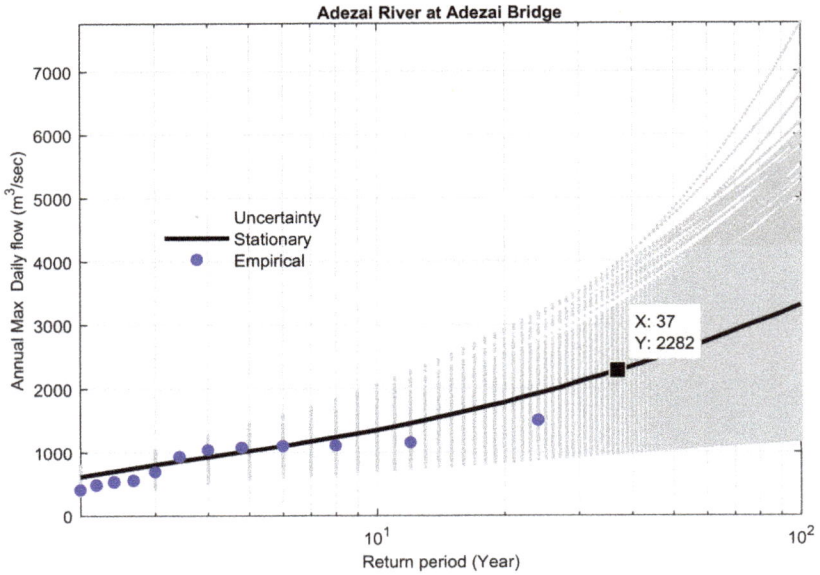

Figure 13. FFCs at site 5, along with stationary uncertainty bounds for the year 2009.

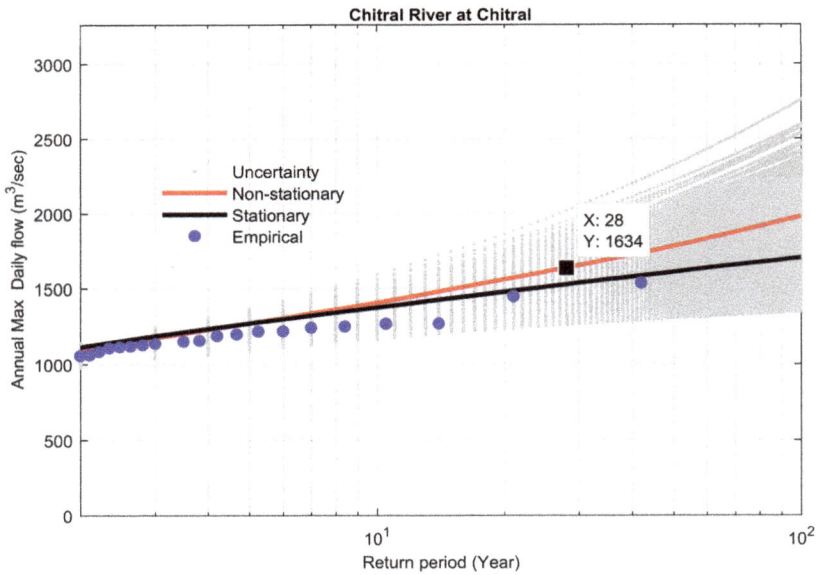

Figure 14. Comparison between non-stationary and stationary FFCs at site 6, along with non-stationary uncertainty bounds for the year 2004.

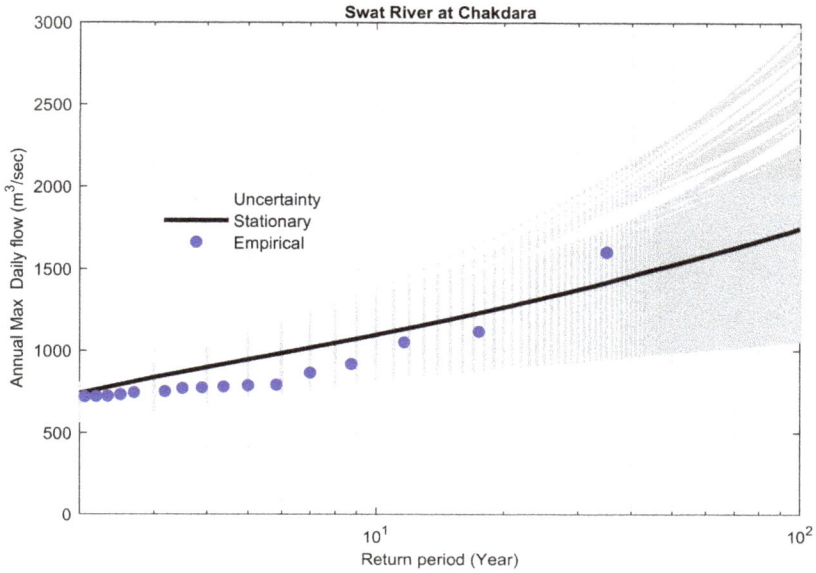

Figure 15. FFCs at site 9, along with stationary uncertainty bounds for the year 1991.

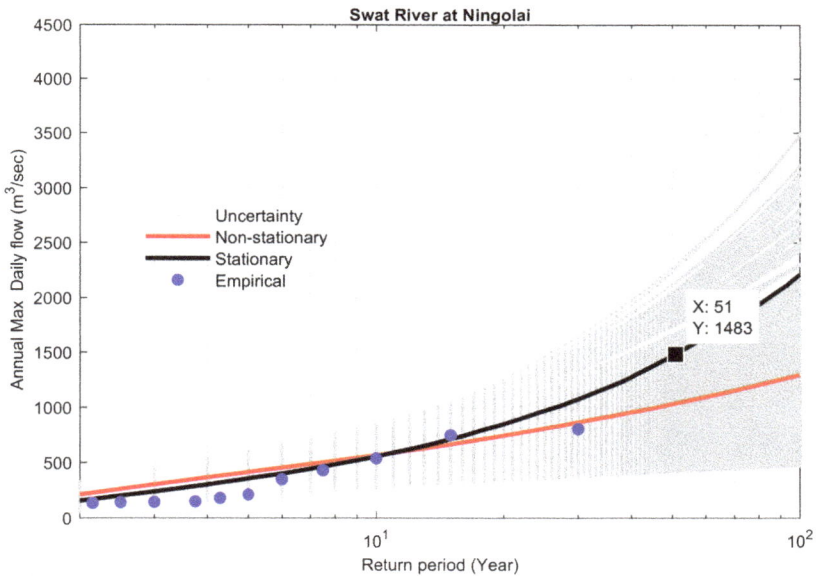

Figure 16. Comparison between non-stationary and stationary FFCs at site 11, along with non-stationary uncertainty bounds for the year 2015.

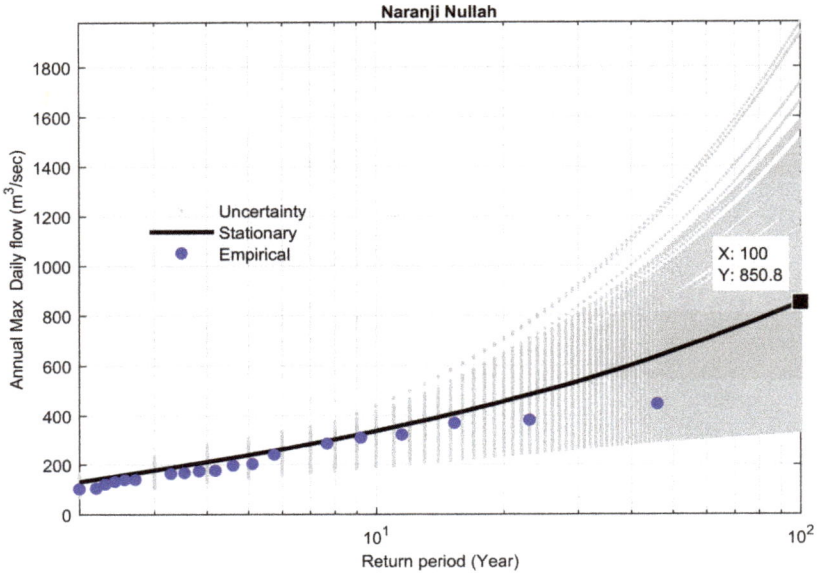

Figure 17. FFCs at site 16, along with stationary uncertainty bounds for the year 2009.

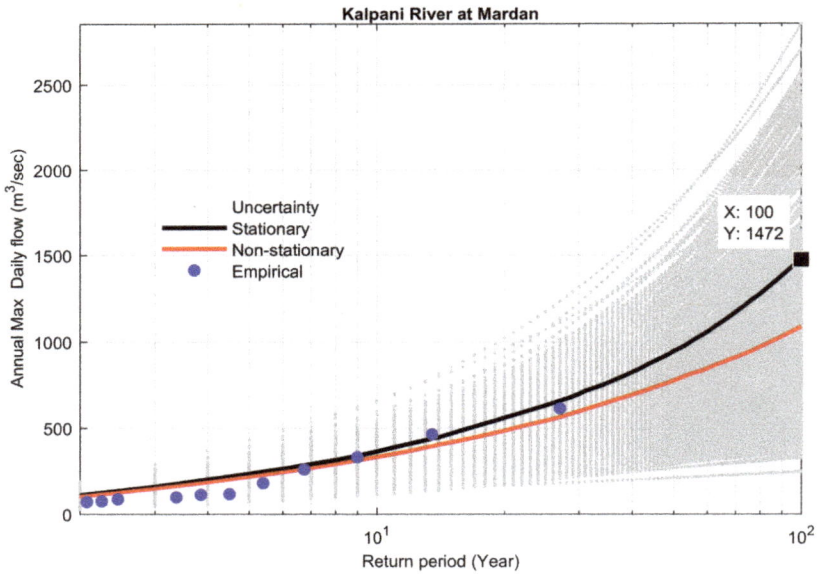

Figure 18. Comparison between non-stationary and stationary FFCs at site 18, along with non-stationary uncertainty bounds for the year 2009.

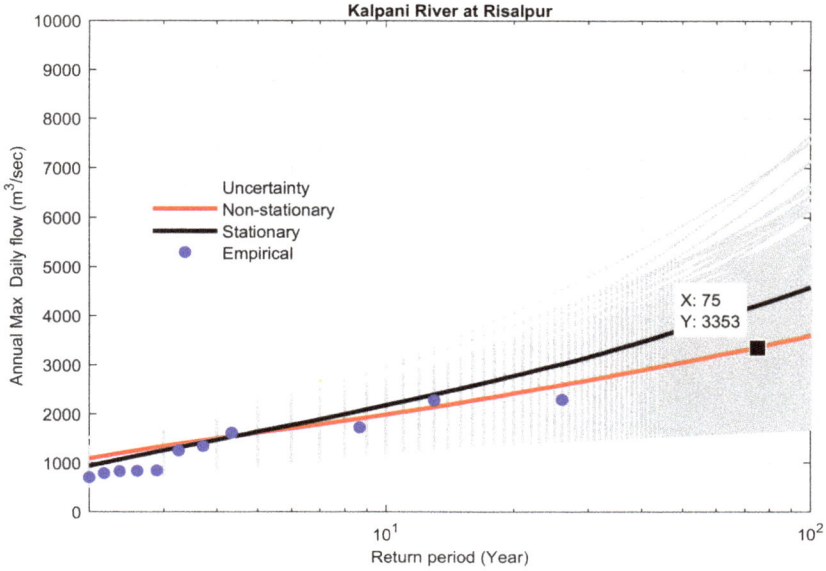

Figure 19. Comparison between non-stationary and stationary FFCs at site 19, along with non-stationary uncertainty bounds for the year 2009.

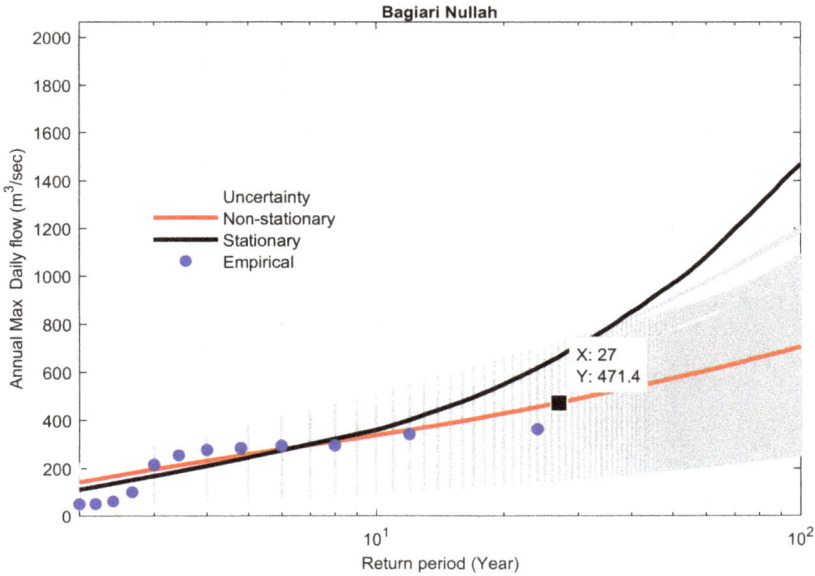

Figure 20. Comparison between non-stationary and stationary FFCs at site 21, along with non-stationary uncertainty bounds for the year 2009.

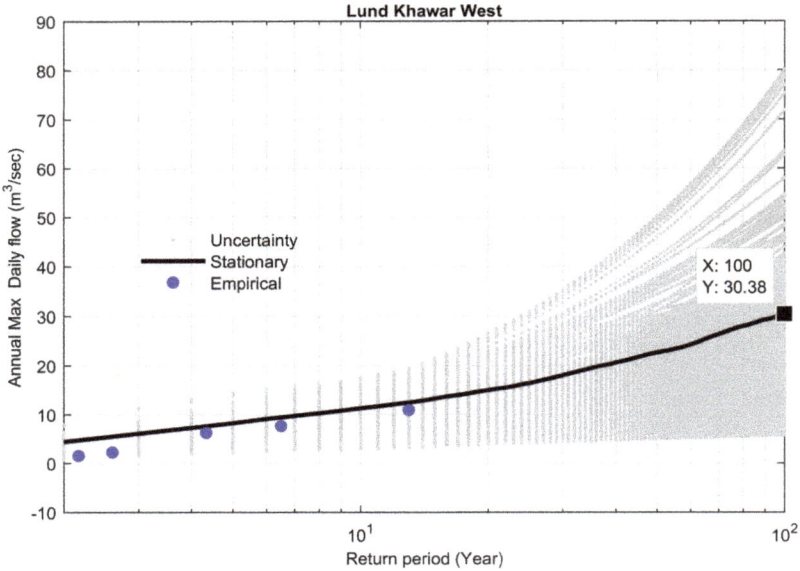

Figure 21. FFCs at site 22, along with stationary uncertainty bounds for the year 1996.

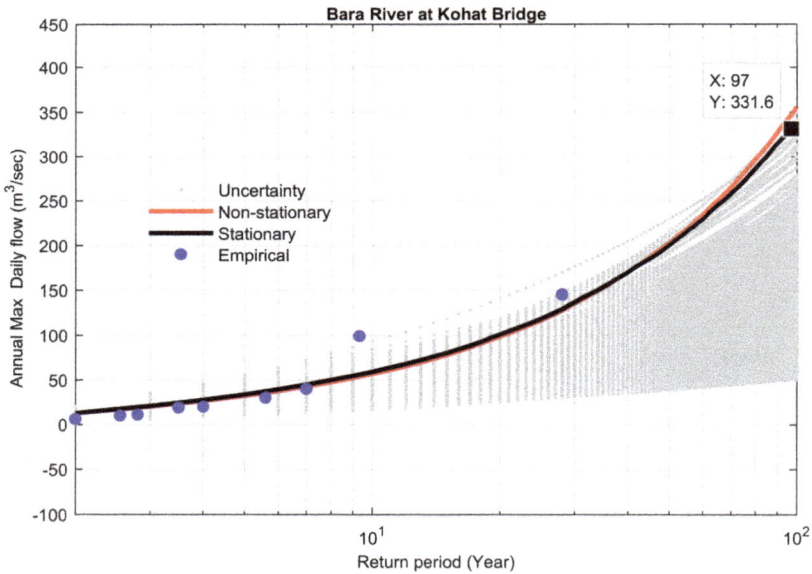

Figure 22. Comparison between non-stationary and stationary FFCs at site 24, along with non-stationary uncertainty bounds for the year 2009.

5. Conclusions

Analyzing the flood regime and its non-stationary modeling in a Bayesian framework for the KRB was the main objective of the present study. To achieve this, a Mann–Kendall trend analysis was performed to explore the flood regime of the KRB in detail, and finally, the stationary and non-stationary Bayesian models with informed priors on shape parameter for GEV distribution were developed, and their results were compared by using the corresponding FFCs, along with their uncertainty bounds. We utilized the annual extreme data of flood series for the study area, with a maximum record of 1962–2016 (55 years) and a minimum of 1987–2016 (30 years). The key findings of the study are described below:

1. Trend analysis showed a mixture of increasing and decreasing trends at different gauges in the KRB at $\alpha = 0.05$. The Chitral River, Kalpani River, Main Swat River, and Bara River basins showed significant increasing trends, and the Panjkora River basin displayed a moderate increasing trend in its annual maximum flood regime. However, the Lower Swat and Kabul sub-basins showed decreasing trends, except for the Adezai River in the Kabul sub-basin, which showed a significant increasing trend.
2. The overall basin was under critical change and signals of clear non-stationarity in the flood regime were evident at various spatial scales throughout the basin.
3. The presence of a significant trend and significant difference in flood estimates for 100-year flood between stationary and non-stationary FFCs were found that represent the clear violation from the so-called stationary assumption.
4. The non-stationary Bayesian model was found to be reliable for the study sites that had a significant trend at $\alpha = 0.05$, while the stationary model overestimated or underestimated the flood risk for these sites. On the other hand, the stationary Bayesian model performed better for the study sites for trends at $\alpha = 0.1$, while the non-stationary Bayesian model overestimated or underestimated the flood risk for such sites.
5. The use of informed priors on the shape parameter based on regional information improved the estimation of flood quantiles and reduced the uncertainty.
6. Proper consideration should be given to identify the outliers while using Bayesian models.
7. The presence of non-stationarity in the flood regime of the KRB has substantial implications for flood management and water resources development. A design with stationary assumption will cause two major concerns: under estimation or overestimation of design for structural and non-structural measures in the KRB. An event-based design may also overestimate or underestimate the risk in hydraulic design that was intended. Some previous studies in other parts of world also provided similar results [1,13,31,84–86].

The study will be helpful for sustainable flood management and provide a reference for studying floods in a changing environment for hydrologists, water resources managers, decision makers, and concerned organizations.

Author Contributions: This research article, A.M. and S.J. formulated research design, plan, organized research flow and manuscript write up. A.M. performed analysis, S.J. supervised research work and contributed in the interpretation of results and discussions, R.M. and M.A. contributed in drafting and map preparations and J.Y. was involved in short listing data sets from 45 flow gauge stations to 29 for the current study. All the authors contributed well to writing at various stages.

Funding: This research was funded by CAS-TWAS President Fellowship Program for doctoral students and the Strategic Priority Research Program of the Chinese Academy of Sciences [XDA20010201].

Acknowledgments: The authors acknowledge SWHP (Surface Water Hydrology Project) WAPDA Pakistan and Irrigation Department Khyber Pakhtunkhwa, Pakistan to provide data for this study.

Conflicts of Interest: The authors declare no conflicts of interests.

References

1. López, J.; Francés, F. Non-stationary flood frequency analysis in continental Spanish rivers, using climate and reservoir indices as external covariates. *Hydrol. Earth Syst. Sci.* **2013**, *17*, 3189–3203. [CrossRef]
2. Khaliq, M.; Ouarda, T.; Ondo, J.-C.; Gachon, P.; Bobée, B. Frequency analysis of a sequence of dependent and/or non-stationary hydro-meteorological observations: A review. *J. Hydrol.* **2006**, *329*, 534–552. [CrossRef]

3. Stedinger, J.R.; Vogel, R.; Foufoula-Georgiou, E. Frequency analysis of extreme events. *Handb. Hydrol.* **1993**, *18*, 68.
4. Salas, J. *Analysis and Modeling of Hydrologic Time Series in Hand Book of Hydrology*; Maidment, D.R., Ed.; McGraw Hill Book Co.: New York, NY, USA, 1993.
5. Council, N.R. *Decade-to-Century-Scale Climate Variability and Change: A Science Strategy*; National Academies Press: Washington, DC, USA, 1998.
6. Norrant, C.; Douguédroit, A. Monthly and daily precipitation trends in the Mediterranean (1950–2000). *Theor. Appl. Climatol.* **2006**, *83*, 89–106. [CrossRef]
7. Mudelsee, M.; Börngen, M.; Tetzlaff, G.; Grünewald, U. No upward trends in the occurrence of extreme floods in central Europe. *Nature* **2003**, *425*, 166–169. [CrossRef]
8. Douglas, E.; Vogel, R.; Kroll, C. Trends in floods and low flows in the United States: Impact of spatial correlation. *J. Hydrol.* **2000**, *240*, 90–105. [CrossRef]
9. Franks, S.W. Identification of a change in climate state using regional flood data. *Hydrol. Earth Syst. Sci.* **2002**, *6*, 11–16. [CrossRef]
10. Milly, P.C.; Dunne, K.A.; Vecchia, A.V. Global pattern of trends in streamflow and water availability in a changing climate. *Nature* **2005**, *438*, 347–350. [CrossRef]
11. Villarini, G.; Serinaldi, F.; Smith, J.A.; Krajewski, W.F. On the stationarity of annual flood peaks in the continental united states during the 20th century. *Water Resour. Res.* **2009**, *45*. [CrossRef]
12. Wilson, D.; Hisdal, H.; Lawrence, D. Has streamflow changed in the nordic countries?–recent trends and comparisons to hydrological projections. *J. Hydrol.* **2010**, *394*, 334–346. [CrossRef]
13. Villarini, G.; Smith, J.A.; Serinaldi, F.; Bales, J.; Bates, P.D.; Krajewski, W.F. Flood frequency analysis for nonstationary annual peak records in an urban drainage basin. *Adv. Water Resour.* **2009**, *32*, 1255–1266. [CrossRef]
14. Vogel, R.M.; Yaindl, C.; Walter, M. Nonstationarity: Flood magnification and recurrence reduction factors in the United States. *J. Am. Water Resour. Assoc.* **2011**, *47*, 464–474. [CrossRef]
15. Hejazi, M.I.; Markus, M. Impacts of urbanization and climate variability on floods in northeastern Illinois. *J. Hydrol. Eng.* **2009**, *14*, 606–616. [CrossRef]
16. Held, I.M.; Soden, B.J. Robust responses of the hydrological cycle to global warming. *J. Clim.* **2006**, *19*, 5686–5699. [CrossRef]
17. Allen, M.R.; Smith, L.A. Monte carlo ssa: Detecting irregular oscillations in the presence of colored noise. *J. Clim.* **1996**, *9*, 3373–3404. [CrossRef]
18. Zaman, C.Q.U.; Mahmood, A.; Rasul, G.; Afzal, M. *Climate Change Indicators of Pakistan*; Report No: PMD-22/2009; Pakistan Meteorological Department: Islamabad, Pakistan, 2009.
19. Ahmad, I.; Tang, D.; Wang, T.; Wang, M.; Wagan, B. Precipitation trends over time using mann-kendall and spearman's rho tests in Swat river basin, Pakistan. *Adv. Meteorol.* **2015**, *2015*, 431860. [CrossRef]
20. Khalid, S.; Rehman, S.U.; Shah, S.M.A.; Naz, A.; Saeed, B.; Alam, S.; Ali, F.; Gul, H. Hydro-meteorological characteristics of Chitral river basin at the peak of the Hindukush range. *Nat. Sci.* **2013**, *5*, 987. [CrossRef]
21. Hartmann, H.; Buchanan, H. Trends in extreme precipitation events in the Indus river basin and flooding in Pakistan. *Atmos. Ocean* **2014**, *52*, 77–91. [CrossRef]
22. Najmuddin, O.; Deng, X.; Siqi, J. Scenario analysis of land use change in Kabul river basin–a river basin with rapid socio-economic changes in Afghanistan. *Phys. Chem. Earth Parts A B C* **2017**, *101*, 121–136. [CrossRef]
23. Qasim, M.; Hubacek, K.; Termansen, M.; Khan, A. Spatial and temporal dynamics of land use pattern in district Swat, Hindu Kush Himalayan region of Pakistan. *Appl. Geogr.* **2011**, *31*, 820–828. [CrossRef]
24. Ullah, S.; Farooq, M.; Shafique, M.; Siyab, M.A.; Kareem, F.; Dees, M. Spatial assessment of forest cover and land-use changes in the Hindu-Kush mountain ranges of northern Pakistan. *J. Mt. Sci.* **2016**, *13*, 1229–1237. [CrossRef]
25. Sajjad, A.; Adnan, S.; Hussain, A. Forest land cover change from year 2000 to 2012 of tehsil Barawal Dir Upper Pakistan. *Int. J. Adv. Res. Biol. Sci.* **2016**, *3*, 144–154.
26. Ahmad, A.; Nizami, S.M. Carbon stocks of different land uses in the Kumrat valley, Hindu Kush region of Pakistan. *J. For. Res.* **2015**, *26*, 57–64. [CrossRef]
27. Yar, P.; Atta-ur-Rahman, M.A.K.; Samiullah, S. Spatio-temporal analysis of urban expansion on farmland and its impact on the agricultural land use of Mardan city, Pakistan. *Proc. Pak. Acad. Sci. B Life Environ. Sci.* **2016**, *53*, 35–46.

28. Raziq, A.; Xu, A.; Li, Y.; Zhao, Q. Monitoring of land use/land cover changes and urban sprawl in peshawar city in khyber pakhtunkhwa: An application of geo-information techniques using of multi-temporal satellite data. *J. Remote Sens. GIS* **2016**, *5*. [CrossRef]
29. Milly, P.C.; Betancourt, J.; Falkenmark, M.; Hirsch, R.M.; Kundzewicz, Z.W.; Lettenmaier, D.P.; Stouffer, R.J. Stationarity is dead: Whither water management? *Science* **2008**, *319*, 573–574. [CrossRef] [PubMed]
30. Delgado, J.M.; Apel, H.; Merz, B. Flood trends and variability in the Mekong river. *Hydrol. Earth Syst. Sci.* **2010**, *14*, 407–418. [CrossRef]
31. Leclerc, M.; Ouarda, T.B. Non-stationary regional flood frequency analysis at ungauged sites. *J. Hydrol.* **2007**, *343*, 254–265. [CrossRef]
32. Olsen, J.R.; Lambert, J.H.; Haimes, Y.Y. Risk of extreme events under nonstationary conditions. *Risk Anal.* **1998**, *18*, 497–510. [CrossRef]
33. McNeil, A.J.; Saladin, T. Developing Scenarios for Future Extreme Losses Using the Pot Method. In *Extremes and Integrated Risk Management*; Embrechts, P., Ed.; CiteseerX: Zurich, Switzerland, 2000; pp. 253–267.
34. Stedinger, J.R.; Crainiceanu, C.M. Climate Variability and Flood-Risk Management. In *Risk-Based Decision Making in Water Resources IX*; ASCE: Reston, VA, USA, 2001; pp. 77–86.
35. Strupczewski, W.; Singh, V.; Mitosek, H. Non-stationary approach to at-site flood frequency modelling. III. Flood analysis of Polish rivers. *J. Hydrol.* **2001**, *248*, 152–167. [CrossRef]
36. He, Y.; Bárdossy, A.; Brommundt, J. Non-Stationary Flood Frequency Analysis in Southern Germany. In Proceedings of the Seventh International Conference on Hydroscience and Engineering, Philadelphia, PA, USA, 10–13 September 2006.
37. Renard, B.; Lang, M.; Bois, P. Statistical analysis of extreme events in a non-stationary context via a bayesian framework: Case study with peak-over-threshold data. *Stoch. Environ. Res. Risk Assess.* **2006**, *21*, 97–112. [CrossRef]
38. Khattak, M.; Anwar, F.; Sheraz, K.; Saeed, T.; Sharif, M.; Ahmed, A. Floodplain mapping using hec-ras and arcgis: A case study of Kabul river. *Arab. J. Sci. Eng. (Springer Sci. Bus. Media BV)* **2016**, *41*, 1375–1390. [CrossRef]
39. Sayama, T.; Ozawa, G.; Kawakami, T.; Nabesaka, S.; Fukami, K. Rainfall–runoff–inundation analysis of the 2010 Pakistan flood in the Kabul river basin. *Hydrol. Sci. J.* **2012**, *57*, 298–312. [CrossRef]
40. Bahadar, I.; Shafique, M.; Khan, T.; Tabassum, I.; Ali, M.Z. Flood hazard assessment using hydro-dynamic model and gis/rs tools: A case study of Babuzai-Kabal tehsil Swat basin, Pakistan. *J. Himal. Earth Sci.* **2015**, *48*, 129–138.
41. Aziz, A. Rainfall-runoff modeling of the trans-boundary Kabul river basin using integrated flood analysis system (ifas). *Pak. J. Meteorol.* **2014**, *10*, 75–81.
42. Ullah, S.; Farooq, M.; Sarwar, T.; Tareen, M.J.; Wahid, M.A. Flood modeling and simulations using hydrodynamic model and aster dem—A case study of Kalpani river. *Arab. J. Geosci.* **2016**, *9*, 439. [CrossRef]
43. Mack, T.J.; Chornack, M.P.; Taher, M.R. Groundwater-level trends and implications for sustainable water use in the Kabul basin, afghanistan. *Environ. Syst. Decis.* **2013**, *33*, 457–467. [CrossRef]
44. Lashkaripour, G.R.; Hussaini, S. Water resource management in Kabul river basin, Eastern Afghanistan. *Environmentalist* **2008**, *28*, 253–260. [CrossRef]
45. Tariq, M.A.U.R.; Van de Giesen, N. Floods and flood management in Pakistan. *Phys. Chem. Earth Parts A B C* **2012**, *47*, 11–20. [CrossRef]
46. Anjum, M.N.; Ding, Y.; Shangguan, D.; Ijaz, M.W.; Zhang, S. Evaluation of high-resolution satellite-based real-time and post-real-time precipitation estimates during 2010 extreme flood event in Swat river basin, Hindukush region. *Adv. Meteorol.* **2016**, *2016*, 2604980. [CrossRef]
47. Rasul, G.; Dahe, Q.; Chaudhry, Q. Global warming and melting glaciers along southern slopes of HKH range. *Pak. J. Meteorol.* **2008**, *5*, 63–76.
48. Mann, H.B. Nonparametric tests against trend. *Econom. J. Econom. Soc.* **1945**, *13*, 245–259. [CrossRef]
49. Kendall, M. *Rank Correlation Methods*; Charles Griffin: London, UK, 1975.
50. Katz, R.W. Statistics of extremes in climate change. *Clim. Chang.* **2010**, *100*, 71–76. [CrossRef]
51. Coles, S.; Bawa, J.; Trenner, L.; Dorazio, P. *An Introduction to Statistical Modeling of Extreme Values*; Springer: Berlin/Heidelberg, Germany, 2001; Volume 208.
52. Smith, R. Extreme value statistics in meteorology and the environment. *Environ. Stat.* **2001**, *8*, 300–357.

53. Shukla, R.K.; Trivedi, M.; Kumar, M. On the proficient use of gev distribution: A case study of subtropical monsoon region in India. *arXiv* **2012**, arXiv:1203.0642.

54. Massey, F.J., Jr. The kolmogorov-Smirnov test for goodness of fit. *J. Am. Stat. Assoc.* **1951**, *46*, 68–78. [CrossRef]

55. Mehrannia, H.; Pakgohar, A. Using easy fit software for goodness-of-fit test and data generation. *Int. J. Math. Arch.* **2014**, *5*, 118–124.

56. Lin, L.; Sherman, P.D. Cleaning Data the Chauvenet Way. In Proceedings of the SouthEast SAS Users Group, Hilton Head Island, SC, USA, 4–6 November 2007; SESUG Proceedings, Paper SA11.

57. Renard, B.; Sun, X.; Lang, M. Bayesian Methods for Non-Stationary Extreme Value Analysis. In *Extremes in a Changing Climate*; Springer: Berlin/Heidelberg, Germany, 2013; pp. 39–95.

58. Meehl, G.A.; Karl, T.; Easterling, D.R.; Changnon, S.; Pielke, R., Jr.; Changnon, D.; Evans, J.; Groisman, P.Y.; Knutson, T.R.; Kunkel, K.E. An introduction to trends in extreme weather and climate events: Observations, socioeconomic impacts, terrestrial ecological impacts, and model projections. *Bull. Am. Meteorol. Soc.* **2000**, *81*, 413–416. [CrossRef]

59. Gilleland, E.; Katz, R.W. New software to analyze how extremes change over time. *Eos Trans. Am. Geophys. Union* **2011**, *92*, 13–14. [CrossRef]

60. Cheng, L.; AghaKouchak, A.; Gilleland, E.; Katz, R.W. Non-stationary extreme value analysis in a changing climate. *Clim. Chang.* **2014**, *127*, 353–369. [CrossRef]

61. Stephenson, A.; Tawn, J. Bayesian inference for extremes: Accounting for the three extremal types. *Extremes* **2004**, *7*, 291–307. [CrossRef]

62. Ragno, E.; AghaKouchak, A.; Love, C.A.; Cheng, L.; Vahedifard, F.; Lima, C.H. Quantifying changes in future intensity-duration-frequency curves using multimodel ensemble simulations. *Water Resour. Res.* **2018**, *54*, 1751–1764. [CrossRef]

63. Martins, E.S.; Stedinger, J.R. Generalized maximum-likelihood generalized extreme-value quantile estimators for hydrologic data. *Water Resour. Res.* **2000**, *36*, 737–744. [CrossRef]

64. Ter Braak, C.J. A Markov chain monte carlo version of the genetic algorithm differential evolution: Easy bayesian computing for real parameter spaces. *Stat. Comput.* **2006**, *16*, 239–249. [CrossRef]

65. Vrugt, J.A.; Ter Braak, C.; Diks, C.; Robinson, B.A.; Hyman, J.M.; Higdon, D. Accelerating markov chain monte carlo simulation by differential evolution with self-adaptive randomized subspace sampling. *Int. J. Nonlinear Sci. Numer. Simul.* **2009**, *10*, 273–290. [CrossRef]

66. Gelman, A.; Shirley, K. Inference from Simulations and Monitoring Convergence. In *Handbook. Markov Chain Monte Carlo*; CRC Press: Boca Raton, FA, USA, 2011; pp. 163–174.

67. Kass, R. Re kass and ae raftery. *J. Am. Stat. Assoc.* **1995**, *90*, 773–795. [CrossRef]

68. Khan, A. Analysis of streamflow data for trend detection on major rivers of the indus basin. *J. Himal. Earth Sci. Vol.* **2015**, *48*, 99–111.

69. Khan, K.; Yaseen, M.; Latif, Y.; Nabi, G. Detection of river flow trends and variability analysis of Upper Indus basin, pakistan. *Sci. Int.* **2015**, *27*, 1261–1270.

70. Sharif, M.; Archer, D.; Fowler, H.; Forsythe, N. Trends in timing and magnitude of flow in the Upper Indus basin. *Hydrol. Earth Syst. Sci.* **2013**, *17*, 1503–1516. [CrossRef]

71. Rosner, A.; Vogel, R.M.; Kirshen, P.H. A risk-based approach to flood management decisions in a nonstationary world. *Water Resour. Res.* **2014**, *50*, 1928–1942. [CrossRef]

72. Sun, X.; Thyer, M.; Renard, B.; Lang, M. A general regional frequency analysis framework for quantifying local-scale climate effects: A case study of enso effects on southeast Queensland rainfall. *J. Hydrol.* **2014**, *512*, 53–68. [CrossRef]

73. Halbert, K.; Nguyen, C.C.; Payrastre, O.; Gaume, E. Reducing uncertainty in flood frequency analyses: A comparison of local and regional approaches involving information on extreme historical floods. *J. Hydrol.* **2016**, *541*, 90–98. [CrossRef]

74. Kyselý, J.; Gaál, L.; Picek, J. Comparison of regional and at-site approaches to modelling probabilities of heavy precipitation. *Int. J. Climatol.* **2011**, *31*, 1457–1472. [CrossRef]

75. Viglione, A.; Merz, R.; Salinas, J.L.; Blöschl, G. Flood frequency hydrology: 3. A bayesian analysis. *Water Resour. Res.* **2013**, *49*, 675–692. [CrossRef]

76. Kuczera, G. Combining site-specific and regional information: An empirical bayes approach. *Water Resour. Res.* **1982**, *18*, 306–314. [CrossRef]

77. Sun, X.; Lall, U.; Merz, B.; Dung, N.V. Hierarchical bayesian clustering for nonstationary flood frequency analysis: Application to trends of annual maximum flow in Germany. *Water Resour. Res.* **2015**, *51*, 6586–6601. [CrossRef]

78. Katz, R.W.; Parlange, M.B.; Naveau, P. Statistics of extremes in hydrology. *Adv Water Resour.* **2002**, *25*, 1287–1304. [CrossRef]

79. Lima, C.H.; Lall, U.; Troy, T.; Devineni, N. A hierarchical bayesian gev model for improving local and regional flood quantile estimates. *J. Hydrol.* **2016**, *541*, 816–823. [CrossRef]

80. Kwon, H.H.; Brown, C.; Lall, U. Climate informed flood frequency analysis and prediction in Montana using hierarchical bayesian modeling. *Geophys. Res. Lett.* **2008**, *35*. [CrossRef]

81. Steinschneider, S.; Lall, U. A hierarchical bayesian regional model for nonstationary precipitation extremes in northern california conditioned on tropical moisture exports. *Water Resour. Res.* **2015**, *51*, 1472–1492. [CrossRef]

82. Lima, C.H.; Lall, U.; Troy, T.J.; Devineni, N. A climate informed model for nonstationary flood risk prediction: Application to negro river at Manaus, Amazonia. *J. Hydrol.* **2015**, *522*, 594–602. [CrossRef]

83. Machado, M.J.; Botero, B.; López, J.; Francés, F.; Díez-Herrero, A.; Benito, G. Flood frequency analysis of historical flood data under stationary and non-stationary modelling. *Hydrol. Earth Syst. Sci.* **2015**, *19*, 2561–2576. [CrossRef]

84. Šraj, M.; Viglione, A.; Parajka, J.; Blöschl, G. The influence of non-stationarity in extreme hydrological events on flood frequency estimation. *J. Hydrol. Hydromech.* **2016**, *64*, 426–437. [CrossRef]

85. Hounkpè, J.; Diekkrüger, B.; Badou, D.F.; Afouda, A.A. Non-stationary flood frequency analysis in the Ouémé river basin, Benin Republic. *Hydrology* **2015**, *2*, 210–229. [CrossRef]

86. Xiong, L.; Du, T.; Xu, C.-Y.; Guo, S.; Jiang, C.; Gippel, C.J. Non-stationary annual maximum flood frequency analysis using the norming constants method to consider non-stationarity in the annual daily flow series. *Water Resour. Manag.* **2015**, *29*, 3615–3633. [CrossRef]

water

MDPI

Article

Characteristics of Heavy Storms and the Scaling Relation with Air Temperature by Event Process-Based Analysis in South China

Cuilin Pan [1,2], Xianwei Wang [1,3,4,5,*], Lin Liu [6,7,*], Dashan Wang [1] and Huabing Huang [1,3,5]

[1] School of Geography and Planning, Sun Yat-sen University, Guangzhou 510275, China; pancuil@mail2.sysu.edu.cn (C.P.); wangdash@mail2.sysu.edu.cn (D.W.); huanghb7@mail.sysu.edu.cn (H.H.)
[2] Department of Information Technology & Automation, The Pearl River Hydraulic Research Institute, Guangzhou 510611, China
[3] Guangdong Key Laboratory for Urbanization and Geo-simulation, Sun Yat-sen University, Guangzhou 510275, China
[4] Southern Laboratory of Ocean Science and Engineering (Guangdong, Zhuhai), Zhuhai 519000, China
[5] Guangdong Provincial Engineering Research Center for Public Security and Disasters, Guangzhou 510275, China
[6] Department of Geography, Guangzhou University, Guangzhou 510006, China
[7] Department of Geography, University of Cincinnati, Cincinnati, OH 45221, USA
[*] Correspondence: wangxw8@mail.sysu.edu.cn (X.W.); liulin2@mail.sysu.edu.cn (L.L.); Tel.: +86-20-841-146-23 (X.W.)

Received: 29 December 2018; Accepted: 18 January 2019; Published: 22 January 2019

Abstract: The negative scaling rate between precipitation extremes and the air temperature in tropic and subtropic regions is still a puzzling issue. This study investigates the scaling rate from two aspects, storm characteristics (types) and event process-based temperature variations. Heavy storms in South China are developed by different weather systems with unique meteorological characteristics each season, such as the warm-front storms (January), cold-front storms (April to mid-May), monsoon storms (late May to June), convective storms, and typhoon storms (July to September). This study analyzes the storm characteristics using the hourly rainfall data from 1990 to 2017; compares the storm hyetographs derived from the one-minute rainfall data during 2008–2017; and investigates the interactions between heavy storms and meteorological factors including air temperature, relative humidity, surface pressure, and wind speed at 42 weather stations in Guangzhou during 2015–2017. Most storms, except for typhoon and warm-front storms, had a short duration (3 h) and intense rates (~13 mm/h) in Guangzhou, South China. Convective storms were dominant (50%) in occurrence and had the strongest intensity (15.8 mm/h). Storms in urban areas had stronger interactions with meteorological factors and showed different hyetographs from suburban areas. Meteorological factors had larger variations with the storms that occurred in the day time than at night. The air temperature could rise 6 °C and drop 4 °C prior to and post-summer storms against the diurnal mean state. The 24-h mean air temperature prior to the storms produced more reliable scaling rates than the naturally daily mean air temperature. The precipitation extremes showed a peak-like scaling relation with the 24-h mean air temperature and had a break temperature of 28 °C. Below 28 °C, the relative humidity was 80%–100%, and it showed a positive scaling rate. Above 28 °C, the negative scaling relation was likely caused by a lack of moisture in the atmosphere, where the relative humidity decreased with the air temperature increase.

Keywords: heavy storm; hyetograph; temperature; clausius-clapeyron scaling

1. Introduction

Heavy storm rainfall is the driving force of urban pluvial flooding. Mega cities, especially in the developing countries, such as China and India, suffered frequent flooding disasters in recent years in the context of global warming and fast urbanization [1–3]. Urban pluvial flooding or waterlogging is a common problem in many mega cites of China, such as Nanjing, Wuhan, and Guangzhou [4]. Guangzhou faces severe challenges for its over-stressed storm water drainage systems due to the heavy tropical storms and rapid urbanization in the past 30 years [5]. The impervious urban areas have a complicated impact on local weather systems, resulting in the phenomena of a heat and rain island [6]. Numerical modeling studies found that the increase of urban areas would significantly intensify the local extreme rainfall [1,7,8]. Experimental observations reported that the precipitation down-wind of large cities could increase 5%–25% from the background values [1,6]. There are urgent needs to study and update the heavy rainfall characteristics for better storm water management and emergency response in the metropolitan areas of Guangzhou, South China.

Heavy storms in South China are developed by different weather systems each season. They have dynamically unique environment structures largely controlled by three-dimensional meteorological factors, such as air temperature, humidity, pressure, and wind speed and direction, leading to different storm types and forming mechanisms. Four types of warm season storms are reported in the literature [9,10], that is, cold-front storms (April to mid-May), monsoon storms (late May to June), convective storms, and typhoon storms (July to September). Most warm season storms have a short duration and intense rates in Guangzhou, except for typhoon storms [5,11].

Heavy storms have complicated interactions with air temperature. The impact of air temperature on precipitation extremes have been extensively investigated after the pioneering work of Lenderink and van Meijgaard [12] in the Netherlands. The ideal gas law and Clausius–Clapeyron (CC) equation is the theoretical basis for such studies. The water-holding capacity of the atmosphere increases with the air temperature by about 7% $°C^{-1}$ globally for a given relative humidity, thus the precipitation extreme is proposed so as to scale with the precipitable water content in the atmosphere [12,13]. Many studies have investigated the scaling rate using numerical models and field observations at regional and global scales. Overall, five types of scaling rates between surface daily mean air temperature and precipitation extremes were reported, namely sub-CC (~3% $°C^{-1}$), close-CC (~7% $°C^{-1}$), super-CC (~14% $°C^{-1}$), peak-like CC (positive and negative), and negative CC [14–17].

The apparent scaling rates are mostly affected by the regional climatic settings, namely air temperature variation ranges and available water vapor. Sub-CC, close-CC, and super-CC were reported in mid and high latitude regions with a daily mean air temperature below ~20 °C, such as in the Netherlands [12], Germany [18], France [13], and Canada [16], and in the winter time of mid-latitude regions, such as the United States [14], southern Australia [19], and China [20]. Peak-like CC were reported in the mid latitude regions (20–55° N and 20–55° S), with the upper range of daily mean temperature above 25 °C [14,21,22], such as in Central Australia [19], South China [20], and Southern France [13]. The negative CC were reported in the tropic regions and the summer of the subtropical regions with a daily mean temperature above 25 °C [14], such as in Brazil [15], Northern Australia [19,23], South China [20], and Hong Kong [24].

Other factors affecting the scaling rates include the available moisture source (humidity), percentiles, and durations used to quantify the precipitation extremes. Higher percentiles and a shorter duration display a better close-CC or supper-CC [13,14,16]. The negative part of the peak-like scaling was explained by the lack of a moisture source, such as in Southern France [13]. This was supported by the fact that there was a general decrease in the relative humidity with a temperature increase at most stations in Australia, which suggests that the precipitation extremes were not only associated with how much moisture the atmosphere can hold, but also with how much moisture was available in the first place [15,19,23,25,26].

In summary, the current studies are mostly aimed at how global/regional warming intensifies the precipitation extremes conditional to the rainfall occurring with an available moisture source.

The orographic and other meteorological factors influencing rainfall occurrence are also important in constraining the changes of the precipitation extremes. However, few studies investigate the feedback and interactions of precipitation extremes with air temperature and other meteorological factors prior to and after a storm, especially in the tropical and subtropical regions. The behavior and mechanisms of tropical and subtropical heavy storms are worthy of further investigations.

The primary objectives of this study are (1) to analyze the characteristics of the different types of heavy storms in the metropolitan areas of Guangzhou, South China (subtropical, 23° N), and (2) to reveal the interactions of the heavy storms with air temperature and other meteorological factors, including relative humidity, surface pressure, and wind, using event process-based analysis.

2. Study Area and Data

2.1. Study Area

The City of Guangzhou is located in the upper Pearl River Delta in Southern China (Figure 1a). It has a sub-tropic climate controlled by the Indian summer monsoon and the South China Sea monsoon later in the year, with an annual mean air temperature of 22 °C and precipitation of 1700 mm [27,28]. The warm and wet rainy season starts from April through to September, and falls over 80% of the annual precipitation [4,29]. The rainy season is usually divided into three periods [30]. From April to mid-May, rainfall is dominated by frontal systems, being affected by the large-scale cold air south down from the mid-latitudes and the southwest warm air along the west flank of the western North Pacific subtropical high [31]. From late May to June, after the summer monsoon onset over the South China Sea, the monsoonal rain band advances up to the Pearl River Delta areas (Guangzhou), and the rainfall mainly results from a southeasterly direction, which transports water vapor into Guangzhou [9,10,32]. From July to September, monsoon rainfall becomes relatively weakened, and convective thunderstorms and tropical cyclones contribute appreciably to the rainfall in Guangzhou [33,34]. The first two periods are also called the first rainy season, while the third period is called the second rainy season [11]. The warm season storms in Guangzhou can be classified into four classes, mostly based on the location of the subtropical high (i.e., the cold-front storms, monsoon storms, convective storms, and typhoon storms) [30].

At present, the administration area of Guangzhou is 7434 km^2. It includes 11 districts—Yuexiu, Haizhu, Liwan, Tianhe, Baiyun, Huangpu, Huadu, Panyu, Nansha, Chonghua, and Zengcheng [5]. The metropolitan area has undergone fast urbanization during the past 30 years, and the built-up area ratio increased from 3% to 24% from 1990 to 2013, according to Landsat images [5]. There are 42 standard automatic weather stations in Guangzhou. These stations are divided into two groups of urban and suburban, so as to examine the generic characteristic of the meteorological factors and their variations with storms in this study (Figure 1b). In addition, six stations in the Tianhe (Site 2/rain gauge) and Panyu (Site 1, 3–6) Districts had a one-minute record of rainfall and water depth data, which were used to develop the rain hyetograph (Figure 1c).

2.2. Rain Depth and Other Meteorological Data

The 42 automatic weather stations contained data on the rainfall accumulation, air temperature, relative humidity, surface pressure, wind speed, and direction. All of the data were processed and archived in an hourly interval. Their precisions were 0.1 mm for precipitation, 0.1 °C for air temperature, 0.1% for relative humidity, 0.1 m/s for wind speed data, and 1° for wind direction. The data duration was 28 months, from July 2015 to October 2017. All of the climate data were validated by using quality control procedures [35–37].

Two sources of rainfall data from six automatic gauges were used to develop the storm hyetograph. The first one was from the national standard weather stations (Sites 1 and 2) of China, where the rainfall data were automatically recorded at one-minute intervals with a precision of 0.1 mm (Figure 1c). Site 2 is in the downtown area of the Tianhe District, Site 1 is in the suburban Panyu District, and both sites are 25 km apart. In addition, the processed hourly-interval rainfall data from 1990–2017 at Sites 1 and 2, and the hourly data at Sites 3–6 from 2014–2017, were used to analyze the storm features separately for the suburban and urban stations at the climatic time scale. Sites 3–6 were set up in the summer of 2014 at the Panyu District by our own research team. The rainfall data were recorded at one-minute intervals with a precision of 1 mm, which aimed to record the heavy storm rainfall. The five-year rainfall data (one-minute interval) from 2008 to 2012 at Sites 1 and 2, and the three-year rainfall data from 2014–2017 at Sites 3–6, were obtained in order to develop the rain hyetographs, respectively.

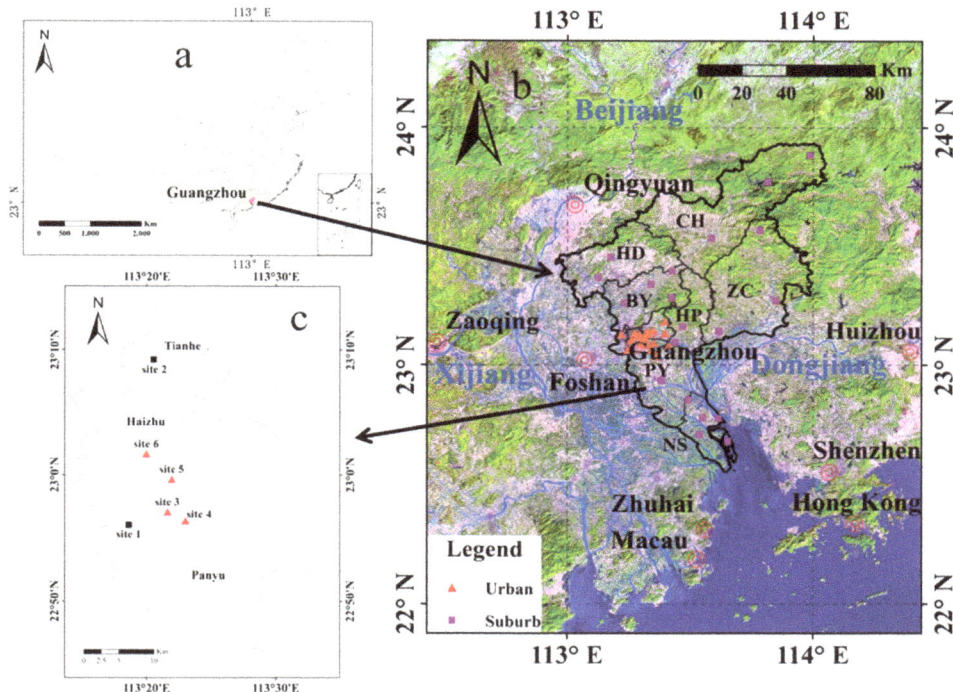

Figure 1. Meteorological sites (urban: 23 triangles; suburban: 19 squares) in the administration areas of Guangzhou (**b**), South China. The urban areas include the four districts of Haizhu, Liwan, Tianhe, and Yuexiu, and the suburban areas comprise the seven districts of Baiyun (BY), Huadu (HD), Conghua (CH), Zengcheng (ZC), Huangpu (HP), Panyu (PY), and Nansha (NS). Map (**c**) shows the meteorological sites in the Panyu District, where Sites 3–6 are maintained by our research team, and Sites 1 and 2 are the national standard meteorological sites.

2.3. China Hourly Merged Precipitation Analysis (CMPA)

The China Hourly Merged Precipitation Analysis (CMPA) data merged the hourly precipitation products with $0.1° \times 0.1°$ spatial resolution [38] (http://cdc.nmic.cn/home.do), and are available from 2008 to present. They show a much better performance in quantifying the extreme rainfall than the other satellite and reanalysis precipitation data in China [11,38,39]. The CMPA data are used to illustrate the spatial distributions of five typical storms for the peak intensity and event total precipitation.

3. Methodology

3.1. Storm Events Classification

This study does not analyze all of the rain events and only focuses on heavy storms, as they can produce a severe impact on meteorological factors and cause surface flooding. Storm events are identified at the individual stations based on the following criteria: (a) rain duration >20 min for one-minute data or one hour for hourly data [40], (b) rain depth in a one-hour moving window >20 mm, and (c) storm event separation with an hourly rain depth <1 mm for at least for three hours [41]. According to these criteria, there were 2611 storms at Sites 1 to 6 during 1990–2017, which were used to analyze the storm features. Among them, there were 214 storms recorded at Sites 1–2 from 2008 to 2012 and at Sites 3–6 from 2014 to 2017, using the one-minute interval. There were another 1454 storms at the 42 weather stations from July 2015 to October 2017. The 1454 storms were not physically separate storm events defined in meteorology, and occurred in Guangzhou. Some of them were actually the same storm events that occurred at the same or at a slightly later time in the metropolitan areas of Guangzhou, but were recorded at different weather stations. Those storm events at the 42 weather stations were mainly used to analyze the variations of the meteorological factors along the process of storm development and evolution.

In order to analyze the interactions between the storm (rainfall) and meteorological factors (air temperature, relative humidity, surface pressure, and wind speed), the 42 weather stations were first divided into urban and suburban groups using the K-means cluster analysis, while considering their location and neighboring land use/cover (Figure 1b). The K-means cluster algorithm set the initial center values of the meteorological variables for the two clusters of urban and suburban, and then calculated their minimum squared distance from the samples to their centers iteratively [11,42]. Finally, all of the stations were classified into the two clusters by the K-means cluster analysis using the time series of the hourly observations of the five meteorological factors for each storm event in this study. There were 23 urban stations (55%) and 19 suburban (Figure 1b) stations. All of the heavy storm events at both the urban and suburban clusters were generally classified into five types according to the season or the locations of the subtropical high, which determines the vapor source and forming mechanisms of heavy storms [9,10]. They are (a) warm-front storms (occurred in January), (b) cold-front storms (April to mid-May), (c) monsoon storms (late May to June), (d) convective storms (July to September), and (e) typhoon storms (July to September). The typhoon storms were precisely identified.

The cold-front storms, monsoon storms, and convective storms were further divided into three groups by occurrence time (i.e., 8:00–12:00, 13:00–18:00, and 19:00–0:00–7:00), so as to assess the impact of the heavy storms on the meteorological factors during the storm process in different periods/solar radiation, and thus could better analyze their interaction with storms. Warm-front and typhoon storms had a limited storm count and did not have such an analysis carried out.

3.2. Anomaly Curves

After the storm events were classified, anomaly curves 36 h prior to and post the storm peak hour were generated so as to analyze the impact on and the interaction of the storms with meteorological factors. The reference values are the diurnal mean of each factor during two weeks centered on the storm time, excluding their values during the 72-h period affected by the storm. The anomalies are the residuals between the actual meteorological factors' value and their reference value during the 72 h centered at the storm peak intensity hour.

3.3. Rain Hyetograph

The rain hyetographs in this study are derived by the Improved Huff curve model reported by Pan et al. [5]. The Huff curve is a dimensionless hyetograph initially developed by Huff for characterizing rainfall temporal distributions in an area, and has been widely applied to describe the hyetograph and to predict the runoff in a catchment [43–46]. In traditional analysis, the storm events are first classified into four quartiles according to their normalized time of peak rain intensity. Next, a quartile curve is developed at a certain provability, normally varying from 10% to 90%, by a 10% increment. Then, a series of Huff curves are developed at different probabilities within each quartile [46]. The 50% probability (median) curve is the most representative in each quartile.

The improved Huff curve method does not separate storms into the four quartiles as usual, but divides each storm into the rising and falling limbs, according to the occurrence time of the peak rain intensity [5]. Then, the dimensionless hyetographs are developed by the Huff curve method based on the normalized rain intensity and the time in the rising and falling limbs separately. Finally, both of the hyetographs are combined to form an Improved Huff curve. The Improved Huff curves in this study were developed at the probability of 50% in both the rising and falling limbs, based on the one-minute rainfall data of Sites 1–6 from 2008 to 2017.

3.4. Precipitation Extremes and Temperature Scaling

The approach of Clausius–Clapeyron (CC) scaling is applied in order to assess the impact of air temperature on precipitation extremes in the subtropical Guangzhou, based on all of the available hourly precipitation and temperature data [12]. Only the hourly precipitation data are analyzed. The daily mean air temperature is computed from the hourly temperature data during the 24-h period prior to the storm, as well as the natural daily mean temperature. The precipitation data were stratified based on the 24-h and daily mean air temperature in bins of 2 °C widths, within which the precipitation extremes were computed from the 75th, 90th, 99th, and 99.9th percentiles. Only the 75th and 99th percentiles have been presented for graph clarity [47,48].

4. Results and Discussions

4.1. Characteristics of Meteorological Factors

The administration area of Guangzhou is located in the upper Pearl River Delta (PRD), facing the low-lying delta plain in West and South China Sea in the southeast, and surrounded by hills in the North and East (Figures 1b and 2a). At the 42 weather stations from July 2015 to October 2017, the main wind direction during the storm duration was from the south (42%) and east (28%), followed by the west (19%) and a few (11%) from the north. The suburban districts of Baiyun, Huangpu, Zengcheng, and Conghua had much larger precipitation than the downtown area of Guangzhou (Figure 2b). Meanwhile, attention must be paid to the big orange area, which has less precipitation as a result of the statistical artifacts caused by lack of weather stations in the hills, and thus its actual annual precipitation could be larger. In the downtown areas of Guangzhou with more weather stations, the urban stations showed distinct patterns of meteorological factors from the suburban stations, that is, less precipitation (Figure 2b), higher air temperature (Figure 2c), lower relative humidity (Figure 2d), lower surface pressure (Figure 2e), and smaller wind speed (Figure 2f) at the urban stations compared with the suburban stations. Considering the short duration of the records, they were just the typical mean states for this area. The precipitation extremes were also found to be positively associated with the urban extent in the Pearl River Delta [39].

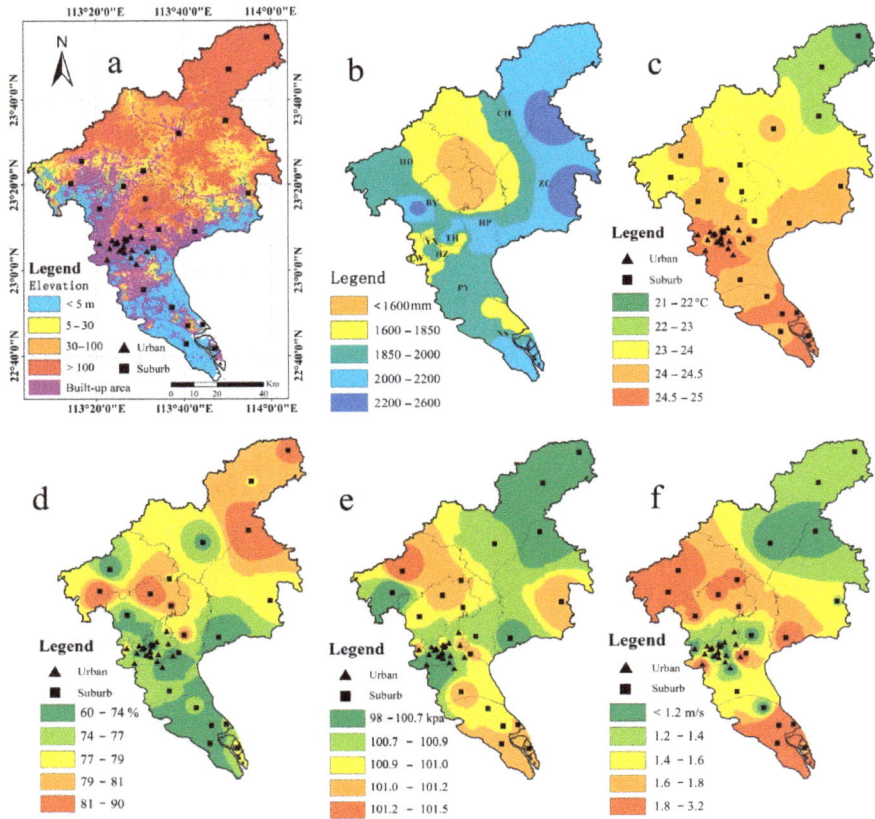

Figure 2. The elevation and built-up areas of Guangzhou and the locations of the meteorological sites (**a**), annual total rainfall (**b**), annual mean air temperature (**c**), relative humidity (**d**), air pressure (**e**), and wind speed (**f**) for two complete years from July 2015 to June 2017.

Besides the annual scale, meteorological factors also demonstrated different patterns for urban and suburban stations at the seasonal and diurnal scales (Figure 3). The urban stations had less rainfall in the first rainy season, from April to June, than the suburban stations, while they had larger rainfalls in the second rainy season of July, September, and October (Figure 3a). At the diurnal cycle, all of the stations showed two peaks of storm events in the morning and afternoon. The urban stations had a shorter duration in the morning peak and a longer duration in the afternoon peak than the suburban stations (Figure 3b). The surface air temperature and pressure showed an inverse temporal pattern at the seasonal scale (Figure 3c,g), while the temperature and relative humidity had an inverse temporal pattern at the diurnal scale (Figure 3d,f). The air pressure also showed a semidiurnal pattern (Figure 3h). Similar to the annual scale, the urban stations generally had a higher temperature, lower humidity and pressure, and much smaller wind speed than the suburban stations (Figure 3).

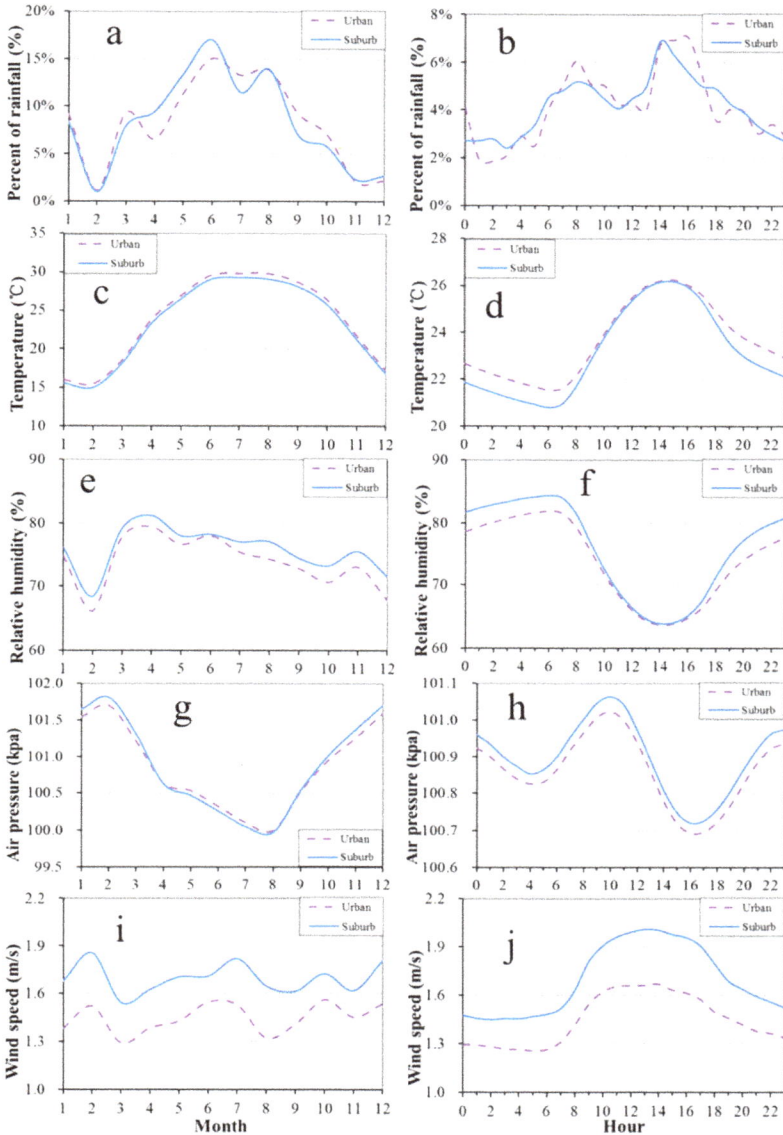

Figure 3. The seasonal (monthly) and diurnal (hourly) distributions of (**a,b**) rainfall, (**c,d**) surface temperature, (**e,f**) relative humidity, (**g,h**) air pressure, and (**i,j**) wind speed at the urban and suburban sites in Guangzhou from July 2015 to October 2017.

4.2. Characteristics of Storms

The storm features displayed little differences between the urban and suburban stations at an event scale, with a similar storm duration, event total, and intensity (Table 1). Half of the storms had an event duration of less than three hours, nearly a quarter of them were three to five hours, and over another quarter were longer than five hours, based on all of the hourly storm rainfall data from 1990 to 2017 (Table 1).

Table 1. Statistics of storm events at Sites 1–6 for the hourly rainfall data from 1990 to 2017.

Stations	Storm Count	Duration (h)				Mean (h) Duration	Mean (mm) Event Total	Mean (mm/h) Rain Intensity
		≤1	1–3	3–5	>5			
Urban	1327	165	505	297	360	4.4	46	10.5
		12%	38%	23%	27%			
Suburb	1284	144	493	292	355	4.6	47	10.3
		11%	38%	23%	28%			

The occurrence time of the peak rainfall plays a crucial role in determining the temporal distribution of the storm rainfall, that is, the rain hyetograph, which further impacts on the design storm, local drainage planning/design, and flooding risk. Figure 4 illustrates the rain hyetographs for the four types of summer storms at urban and suburban stations using the improved Huff curve established by Pan et al. [5]. Table 2 summarizes the statistics of these storms used in Figure 4. The urban stations had similar hyetographs, for example, having a similar peak rainfall occurrence time (29%–32% of event duration) and peak rainfall percentage (52%–57% of total rainfall) during a 0.5-h peak rainfall time. In contrast, the suburban stations had a wider range and later peak rainfall occurrence time (30%–41%) and a larger range (45%–61%) of peak rainfall percentage. The difference in the storm hyetograph will generate a different peak runoff, requiring a different drainage capability even for a same scale storm event. This indicates that different rain hyetographs are required in the storms for the design of drainage planning and flooding infrastructure in urban and suburban areas, even in the same administration area of Guangzhou [5].

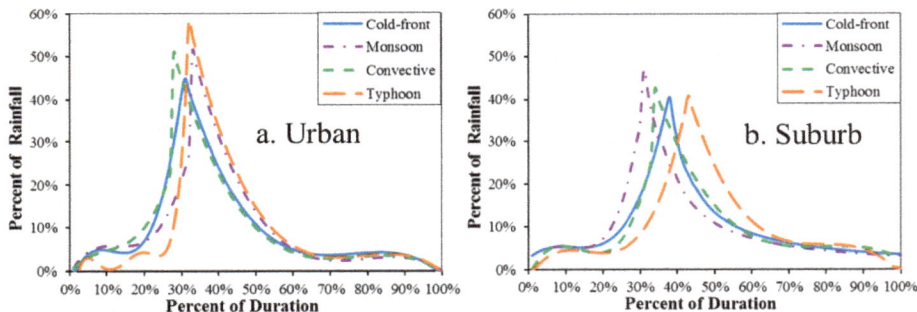

Figure 4. Storm hyetographs derived from the Improved Huff curve model at a probability of 50% from the one minute-interval rainfall data at Sites 1–2 during 2008–2012, and Sites 3–6 from 2014–2017 for the cold-front storms, monsoon storms, convective storms, and typhoon storms in (**a**) urban and (**b**) suburban areas.

Table 2. Statistics of storm events for the one minute-interval data during 2008–2012 (Sites 1–2) and 2014–2017 (Sites 3–6). The peak rainfall time is determined using the maximum five-minute rainfall accumulation. The rainfall time is normalized using the total rainfall duration.

Stations	Storm Types	Event Mean Rainfall (mm)	Percent of Max 0.5h Rainfall	Peak Rainfall Time	Rainfall Depth		Intensity (mm/min)	
					Rising	Falling	Rising	Falling
Urban	Cold-front	45	52%	29%	40%	60%	0.64	0.31
	Monsoon	39	56%	32%	46%	54%	0.64	0.27
	Convective	41	57%	28%	41%	59%	0.77	0.39
	Typhoon	29	57%	32%	31%	69%	0.17	0.20
	Mean	41	55%	29%	42%	58%	0.68	0.33
Suburban	Cold-front	55	52%	34%	40%	60%	0.76	0.54
	Monsoon	52	49%	30%	37%	63%	0.64	0.40
	Convective	36	61%	33%	42%	58%	0.67	0.44
	Typhoon	34	45%	41%	50%	50%	0.32	0.21
	Mean	46	54%	32%	40%	60%	0.69	0.45

Table 3 summarizes the storm information recorded at the 42 weather stations from July 2015 to October 2017. The coastal zone had more rainfall than the inner land for the warm-front events (Figure 5b). There were only two actual warm-front events that occurred during 27–29 January 2016. For example, one, which occurred on 28 January 2016, was a wide spread and long-duration storm (Figure 5a,b), and most of the stations reported this storm. Another storm on 27 January 2016 had a smaller intensity, and only a few stations reported it as a storm event. This explained why there were 50 storm events at the 42 weather stations for the actual two events.

Table 3. Statistics of the storms at the 42 weather stations from July 2015 to October 2017.

Items	Stations	All	Warm-Front Storm	Cold-Front Storm	Monsoon Storm	Convective Storm	Typhoon Storm
Storm Count	Urban (23)	752	*3%	13%	26%	50%	8%
	Suburb (19)	702	*4%	23%	29%	39%	7%
Storm duration (h)	Urban	3.2	7.8	3.3	3	2.6	5.9
	Suburb	3.6	11.9	3.2	3.3	2.9	6
Mean rainfall (mm/event)	Urban	41.3	82.1	37.9	37.4	41	43.8
	Suburb	44.0	83.1	37.2	37.9	43.3	76.7
Mean rain rate (mm/h)	Urban	12.8	10.5	11.5	12.5	15.8	7.4
	Suburb	12.2	7.0	11.6	11.5	14.9	12.8
Storm count 8:00–12:00	Urban	#25%	52%	19%	26%	24%	19%
	Suburb	#19%	40%	16%	17%	20%	15%
Storm count 13:00–18:00	Urban	#40%	0%	38%	30%	51%	28%
	Suburb	#41%	12%	42%	43%	42%	34%
Storm count 19:00–7:00	Urban	#35%	48%	43%	44%	25%	52%
	Suburb	#40%	48%	41%	40%	37%	51%
**Mean storm rainfall (mm)	Urban	926	88	235	161	414	28
	Suburb	1056	117	374	217	312	36
##Mean total rainfall (mm)	Urban	2708	296	530	734	1095	53
	Suburb	2822	289	609	860	1014	50

Note: * = count percentage of each storm type against all of the storm events. # = count percentage of the morning, afternoon, and night for each storm type, against those that occurred all day. ** = mean storm rainfall at each site. ## = mean total rainfall, including storms and no storm, at each site.

The cold-front storms were fast moving and wide spreading (Figure 5c,d). The suburban stations (158) had more storms than the urban stations (94) (Table 3). The afternoon had more storms than the morning and night on average, for example, a quarter of a day (6 h) in the afternoon had 38% and 42% of all of the storms in the urban and suburban stations, respectively.

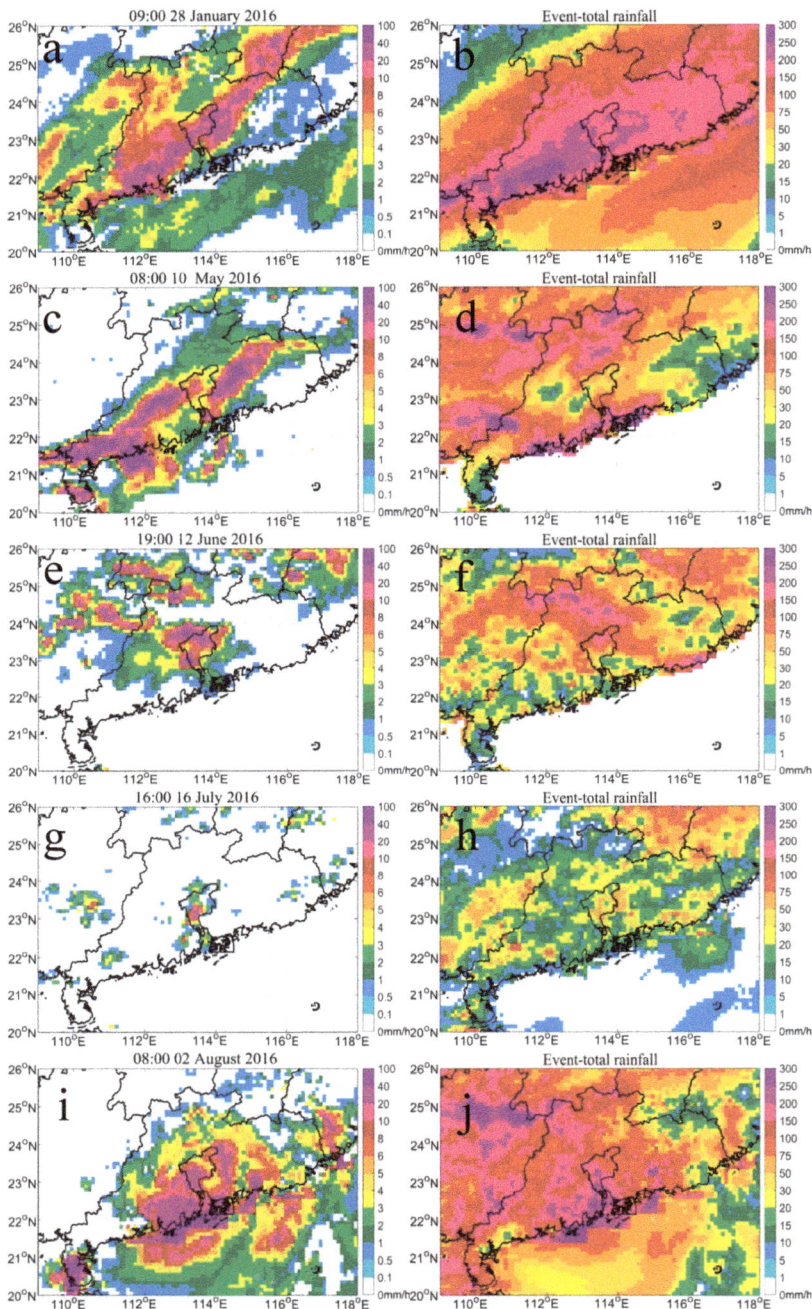

Figure 5. The spatial distribution of the peak rainfall and event-total rainfall plotted from the China Hourly Merged Precipitation Analysis (CMPA) product for five typical storm events (types) that occurred in the metropolitan areas of Guangzhou, China, that is, a warm-front storm (**a,b**), a cold-front storm (**c,d**), a monsoon storm (**e,f**), a convective storm (**g,h**), and the typhoon Nida storm (**i,j**).

Monsoon storms were the second most recorded storms next to convective storms, 26% in urban and 29% in suburban (Table 3). They had a similar rain duration and rain depth, but a smaller rainfall range than the cold-front storms (Figure 5e,f). The urban stations had more storms in the morning and less storms in the afternoon than the suburban stations, plus an overall shorter duration.

Most of the convective storms were localized and small-range events (Figure 5g,h), while they had the shortest duration and the largest mean intensity, resulting in most of the urban waterlogging incidents (Table 3). They were dominant in both the urban (50%) and suburban (39%) stations. The afternoon had the most events, especially in the urban areas (51%), due to strong solar radiation and the urban heat island effect.

The typhoon-brought storms were the most-wide spreading (Figure 5i,j), and had the second longest rain duration following the warm-front storms (Table 3). They were near evenly distributed through the day. The suburban stations had much more rainfall and a stronger rain rate than the urban stations.

4.3. Variations of Meteorological Factors with Storms

The interactions between the storm and meteorological factors were investigated from two aspects. Firstly, five storm events that occurred at a typical station were used to illustrate their specific interactions (Figure 6). Then, the mean conditions of all events were divided into three storm occurrence periods of morning, afternoon, and night time, so as to show the impact of solar radiation on their interactions with cold-front storms, monsoon storms, and convective storms (Figures 7–9). Warm-front and typhoon storms were not separated into these three periods because of their limited storm count.

4.3.1. Warm-Front Storms (in January)

The development of warm-front storms was mainly caused by the El Nino effect, a special case in Guangzhou and Southern China in January 2016. They were controlled by the cold air in the winter time, and then encountered the warm moist air that moved up from the Bengal Bay and the South China Sea. During the two weeks centered on 27–29 January 2016, the mean diurnal air temperature varied between 10 °C and 13 °C one week prior to and post storm, while it decreased to 5 °C prior to the storm and increased to 18 °C after the storm, resulting in a mean storm total of 90 mm (Figure 6a). The warming effect lasted over three days after the storm. As the warm and moist air moved up and the temperature increased, the relative humidity dramatically increased from 25% to above 80% (Figure 6b), the surface pressure was lower than the normal mean prior to and during the storm, and was higher than the normal after the storm (Figure 6c). The wind speed had a much larger variation than the normal mean (Figure 6d).

4.3.2. Cold-Front Storms (April to Mid-May)

The cold-front storms are controlled by the southwesterly wind (northeasterly) in South China, before the South China Sea summer monsoon is formed [9,10]. It generates a heavy storm center in Qingyuan and Shaoguan, the northern Guangzhou, mainly because of the uplifting effect of the topography (Figures 1b and 5d). The suburban stations (158 events) in Northern Guangzhou had much more cold-front storms than the urban stations (94 events) (Table 3).

Taking the storm on 9–11 May 2016 as an example, the air temperature rose by 2 °C above the two-week diurnal mean before the cold front arrived, and it dramatically decreased from 32 °C to 23 °C within 20 h as the cold front was approaching and the storm was formed (Figure 6e). It returned to the normal diurnal variations about 24 h after the storm ended. Relative humidity always accompanied the air temperature changes in an inverse pattern, that is, a lower and higher relative humidity than the diurnal mean immediately before and after the storm (Figure 6f). The surface pressure was lower than the diurnal mean 24 h before the storm, and the difference was reduced after the storm. It returned to the normal variation 12 h after the storm (Figure 6g). There was a larger wind speed about 20 h prior to the storm, and it fell back to the normal variations during and after the storm (Figure 6h).

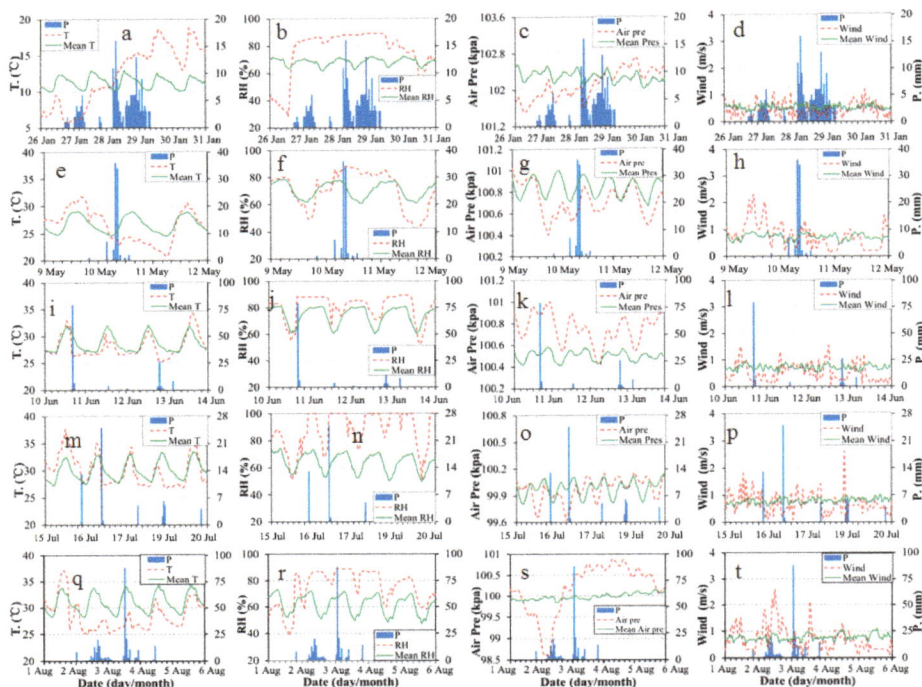

Figure 6. The fluctuations of air temperature, relative humidity, surface pressure, and wind speed at an urban station for a warm-front storm on 27–29 January 2016 (**a–d**), a cold-front storm on 9–11 May 2016 (**e–h**), a monsoon storm on 9–15 June 2016 (**i–l**), a convective storm on 14–20 July 2016 (**m–p**), and the typhoon Nida storm on 1–4 August 2016 (**q–t**). The green lines are the diurnal average during a two-week period centered at but excluded from the storm-affecting duration. This urban station (G3221) is located in the Tianhe District downtown of Guangzhou.

The impact of the storms on the meteorological factors were investigated by dividing the occurrence time into morning 8:00–12:00 (Figure 7a–d), afternoon 13:00–18:00 (Figure 7e–h), and night time 19:00–7:00 (Figure 7i–l). Prior to and after the storm, there were overcast clouds, which blocked the shortwave solar radiation and retained the Earth's surface long wave radiation. The storms disturbed the normal diurnal variation of the meteorological factors, among which the air temperature was most impacted. When the storm occurred in the morning, the air temperature was +6 °C higher than the diurnal mean about 24 h prior to the storm, lasted about 8 h at that anomaly high status, and then dramatically decreased to −4 °C after the storm (Figure 7a). It decreased at a larger magnitude and longer duration at the urban stations than at the suburban stations.

When the storm occurred in the afternoon, the air temperature was +2 °C higher than the diurnal mean about 18 h prior to the storm, lasted about 6 h at that positive status, and then dramatically decreased to −4 °C immediately before the storm (Figure 7e). The cooling effect was reduced quickly after the storm, and the urban stations had a much larger cooling magnitude than the suburban stations.

When the storm occurred in night, the air temperature was +2 °C higher than the diurnal mean about 12 h prior to the storm, lasted about 12 h at that positive status and then immediately decreased to −2 °C during the storm (Figure 7i). The cooling impact lasted 12 h after the storm. The mean wind speed was higher than the diurnal mean 12 h prior to, and after the storm at the urban stations (Figure 7l).

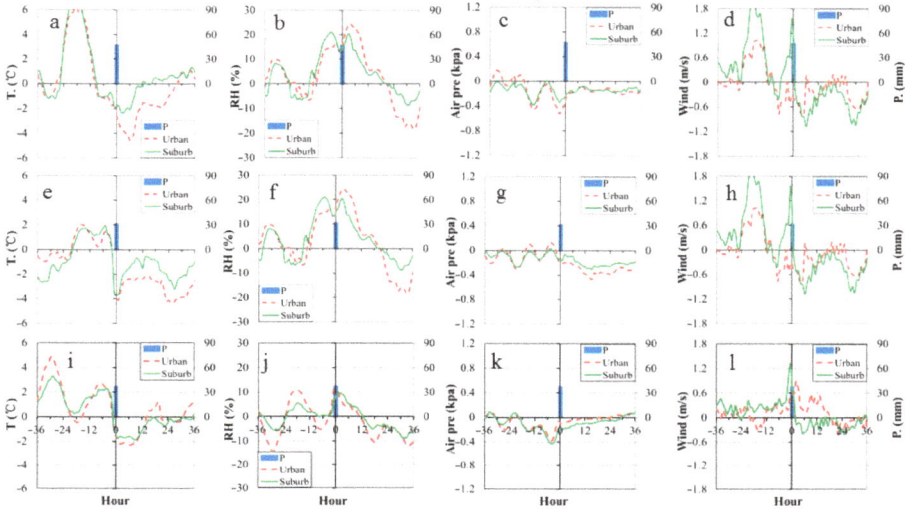

Figure 7. Anomalies of air temperature, relative humidity, surface pressure, and wind speed 36 h prior to, and after the cold-front storms that occurred in the morning (8:00–12:00; **a–d**), afternoon (13:00–18:00; **e–h**), and night (19:00–7:00; **i–l**). The references are the diurnal mean during a two-week period centered at but excluded from the 72 h of storm period from July 2015 to October 2017. The precipitation in the right-hand axis is the mean event-total rainfall.

4.3.3. Monsoon Storms (Late May to June)

When the southwesterly (northeasterly) wind weakened in the mid-May, the southeasterly strengthened and then dominated the monsoon rains in late May and June in South China [9]. One important feature of the South China Sea summer monsoon onset is that the upper tropospheric (100 hPa) zonal wind shifts from westerly to easterly, corresponding to the northward move of the South Asia High [10]. Thus, the storms occurring in late May and June are caused mainly by the warm and moist South China Sea summer monsoon. Both the cold frontal and monsoonal rain are the dominant rain sources in the first rainy season, from April to June. They normally form a storm center in the southeast coast during the monsoon rain period (Figure 5f). The monsoon rain decreased from the southeastern coast to the northwestern inland [9]. The monsoon storms had similar storm durations, event total rainfall, and mean rain rates to the cold-front storms (Table 3). There were more monsoon storms that occurred in the afternoon, especially at the suburban stations.

The monsoon storms did not form an evident front, such as the storm event that occurred on 12 June 2016 (Figure 5e,f). The air temperature did not show obviously changes before the storm, but immediately decreased during the storm (Figure 6i). Both the relative humidity and surface pressure were higher than the two-week diurnal mean for a few days after the storm (Figure 6j,k). The wind speed was larger and smaller than the diurnal mean several hours prior to and after the storm, respectively (Figure 6l).

When storms occurred in the morning, the air temperature was +3 °C higher than the diurnal mean about 20 h prior to the storm, lasted about 6 h at that positive status, and then decreased to −3 °C two hours prior to the storm. This negative value reduced slowly, and it returned to the normal variations 10 h after the storm at the urban stations (Figure 8a). The cooling impact lasted about eight hours longer at the suburban stations than at the urban stations.

When the storm occurred in the afternoon, the air temperature was +2 °C higher than the diurnal mean, about six hours prior to the storm, lasted about three hours at that positive status, and then dramatically decreased to −4 °C (anomaly) during the storm (Figure 8e). The cooling impact quickly reduced after the storm and lasted about 10 h.

When the storms occurred at night, the impact of the storms on the air temperature and other meteorological factors were reduced compared with the morning and afternoon storms (Figure 8i–l). The air temperature was +2 °C higher than the diurnal mean about 12 h prior to the storm, lasted about three hours at that positive status and then decreased to −2 °C (anomaly) during the storm (Figure 8i). The cooling impact lasted 10 h after the storm. The mean wind speed was higher than the diurnal mean after the storm at the urban stations, while it was lower than the diurnal mean at the suburban stations (Figure 8l).

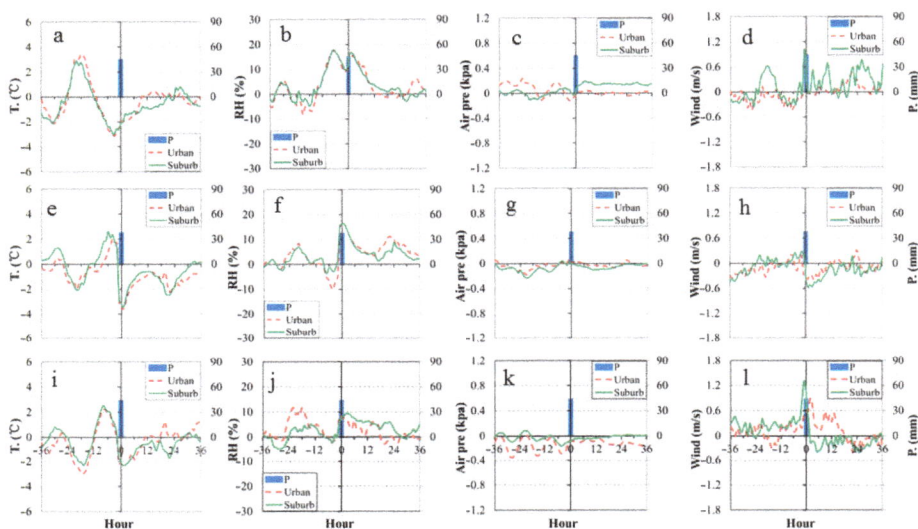

Figure 8. Anomalies of air temperature, relative humidity, surface pressure, and wind speed during 36 h prior to and post the monsoon storms that occurred in the morning (8:00–12:00; a–d), afternoon (13:00–18:00; e–h), and night (19:00–7:00; i–l). The references are the diurnal mean during a two-week period centered at but excluding the 72 h of storm period from July 2015 to October 2017. The precipitation in the right-hand axis is the mean event-total rainfall.

4.3.4. Convective Storms (July to September)

There is strong solar radiation and intense surface heating in Guangzhou (~N23°) from 22 June to 23 September each year, when the sun can vertically shed light on the Tropic of Cancer, and then moves southward back to the equator. Such a surface heating causes intense convection, resulting in localized convective storms or thunderstorms at local (micro) scales, especially in urban areas. These storms have unique dynamical structures largely controlled by the three-dimensional air temperature, humidity, pressure, and wind in the environment of the convection developing. One example was the convective thunderstorm that occurred in Guangzhou on the morning of 16 July 2016 (Figure 5g,h). The air temperature was +3 °C above the diurnal mean 18 h prior to the storm, and then decreased to −3 °C below the mean during the storm (Figure 6m). The relative humidity was much higher than the diurnal mean prior to and after the storms (Figure 6n).

The convective storms had a dominant occurrence frequency in all of the storm types at both the urban (50%) and suburban (39%) stations (Table 3). The afternoon had the largest share on average, while night had the least possibility, especially at the urban stations, with 51% count in the afternoon

(13:00–18:00) and only 25% in the night. When storms occurred in the morning, the air temperature was +5 °C higher than the diurnal mean about 20 h prior to the storm, lasted about four hours at that anomaly high status, and then dramatically decreased to −4 °C (anomaly) six hours prior to the storm. That negative value reduced slowly and returned to the normal variations 12 h after the storm (Figure 9a). The wind speed was 0.6 m/s higher than the mean 18–24 h prior to the storm, then reduced to −0.4 m/s lower than the mean 10 h prior to the storm, and returned to the normal variation after the storm (Figure 9d).

When the convective storms occurred in the afternoon, the air temperature did not show an obvious change until several hours prior to the storm, and then dramatically decreased to −4 °C (anomaly) during the storm (Figure 9e). The cooling impact quickly reduced, and it returned to the normal variations 12 h after the storm. Wind speed was +0.4 m/s larger than the mean 6–12 h prior to the storm, and −0.4 m/s smaller 0–6 h after the storm (Figure 9h).

When convective storms occurred at night, the impact of the storms on the air temperature and other meteorological factors were reduced compared to the morning and afternoon-occurring storms (Figure 9i–l). The air temperature was +2 °C higher than the diurnal mean about 10 h prior to the storm, and then decreased to −2 °C (anomaly) during the storm (Figure 9i). The cooling impact lasted 12 h after the storm.

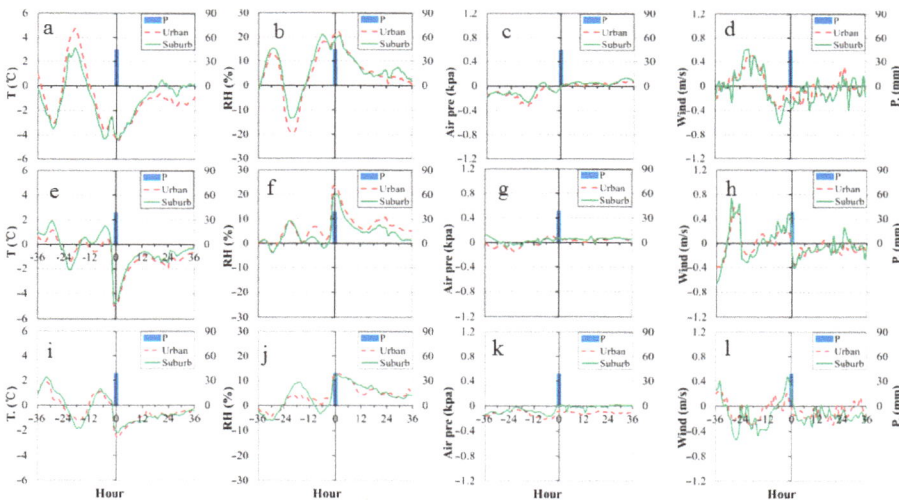

Figure 9. Anomalies of air temperature, relative humidity, surface pressure, and wind speed 36 h prior to, and after the convective storms that occurred in the morning (8:00–12:00; **a–d**), afternoon (13:00–18:00; **e–h**), and night (19:00–7:00; **i–l**). The references are the diurnal mean during a two-week period centered at but excluding the 72 h of the storm period from July 2015 to October 2017. The precipitation in the right-hand axis is the mean event-total rainfall.

4.3.5. Typhoon Storms (July to September)

Tropical cyclones are rapidly rotating storm systems featuring a low-pressure center, a closed low-level atmospheric circulation, strong winds, spiraling storms, and heavy rain. They are called typhoons in the northwestern Pacific Ocean and hurricanes in the Atlantic Ocean and northeastern Pacific Ocean. Most tropical cyclones that made landfall in South China were formed in the South China Sea and Philippine Sea in the northwestern Pacific Ocean, and the winds blew counterclockwise. There were 2.8 landfall typhoons on average in South China during 1957–1996, contributing 20%–30% to the annual rainfall [49].

Typhoon storms are mesoscale weather systems (Figure 5i,j). They were accurately identified and consisted 8% of all storms in the three years examined (Table 3). They occurred evenly within the three periods, with no obvious difference between the urban and suburban stations. Meteorological factor variations were illustrated as an example during the Typhoon Nida, which made landfall in the east of Shenzhen at 04:00 on 2 August 2016 (Figure 6q–t). The surface pressure declined to 98 kPa, and the hourly mean wind speed rose to 3 m/s during landfall at Tianhe in Guangzhou. The air temperature was more than 6 °C lower than the diurnal mean 12 h prior to landfall, and the cooling impact lasted three days after the landfall. It brought 100–200 mm rainfall in Guangzhou, and the peak rain intensity was 90 mm/h 24 h after the landfall.

Generally, heavy storms are developed by different weather systems each season, and have unique and dynamically environment structures largely controlled by the three-dimensional air temperature, humidity, pressure, and wind (Figures 5 and 6). Prior to and after a storm, there is usually overcast clouds, which scatters back the shortwave solar radiation and blocks in the Earth's surface long wave radiation. It disturbs the normal diurnal variation of the meteorological factors, thus heavy storms had different impact on meteorological factors when they occurred in the morning, afternoon, and night (Figures 8–10). Meanwhile, in the formation of clouds and storms, the condensation of water vapor releases a large latent energy into atmosphere, resulting in an abnormal rise of air temperature. Subsequently, the rainfall brings down cool water, and the evaporation of the surface rain water absorbs the heat, resulting in a cooling effect on both the Earth's surface and on the lower atmosphere [23]. Thus, the air temperature could rise several degrees above the normal range during the 24 h prior to the storms, and immediately dropped several degrees below the normal range during and after the storm, resulting in an approximately 6–10 °C air temperature difference before and during the storms (Figures 6–9, Table 4). The 24-h mean air temperature prior to the storms could be a better indicator for computing the scaling rates of the precipitation extremes with the surface air temperature.

Table 4. Mean air temperature (T. = °C) 24 h prior to the storms and during the storms, and the break air temperature of the scaling rates using the 24-h mean and natural daily mean temperature at the 42 automatic weather stations from July 2015 to October 2017. *All storms only include the three types of storms.

Storms	24 h Mean T. Prior to Rain	Mean T. in Rain Hours	T. Difference	Break T. 24 h Mean	Break T. Daily Mean
*All Three	32.5	23.9	8.6	28	26
Cold-front	31.4	22.9	8.5	28	24
Monsoon	31.9	25.3	6.7	28	28
Convective	32.7	25.8	7.0	28	26

4.4. Scaling Rates

When all of the storms were considered except for the warm-front and typhoon storms, it showed a peak-like scaling with a break temperature of 28 °C and a peak precipitation intensity of 67 mm/h in the 99th percentile (Figure 10a). The hourly precipitation extremes in the 75th and 99th percentiles increased at a close-CC rate (~7% °C^{-1}), with air temperature below 28 °C, while a negative scaling existed when it was above 28 °C (Figure 10a). The break temperature was 26 °C in the 75th percentile for the cold-front and monsoon storms. The scaling rate of the cold-front storms was overall similar to that of all of the storms, but with a smaller peak intensity of 57 mm/h in the 99th percentile (Figure 10b). It displayed a super CC rate for monsoon and convective storms when the 24-h mean air temperature was below 28 °C and a negative scaling rate when it was above 28 °C (Figure 10c,d). Their 24-h mean air temperature varied from 24 °C to 34 °C prior to the storm. The hourly peak precipitation intensities in the 99th were 57, 71, and 69 mm/h for the cold-front, monsoon, and convective storms, respectively. Meanwhile, the scaling rates were generally similar for the precipitation extremes at the urban and suburban stations (Figure 10e,f).

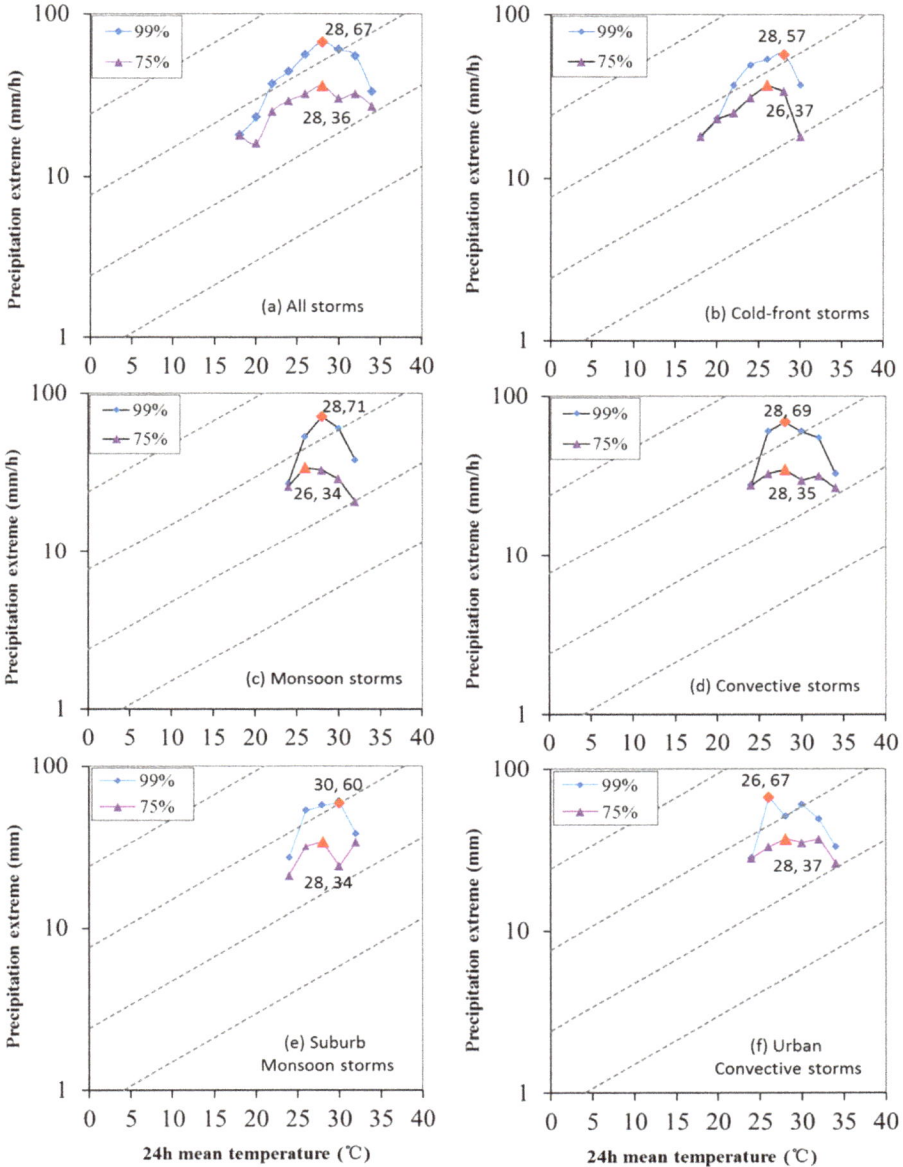

Figure 10. The scaling rates between the 24-h mean air temperature prior to the storms and the hourly precipitation extremes at the 75th and 99th percentile for (**a**) all of the storms from April to September, (**b**) cold-front storms from April to mid-May, (**c**) monsoon storms from mid-May to June, (**d**) the convective storms from July to September, (**e**) monsoon storms at the suburb stations, and (**f**) convective storms at the urban stations. The marked values on each plot are the break temperature and precipitation extremes. The gray dashed lines are the standard Clausius–Clapeyron (CC) scaling rate. The Y-axis scale is in logarithm.

In contrast, when using the natural daily mean air temperature as in literature, the peak-like scaling structure almost disappeared and only the negative scaling existed, especially in the 75th percentile for the monsoon storms (Figure 11c). The break air temperature dropped to 24 °C and 26 °C for the cold-front and convective storms, resulting in a break air temperature of 26 °C for all three types of storms. A negative scaling rate existed for most of the convective storms (Figure 11d). When the urban and suburban stations were separated, a negative CC scaling rate appeared for the cold-front storms at the suburban stations (Figure 11e) and for convective storms at the urban stations (Figure 11f). Meanwhile, the peak-like scaling structure still existed and was similar at the urban and suburban stations for both the monsoon and convective storms when the 24-h mean air temperature prior to the storms was used in computing the scaling rates (Figure 10e,f). On the one hand, although the air temperature affects extreme precipitation, the atmospheric conditions and precipitation affect the surface air temperature as well. The cooling effect of the storms on the air temperature disturbs the scaling rate between the precipitation extremes and the air temperature. A lower temperature during the storms is widely related to the local saturated downdraughts, rain evaporative cooling, and the synoptic atmospheric properties of colder air in low-pressure systems [23]. This indicates that the 24-h mean air temperature could produce more reliable scaling rates than the naturally daily mean air temperature used in literature.

The peak-like structure of the scaling rates between the precipitation extremes and air temperature revealed in this study are similar to those reported in the literature [13,14,16,17], but are slightly different from those who found negative scaling rates in the tropic and subtropical regions when daily mean air temperature was above 25 °C, such as in Brazil [15], Northern Australia [19], Southern China [20], and Hong Kong [24]. Such negative scaling rates were also identified by using the daily mean temperature for the cold-front and convective storms (Figure 11e,f). Using the 24-h mean air temperature prior to the storms, this study presents a break temperature of 28 °C, above which there was a negative scaling rate for the warm season storms. Figures 6–9 demonstrate that the air temperature normally increased by several degrees 4–24 h prior to the storm, while it decreased by several degrees immediately during the storm. The transient cooling effect of the tropical storms could be up to 4 °C in Northern Australia [23]. Table 4 shows that the 24-h mean air temperature prior to the storms was about 7–8 °C higher than that during the rain hours. This suggests that the 24-h mean air temperature prior to the storms could be a better indicator than the natural daily mean air temperature in the scaling rate computation, especially for the sub-tropic and tropic storms [15,23,24].

The break air temperature acts like an atmospheric threshold of water vapor availability in the subtropical Guangzhou. The values of the mean temperature 24 h prior to the storms were 31.4, 31.9, and 32.7 °C for the cold-front, monsoon, and convective storms, respectively, and their mean temperatures during the rain hours were 22.9, 25.3, and 25.8 °C (Table 4). Although both the 24-h and rain-hour mean air temperature were different for the cold-front, monsoon, and convective storms, they all showed a similar break temperature of 28 °C in the 99th percentile (Figure 10). This was 4 °C lower than the 24-h mean air temperature prior to the storms and 2–5 °C higher than those during the rain hours. When the 24-h mean air temperature was lower than 28 °C, the relative humidity was 80%–100%, and showed positive scaling rates (Figures 6–10). In contrast, when it was higher than 28 °C, the relative humidity and precipitation extremes had a negative relationship with the air temperature (Figure 12). This further confirms a previous explanation that the negative scaling rates were mainly caused by a lack of moisture [15,20].

The primary mechanism of moisture lack for the high temperature range is likely caused by the delay of evapotranspiration following the rapid increase of air temperature, rather than by the absolute lack of water resource, especially in the humid subtropical area of Guangzhou. As the air temperature rises above 28 °C, the atmosphere is more dynamic. A further rising temperature is potentially associated with different synoptic systems, atmospheric circulation, and moisture advection, thus resulting in different meteorological or precipitation regimes [19,50]. Meanwhile, the spatial variability of the mean air temperature is much smaller than the precipitation extreme, and it might contribute to the negative scaling rates when the scaling rates were computed using all of the precipitation extremes at each of the weather stations [14–17]. In other words, such negative scaling rates might be partially related to the analytical method [51], such as those showed in Figures 10 and 11. They showed some positive scaling rates in the tropical regions of Australia, by conditioning the precipitation intensity and storm duration. Nevertheless, given enough time and moisture sources, the scaling rate is still appropriate to project the future rainfall extremes in the context of climate change and global warming [17,23].

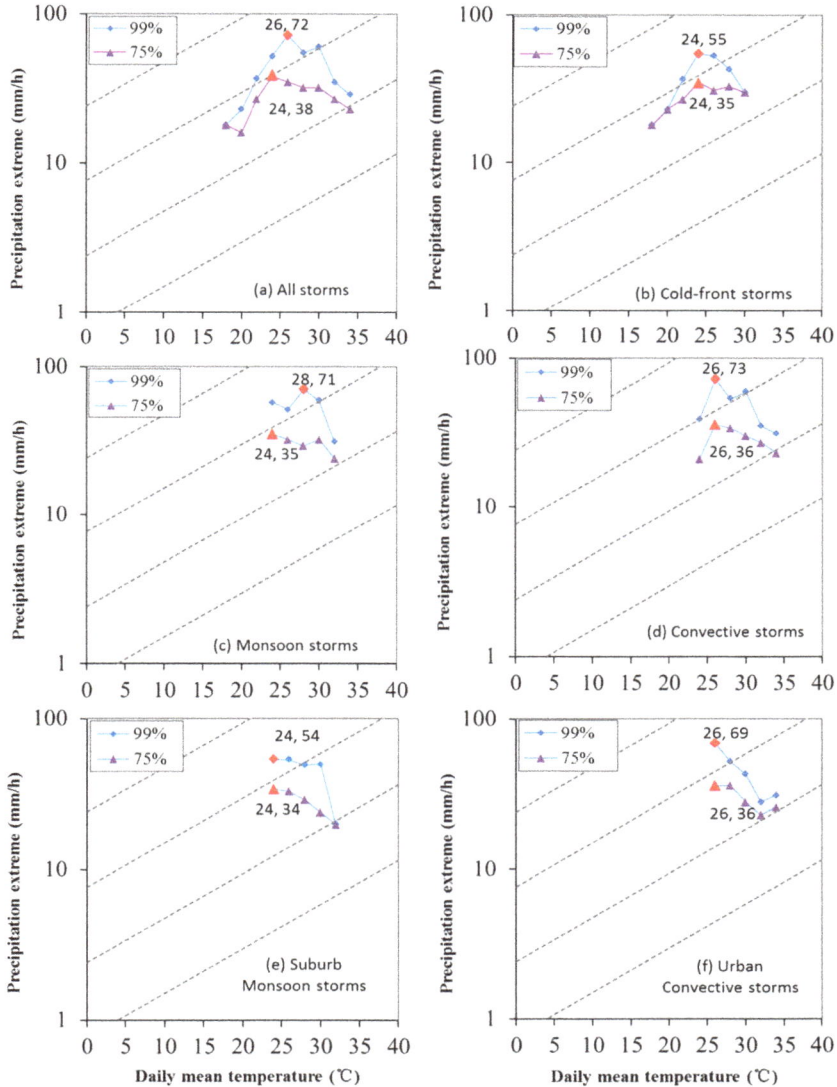

Figure 11. The same as Figure 10, but using the natural daily mean air temperature.

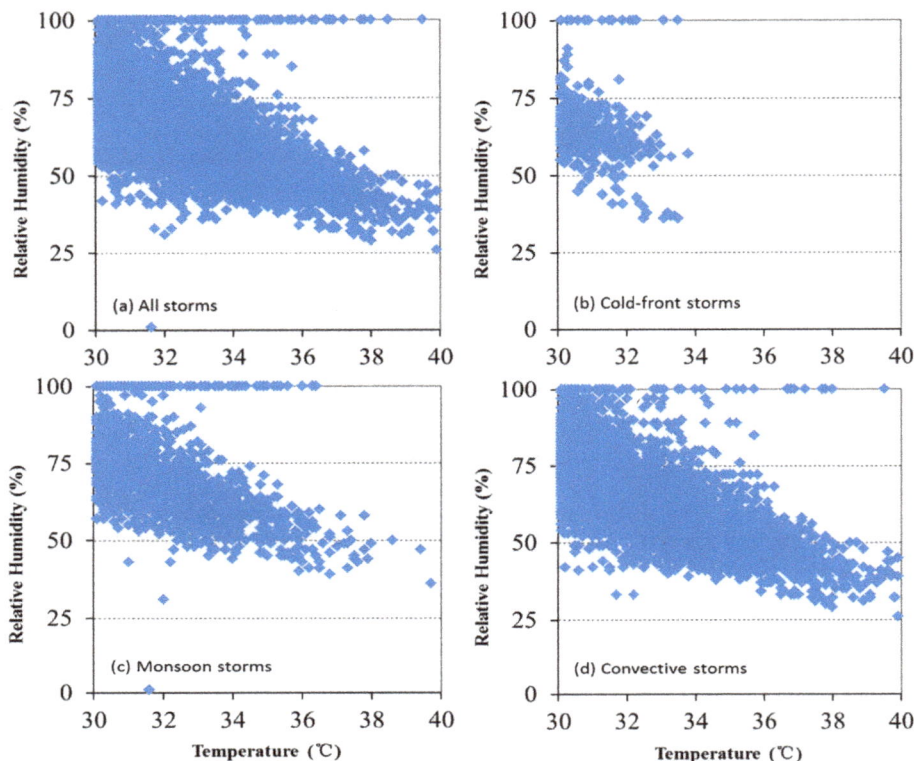

Figure 12. The scatter plots between the relative humidity (%) and hourly mean air temperature during the periods prior to the storms for (**a**) all of the storms from April to September, (**b**) cold-front storms from April to mid-May, (**c**) monsoon storms from late May to June, and (**d**) the convective storms from July to September.

5. Summary and Remark

There is an ongoing debate on the negative scaling rates between precipitation extremes and surface air temperature in tropic and subtropic regions. A lack of moisture resource was mainly applied to explain the negative scaling rates. However, heavy storms are developed by different weather systems each season in the Southern China. They have complicated interactions with meteorological factors and are the driving force of urban pluvial flooding. This study analyzes the characteristics of heavy storms in the administration areas of Guangzhou, South China, and investigates the variations of meteorological factors with different types of storms, and quantifies the scaling rates between the hourly precipitation extremes with the surface air temperature (i.e., the naturally daily mean temperature and the 24-h mean values prior to the storms).

Except for the warm-front and typhoon storms, the warm season storms have a short duration and intense rates in Guangzhou. Half of the storms had rain duration shorter than three hours, a quarter were in the range of three to five hours, and another quarter were longer than five hours, respectively. The convective storms were dominant by 50% in urban, followed by monsoon storms and cold-front storms. Urban and suburban areas had different storm hyetographs.

The air temperature showed a different magnitude of fluctuations prior to and after the different types of storms, while the storm types had little influence on the scaling rates between the precipitation extremes and the temperature. Air temperature is one of the leading meteorological factors that interacts with heavy storms. It could rise by 6 °C and drop by 4 °C prior to and after summer storms. The precipitation extremes showed peak-like scaling rates with the 24-h mean air temperature prior to the storms. For the cold-front, monsoon, and convective storms, they all showed the same break temperature of 28 °C in the 99th percentile, which was 4 °C lower than the 24-h mean air temperature prior to the storms and 2–5 °C higher than those during the storms. Below 28 °C, the relative humidity was 80%–100%, and it showed a positive scaling. In contrast, above 28 °C, the relative humidity decreased with the air temperature increase, which suggests that the negative scaling rates were likely caused by lack of moisture in the atmosphere, instead of by the atmospheric water vapor-holding capacity. Meanwhile, when using the natural daily mean air temperature as in the literature, a lower break temperature appeared for all of the summer storms, partially due to the transient cooling effect, and even purely negative scaling rates appeared for the monsoon storms at the suburban stations and the convective storms at the urban stations. This suggests that the 24-h mean air temperature could be a better variable to use for compute scaling rates rather than the naturally daily mean air temperature.

The storm process-based analysis reveals detailed variations of the meteorological factors prior to, during, and after the storms, especially for the cold-front, monsoon, and convective storms. For large scale storms, such as the winter warm-front storms and typhoon storms, it is limited to analyzing the interactions between the storms and the meteorological factors by using the local weather observations. Fine atmospheric models are needed in order to investigate their full interactions and feedbacks. The accurate forecasting of localized heavy storms is still a sever challenge in present-day climate and weather forecasting models. This study paves a path towards a greater storm-process understanding of the scaling relation between precipitation extreme and air temperature, and offers some suggestions to the forecast of local heavy storms and the urban drainage management in the Southern China.

Author Contributions: Conceptualization, X.W.; methodology, C.P. and X.W.; software, X.W.; validation, D.W. and H.H.; formal analysis, X.W. and C.P.; investigation, C.P., X.W., and H.H.; resources, L.L. and X.W.; data curation, C.P. and D.W.; writing (original draft preparation), C.P. and X.W.; writing (review and editing), L.L. and X.W.; visualization, C.P.; supervision, L.L. and X.W.; project administration, X.W. and L.L.; funding acquisition, X.W.

Funding: This study is funded by the National Natural Science Foundation of China (#41871085) and the Water Resource Science and Technology Innovation Program of Guangdong Province (#2016-19).

Acknowledgments: We thank all of the researchers and staff for providing and maintaining the meteorological data in Guangdong and the CMPA data from all agencies. The comments and suggestions from two anonymous reviewers greatly improved this manuscript and are highly appreciated by the authors.

Conflicts of Interest: The authors declare no conflict of interest. The founding sponsors had no role in the design of the study; in the collection, analyses, or interpretation of data; in the writing of the manuscript; and in the decision to publish the results

References

1. Yang, L.; Tian, F.; Smith, J.A.; Hu, H. Urban signatures in the spatial clustering of summer heavy rainfall events over the Beijing metropolitan region. *J. Geophys. Res. Atmos.* **2014**, *119*, 1203–1217. [CrossRef]
2. Liu, L.; Liu, Y.; Wang, X.; Yu, D.; Liu, K.; Huang, H.; Hu, G. Developing an effective 2-D urban flood inundation model for city emergency management based on cellular automata. *Nat. Hazards Earth Syst.* **2015**, *15*, 381–391. [CrossRef]
3. Shastri, H.; Paul, S.; Ghosh, S.; Karmakar, S. Impacts of urbanization on Indian summer monsoon rainfall extremes. *J. Geophys. Res. Atmos.* **2015**, *120*, 495–516. [CrossRef]
4. Yang, L.; Scheffran, J.; Qin, H.; You, Q. Climate-related flood risks and urban responses in the Pearl River Delta, China. *Reg. Environ. Chang.* **2015**, *15*, 379–391. [CrossRef]
5. Pan, C.; Wang, X.; Liu, L.; Huang, H.; Wang, D. Improvement to the Huff Curve for Design Storms and Urban Flooding Simulations in Guangzhou, China. *Water* **2017**, *9*, 411. [CrossRef]
6. Shepherd, J.M. A Review of Current Investigations of Urban-Induced Rainfall and Recommendations for the Future. *Earth. Interact.* **2005**, *9*, 1–27. [CrossRef]
7. Shem, W.; Shepherd, M. On the impact of urbanization on summertime thunderstorms in Atlanta: Two numerical model case studies. *Atmos. Res.* **2009**, *92*, 172–189. [CrossRef]

8. Pathirana, A.; Denekew, H.B.; Veerbeek, W.; Zevenbergen, C.; Banda, A.T. Impact of urban growth-driven landuse change on microclimate and extreme precipitation—A sensitivity study. *Atmos. Res.* **2014**, *138*, 59–72. [CrossRef]

9. Zheng, B.; Gu, D.; Li, C.; Lin, A.; Liang, J. Frontal Rain and Summer Monsoon Rain during Pre-rainy Season in South China. Part II: Spatial Patterns. *Chin. J. Atmos. Sci.* **2007**, *31*, 495–504. (In Chinese) [CrossRef]

10. Zheng, B.; Liang, J.; Lin, A.; Li, C.; Gu, D. Frontal Rain and Summer Monsoon Rain During Pre-rainy Season in South China. Part I: Determination of the Division Dates. *Chin. J. Atmos. Sci.* **2006**, *30*, 1207–1216. (In Chinese) [CrossRef]

11. Wang, D.S.; Wang, X.; Liu, L.; Wang, D.; Huang, H.; Pan, C. Evaluation of CMPA precipitation estimate in the evolution of typhoon-related storm rainfall in Guangdong, China. *J. Hydroinform.* **2016**, *18*, 1464–7141. [CrossRef]

12. Lenderink, G.; van Meijgaard, E. Increase in hourly precipitation extremes beyond expectations from temperature changes. *Nat. Geosci.* **2008**, *1*, 511–514. [CrossRef]

13. Drobinski, P.; Alonzo, B.; Bastin, S.; Silva, N.D.; Muller, C. Scaling of precipitation extremes with temperature in the French Mediterranean region: What explains the hook shape? *J. Geophys. Res. Atmos.* **2016**, *121*, 3100–3119. [CrossRef]

14. Utsumi, N.; Seto, S.; Kanae, S.; Maeda, E.E.; Oki, T. Does higher surface temperature intensify extreme precipitation? *Geophys. Res. Lett.* **2011**, *38*, 239–255. [CrossRef]

15. Maeda, E.E.; Utsumi, N.; Oki, T. Decreasing precipitation extremes at higher temperatures in tropical regions. *Nat. Hazards* **2012**, *64*, 935–941. [CrossRef]

16. Panthou, G.; Mailhot, A.; Laurence, E.; Talbot, G. Relationship between Surface Temperature and Extreme Rainfalls: A Multi-Time-Scale and Event-Based Analysis. *J. Hydrol.* **2014**, *15*, 1999–2011. [CrossRef]

17. Wang, G.; Wang, D.; Trenberth, K.E.; Erfanian, A.; Yu, M.; Bosilovich, M.G.; Parr, D.T. The peak structure and future changes of the relationships between extreme precipitation and temperature. *Nat. Clim. Chang.* **2017**, *7*, 268–274. [CrossRef]

18. Berg, P.; Haerter, J.O. Unexpected increase in precipitation intensity with temperature—A result of mixing of precipitation types? *Atmos. Res.* **2013**, *119*, 56–61. [CrossRef]

19. Hardwick Jones, R.; Westra, S.; Sharma, A. Observed relationships between extreme sub-daily precipitation, surface temperature, and relative humidity. *Geophys. Res. Lett.* **2010**, *37*, 1–5. [CrossRef]

20. Sun, W.; Li, J.; Yu, R. Corresponding Relation between Warm Season Precipitation Extremes and Surface Air Temperature in South China. *Adv. Clim. Chang. Res.* **2013**, *4*, 160–165. [CrossRef]

21. Wasko, C.; Parinussa, R.M.; Sharma, A. A quasi-global assessment of changes in remotely sensed rainfall extremes with temperature. *Geophys. Res. Lett.* **2016**, *4324*, 12659–12668. [CrossRef]

22. Ali, H.; Fowler, H.J.; Mishra, V. Global Observational Evidence of Strong Linkage Between Dew Point Temperature and Precipitation Extremes. *Geophys. Res. Lett.* **2018**, *45*, 320–330. [CrossRef]

23. Bao, J.; Sherwood, S.C.; Alexander, L.V.; Evans, J.P. Future increases in extreme precipitation exceed observed scaling rates. *Nat. Clim. Chang.* **2017**, *7*, 128–132. [CrossRef]

24. Lenderink, G.; Mok, H.Y.; Lee, T.C.; van Oldenborgh, G.J. Scaling and trends of hourly precipitation extremes in two different climate zones—Hong Kong and the Netherlands. *Hydrol. Earth Syst. Sci.* **2011**, *15*, 3033–3041. [CrossRef]

25. Ali, H.; Mishra, V. Contrasting response of rainfall extremes to increase in surface air and dewpoint temperatures at urban locations in India. *Sci. Rep.* **2017**, *7*, 1228. [CrossRef] [PubMed]

26. Barbero, R.; Westra, S.; Lenderink, G.; Fowler, H.J. Temperature-extreme precipitation scaling: A two-way causality? *Int. J. Climatol.* **2017**, *38*, e1274–e1279. [CrossRef]

27. Chen, D.; Chen, H.W. Using the Köppen classification to quantify climate variation and change: An example for 1901–2010. *Environ. Dev.* **2013**, *6*, 69–79. [CrossRef]

28. Liu, T.; Zhang, Y.H.; Xu, Y.J.; Lin, H.L.; Xu, X.J.; Luo, Y.; Xiao, J.; Zeng, W.L.; Zhang, W.F.; Chu, C.; et al. The effects of dust–haze on mortality are modified by seasons and individual characteristics in Guangzhou, China. *Environ. Pollut.* **2014**, *187*, 116–123. [CrossRef] [PubMed]

29. Xie, L.; Wei, G.; Deng, W.; Zhao, X. Daily $\delta^{18}O$ and δD of precipitations from 2007 to 2009 in Guangzhou, South China: Implications for changes of moisture sources. *J. Hydrol.* **2011**, *400*, 477–489. [CrossRef]

30. Qin, S.; Chao, F.; Lu, R. Large-scale Circulation Anomalies Associated with Interannual Variation in Monthly Rainfall over South China from May to August. *Adv. Atmos. Sci.* **2014**, *31*, 273–282. [CrossRef]

31. Qin, W.; Sun, Z.; Ding, B.; Zhang, A. Precipitation and circulation features during late-spring to early-summer flood rain in South China. *J. Nanjing Inst. Meteorol.* **1994**, *17*, 455–461. (In Chinese)

32. Chen, L.; Li, W.; Zhao, P.; Tao, S. On the process of summer monsoon onset over East Asia. *Acta Meteorol. Sin.* **2001**, *5*, 345–355. [CrossRef]

33. Lee, M.; Ho, C.; Kim, J. Influence of Tropical Cyclone Landfalls on Spatiotemporal Variations in Typhoon Season Rainfall over South China. *Adv. Atmos. Sci.* **2010**, *27*, 443–454. [CrossRef]

34. Ren, F.; Gleason, B.; Easterling, D. Typhoon Impacts on China's Precipitation during 1957–1996. *Adv. Atmos. Sci.* **2002**, *19*, 943–952.

35. Costa, A.C.; Soares, A. Homogenization of climate data: review and new perspectives using geostatistics. *Math. Geosci.* **2009**, *413*, 291–305. [CrossRef]

36. Xu, W.; Li, Q.; Wang, X.L.; Yang, S.; Cao, L.; Feng, Y. Homogenization of Chinese daily surface air temperatures and analysis of trends in the extreme temperature indices. *J. Geophys. Res. Atmos.* **2013**, *118*, 9708–9720. [CrossRef]

37. Gentilucci, M.; Barbieri, M.; Burt, P.; D'Aprile, F. Preliminary Data Validation and Reconstruction of Temperature and Precipitation in Central Italy. *Geosciences* **2018**, *8*, 202. [CrossRef]

38. Shen, Y.; Zhao, P.; Pan, Y.; Yu, J. A high spatiotemporal gauge-satellite merged precipitation analysis over China. *J. Geophys. Res. Atmos.* **2014**, *119*, 3063–3075. [CrossRef]

39. Wang, D.S.; Wang, X.; Liu, L.; Wang, D.G.; Liang, X.; Pan, C.; Huang, H. Comprehensive evaluation of TMPA 3B42V7, GPM IMERG and CMPA precipitation estimates in Guangdong Province, China. *Int. J. Climatol.* **2018**, 1–18. [CrossRef]

40. Kottegoda, N.T.; Natale, L.; Raiteri, E. Monte Carlo Simulation of rainfall hyetographs for analysis and design. *J. Hydrol.* **2014**, *519*, 1–11. [CrossRef]

41. Azli, M.; Rao, A.R. Development of Huff curves for Peninsular Malaysia. *J. Hydrol.* **2010**, *388*, 77–84. [CrossRef]

42. Hartigan, J.; Wong, M. Algorithm AS 136: A k-means clustering algorithm. *Appl. Stat.* **1979**, *28*, 100–108. [CrossRef]

43. Bonta, J.V.; Rao, A.R. Fitting equations to families of dimensionless cumulative hyetographs. *Trans. ASAE* **1988**, *31*, 756–760. [CrossRef]

44. Terranova, O.G.; Gariano, S.L. Rainstorms able to induce flash floods in a Mediterranean-climate region (Calabria, southern Italy). *Nat. Hazards Earth Syst.* **2014**, *14*, 2423–2434. [CrossRef]

45. Todisco, F. The internal structure of erosive and non-erosive storm events for interpretation of erosive processes and rainfall simulation. *J. Hydrol.* **2014**, *519*, 3651–3663. [CrossRef]

46. Yin, S.Q.; Xie, Y.; Nearing, M.A.; Guo, W.L.; Zhu, Z.Y. Intra-Storm Temporal Patterns of Rainfall in China Using Huff Curves. *Trans. ASAE* **2016**, *59*, 1619–1632. [CrossRef]

47. Fadhel, S.; Rico-Ramirez, M.A.; Han, D. Sensitivity of peak flow to the change of rainfall temporal pattern due to warmer climate. *J. Hydrol.* **2018**, *560*, 546–559. [CrossRef]

48. Hettiarachchi, S.; Wasko, C.; Sharma, A. Increase in flood risk resulting from climate change in a developed urban watershed—The role of storm temporal patterns. *Hydrol. Earth Syst. Sci.* **2018**, *22*, 2041–2056. [CrossRef]

49. Ren, F.; Wu, G.; Dong, W.; Wang, X.; Wang, Y.; Ai, W.; Li, W. Changes in tropical cyclone precipitation over China. *Geophys. Res. Lett.* **2006**, *33*, 131–145. [CrossRef]

50. Blenkinsop, S.; Chan, S.C.; Kendon, E.J.; Roberts, N.M.; Fowler, H.J. Temperature influences on intense UK hourly precipitation and dependency on large-scale circulation. *Environ. Res. Lett.* **2015**, *10*, 054021. [CrossRef]

51. Wasko, C.; Sharma, A.; Johnson, F. Does storm duration modulate the extreme precipitation-temperature scaling relationship? *Geophys. Res. Lett.* **2015**, *42*, 8783–8790. [CrossRef]

water

MDPI

Article

Extreme Precipitation in China in Response to Emission Reductions under the Paris Agreement

Jintao Zhang and Fang Wang *

Key Laboratory of Land Surface Pattern and Simulation, Institute of Geographic Sciences and Natural
Resources Research, Chinese Academy of Sciences, Beijing 100101, China; zhangjt.17s@igsnrr.ac.cn
* Correspondence: wangf@igsnrr.ac.cn

Received: 21 May 2019; Accepted: 31 May 2019; Published: 4 June 2019

Abstract: To avoid more severe impacts from climate change, countries worldwide pledged to implement intended nationally determined contributions (INDCs) for emission reductions (as part of the Paris Agreement). However, it remains unclear what the resulting precipitation change in terms of regional extremes would be in response to the INDC scenarios. Here, we analyzed China's extreme precipitation response of the next few decades to the updated INDC scenarios within the framework of the Paris Agreement. Our results indicate increases in the intensity and frequency of extreme precipitation (compared with the current level) in most regions in China. The maximum consecutive five-day precipitation over China is projected to increase ~16%, and the number of heavy precipitation days will increase as much as ~20% in some areas. The probability distributions of extreme precipitation events become wider, resulting in the occurrence of more record-breaking heavy precipitation in the future. We further considered the impacts of precipitation-related extremes and found that the projected population exposure to heavy precipitation events will significantly increase in almost all Chinese regions. For example, for heavy precipitation events that exceed the 20 year baseline return value, the population exposure over China increases from 5.7% (5.1–6.0%) to 15.9% (14.2–16.4%) in the INDC-pledge scenario compared with the present-day level. Limiting the warming to lower levels (e.g., 1.5 °C or 2.0 °C) would reduce the population exposure to heavy precipitation, thereby avoiding impacts associated with more intense precipitation events. These results contribute to an improved understanding of the future risk of climate extremes, which is paramount for the design of mitigation and adaptation policies in China.

Keywords: INDC pledge; precipitation; extreme events; extreme precipitation exposure

---

1. Introduction

In recent decades, a large number of climate extremes related to precipitation have been observed in conjunction with global warming, for example, flood events in the entire Yangtze River Basin in 1998 [1] and the heaviest rainfall in Beijing in 2012 [2], which caused great damages. Variations in the temperature, atmospheric moisture, precipitation, and atmospheric circulation have been observed; meanwhile, the moisture-holding capacity has been increasing at a rate of ~7%/°C with increasing temperature, which further alters the precipitation extremes as well as the hydrological cycle [3,4]. However, the changes of the water cycle projected for the future, including prominent regional and seasonal differences in response to climate change, are far more complex than projected temperature changes [5]. Precipitation-related extremes are among the most relevant consequences of a warmer climate. Therefore, it is crucial to more accurately project future precipitation extremes at regional scales. In China, agriculture heavily depends on the hydrological cycle; a variety of precipitation changes may affect different regions of China because of the diverse climate types. So, regional assessments of extreme heavy precipitation risks and impacts that could be avoided by limiting the warming to a lower level are critical for the design of adaptation and mitigation policies.

In the Paris Agreement, a goal was set to keep the mean global warming well below 2.0 °C above the preindustrial level and make efforts to limit the warming to 1.5 °C [6,7]. To achieve this goal, countries participating in the Paris Agreement submitted national mitigation plans in the form of Intended Nationally Determined Contributions (INDC). A total of 192 countries have reported their respective INDC mitigation targets to the United Nations, as of Dec. 2018. The bottom-up approach in using national efforts reflecting the willingness of each country to reduce their emissions is easier to implement because it avoids the divergence of different countries from the distribution quota [8].

Future emissions are the key factors in determining the climate impacts of the next few decades. Recently, researchers have paid increased attention to changes in the extreme precipitation at the 1.5 °C and 2.0 °C warming levels and the benefits of limiting the global warming to 1.5 °C rather than 2.0 °C [9–12]. A preliminary synthesis is included in Chapter III of the Intergovernmental Panel on Climate Change (IPCC) 1.5 °C report [13]. However, these studies are based on idealized emission pathways to reach the 1.5 °C and 2 °C global warming targets [14–16] that are hard to achieve. Some studies have also focused on the mean global warming response to INDC emission reduction. For example, Rogelj et al. [15] and CAT [16] evaluated the impact of INDC emission reduction commitments on mean global warming. The "Emissions Gap Report" released by UNEP [17] evaluated the gap between INDC emissions and the 2 °C target. There is still a lack of evaluations of the potential changes in regional precipitation and extreme events under INDC emission pledges. Therefore, the risks associated with future changes in the extreme precipitation are still unclear.

In this study, we analyzed China's extreme precipitation response of the next few decades to emission reductions based on the INDC under the Paris Agreement using an ensemble of state-of-the-art global climate models from the Fifth Coupled Climate Model Intercomparison Project (CMIP5). We further explored the exposure to extreme precipitation events under the INDC emission scenarios by considering the potential socioeconomic impacts of future climate change. The results of the 1.5 °C/2.0 °C scenario are shown as reference.

2. Data and Methods

2.1. Emission Scenarios

We used two categories of emission scenarios: (1) the 1.5 °C and 2.0 °C target scenarios under the Paris Agreement and (2) the INDC scenarios. The 1.5 °C/2.0 °C target scenarios were derived from the AR5 (Fifth Assessment Report) and 1.5 °C special report of the IPCC, respectively [18,19]. The INDC scenarios are based on emissions data submitted by 192 countries according to the Paris Agreement. The INDC dataset is continuously updated and can be obtained from the United Nations Framework Convention on Climate Change (UNFCCC) website [20]. The emission targets reported by different countries include a range of absolute emission targets to those relative to a base year level or emission reduction targets relative to a baseline emission scenario. We analyzed and extracted the emission targets of each country.

Simulations of future emissions based on 28 socioeconomic models were used to extend the INDC scenarios to the end of this century. Several key features (the rate of decarbonization, carbon capture, storage technology (CCS), energy structure improvement, and time to carbon neutralization) were considered. In this study, we considered many possible interpretations of "INDC mitigation actions" based on the IPCC AR5 Scenario Database (https://secure.iiasa.ac.at/web-apps/ene/AR5DB/). We used scenarios that conform to the 2030 GHG emission levels, which are in agreement with the INDC (50–56 Gt CO_2eq/year) scenarios. Considering future difficulties and uncertainties with respect to carbon removal technology, we used a conservative approach regarding the future availability of negative emission technologies and scenarios based on which CCS > 15 Gt CO_2eq/year was eliminated. The emission pathways were classified into six groups based on several key characteristics (e.g., the emission targets for specific years, renewable energy structure, and amount of CCS), as shown in Text

S1 and Figure S1. In this study, we focused on Groups III and IV, which represent the INDC "continued action" scenario, and compared the results with that of the 1.5 °C/2.0 °C target scenario.

Subsequently, we evaluated the global mean warming level under INDC "continued action" scenarios. Based on the 78 climate sensitivity experiments from the earth system models (ESMs) ensemble of CMIP5 [21], we assessed the possible corresponding global mean temperature rise [22,23]; we also integrated several other studies (temperature rise levels for some pathways have been provided) [14,15,17,23]. After a comprehensive assessment, we determined 2.9~3.3 °C (median 3.1 °C) as the most likely range of temperature increase for the "continued action" pathways of INDCs.

2.2. Data Description

To better project precipitation extremes, 14 state-of-art CMIP5 GCMs were adopted in this study [21]. The model details are provided in Text S2. For further analyses, all model data were interpolated to a common $1° \times 1°$ horizontal grid using a bilinear interpolation method. The ability of these CMIP5 models to simulate precipitation extremes was assessed using a daily gridded observation dataset CN05.1 ($0.5° \times 0.5°$) established by the China Meteorological Administration, which was obtained at 2416 observation stations and covers the period of 1961–2013 [24]. The assessment results show the validation of these models for simulations of precipitation. It has been widely used in many studies of precipitation extremes across China [25,26].

To estimate the space pattern of precipitation change, we used a time-slice approach [10,27,28] where the spatial state at a specific warming point related to ΔT_{INDC} (or 1.5 °C, 2.0 °C) was separately derived from decadal time slices with the respective mean warming for each model. For the detailed analysis process, please refer to Text S3, Figures S2 and S3. The multi-model ensemble (MME) was calculated based on equal weights.

The period of 1985–2005 is referred to as the present-day period. The preindustrial period in this study is 1861–1900.

Climate extremes are largely affected by distinct topography. To investigate the characteristics of future regional changes in the precipitation extremes, the country has been divided into eight subregions based on geographical conditions and climatic features (Figure 1). Detailed information regarding these eight subregions is provided in Text S4.

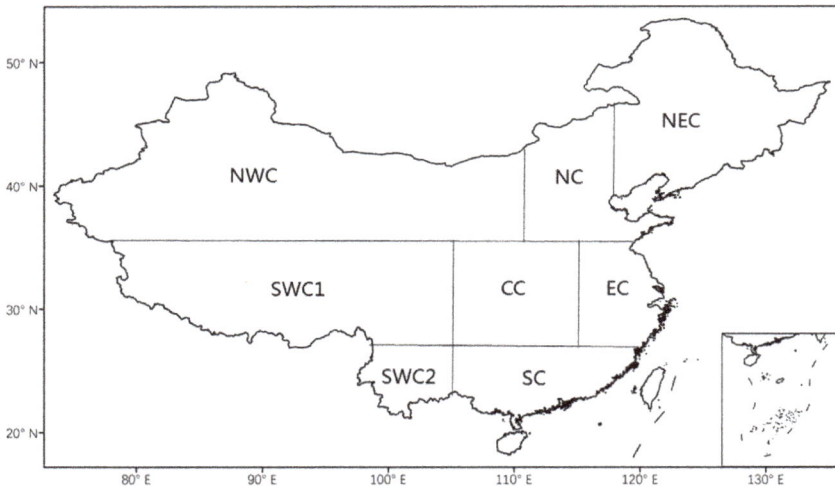

Figure 1. Map of China's eight geographical subregions.

2.3. Extreme Precipitation Indices

It is difficult to provide a universally valid definition of extreme precipitation because of the diversity of climates worldwide. In general, the indices established by the Expert Team on Climate Change Detection and Indices (ETCCDI) are adopted in most studies of global precipitation extremes and the extreme events can be defined using either relative or absolute thresholds [29]. Here, two precipitation indices recommended by the ETCCDI, that is, Rx5day and SDII, which have been widely used in many studies, have been selected to represent the intensities of precipitation extremes. In China, the precipitation can be divided into the following five categories: trace rain (0.1–1.0 mm/day), light rain (1.0–10 mm/day), medium rain (10–25 mm/day), large rain (25–50 mm/day), and heavy rain (≥50 mm/day); these categories have been widely used in previous research in the country [30,31]. Heavy and severe rainstorms in China are traditionally defined as events with daily precipitation values larger than 25, 50, or 100 mm [32]. In this study, the absolute threshold of 25 mm/day was chosen to define a precipitation extreme. The R25 represents the cumulative number of days during which the daily precipitation exceeds the 25 mm threshold.

Thus, three extreme precipitation indices have been adopted in this study to investigate the changes in the frequency and intensity of heavy rainfall in China. The detailed definitions of these indices are provided in Table 1. The evaluation of 14 CMIP5 models based on these indices is summarized in Text S2.

Table 1. Definitions of extreme precipitation indices used in this study.

Extreme Precipitation Indices	Definition (Unit)
Rx5day	Maximum consecutive 5 day precipitation (mm)
SDII	Simple precipitation intensity index. Let PR_{wj} be the daily precipitation amount on wet days, $PR \geq 1$ mm in period j. If W represents the number of wet days in period j, then $SDII_j = \left(\sum_{w=1}^{W} PR_{wj} \right)/W$ (mm/day).
R25	Cumulative number of precipitation days during which the daily precipitation exceeds 25 mm per year (days)

To quantitatively express the occurrence probability of extreme events, we used the following risk ratio (RR):

$$RR = \frac{P_1}{P_0} \tag{1}$$

where P_0 is the probability of reaching a specific present-day intensity (5% is used in this paper), P_1 is the corresponding probability of reaching this temperature intensity in future scenarios, and $RR > 1$ indicates an increased risk of extreme events.

The Wilcoxon rank sum test was applied to identify if there is a statistical significance of differences between two warming levels based on multi-model results. We further assessed the signal-to-noise ratio (SNR), expressed as the significance of the change compared with the internal variability.

2.4. Exposure and Avoided Impacts

Extreme events that substantially deviate from their climatology can result in the greatest losses. Climate change risks are determined based on the hazards, vulnerability, and exposure of the human society and natural ecosystems according to IPCC reports [33]. Climate extreme indices (e.g., Rx5day) are used to characterize the hazard intensity. For further analyses, we defined dangerous extreme events as those exceeding specific return values (RV) compared with the 1961–2005 baseline and quantified the changes in the exposure to dangerous extreme precipitation under different scenarios.

We fitted a generalized extreme value (GEV) distribution to the RX5day in 1961–2005 on the native grids of each model using the method of maximum likelihood [34]. The cumulative distribution function of GEV is given by

$$F(x; \mu, \sigma, \xi) = \begin{cases} \exp\left[-\exp\left(-\frac{x-\mu}{\sigma}\right)\right], & \xi = 0 \\ \exp\left[-\left(1 + \xi\frac{x-\mu}{\sigma}\right)^{-\frac{1}{\xi}}\right], & \xi \neq 0, \ 1 + \xi\frac{x-\mu}{\sigma} > 0 \end{cases} \tag{2}$$

where μ is the location parameter, σ is the (positive) scale parameter, and ξ is the shape parameter. These parameters are estimated by the method of maximum likelihood. Considering that the noise in changing patterns of extreme precipitation stems from sampling, the GEV parameter estimates of the extreme precipitation are smoothed spatially. This is done by smoothing the estimated GEV parameters at each grid point by its eight surrounding neighbors [35]. Then, the return values (RV) are obtained by inverting the fitted GEV distributions derived from the smoothed parameters. The different level RVs from the baseline are derived on the native grid points for each model. Finally, exposures to these dangerous extremes under different scenarios are estimated, and the area (population) that experiences RX5day events exceeding the threshold for danger is aggregated spatially to represent the total area (population) exposed. Population exposure is estimated on the population distribution projected under different socioeconomic development scenarios of Shared Socioeconomic Pathways (SSPs) [36]. The vulnerability of socioeconomical systems is not discussed in this study, thus the results of exposure only reflect the risk of physical climate change.

In addition, we investigated the avoided impacts at different levels of warming, which correspond to different mitigation policies. The impacts in terms of exposure induced by warming are quantified using the 1985–2005 present-day levels:

$$\text{Impacts (k)} = \frac{E_k - E_{present}}{E_{present}} \tag{3}$$

where E stands for the exposure and the subscript k indicates different warming levels based on different scenarios (1.5 °C, 2.0 °C, and INDC). Thus, the impacts avoided by less warming can be derived as the difference between the impacts at the two levels.

3. Results

3.1. Changes in the Extreme Precipitation Indices

We first evaluated the model performance; while the general features of the observed temperature indices were reasonably reproduced by the model, moderate biases were evident (Text S5, Figures S4 and S5).

The changes in the three annual extreme precipitation indices over China and eight subregions (Text S4) until the end of the century exhibit contrasting patterns in terms of the signal strength and robustness (Figure 2 for Rx5day, Figure 3 for R25, and Figure 4 for SDII). The national averages and relative differences between the various sets of scenarios are shown in Figure 5. These results reveal similar increases for all indices and indicate that heavier precipitation will be more frequent and intense when higher warming thresholds are crossed.

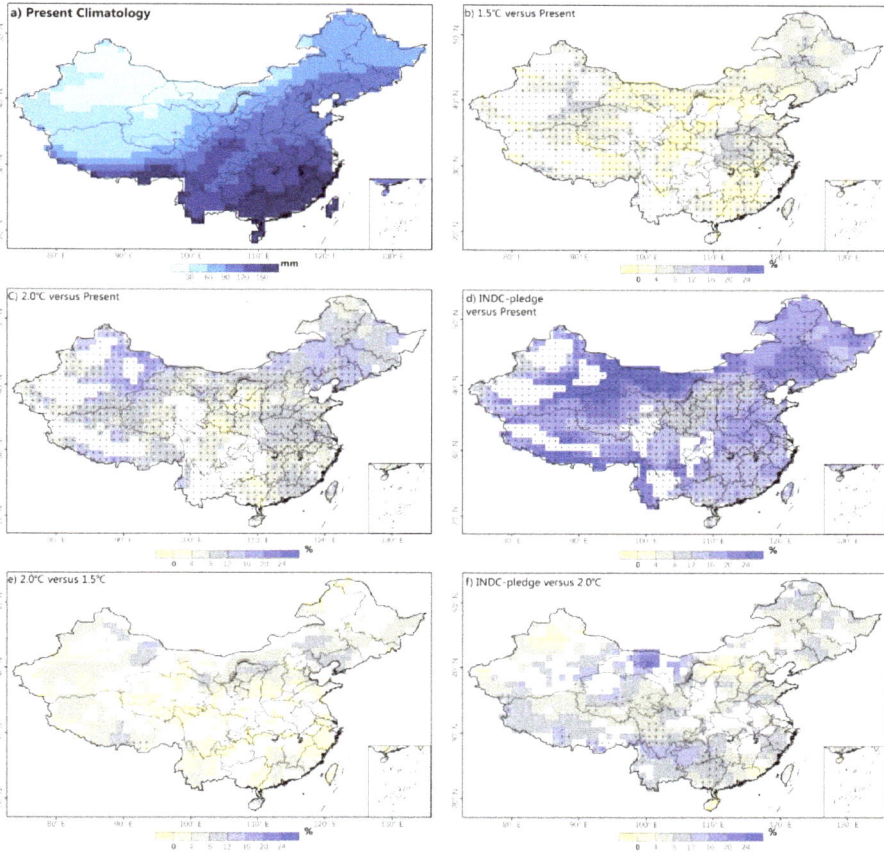

Figure 2. Present climatology of the annual Rx5day over Asia (**a**) and changes in the annual Rx5day under different scenarios (**b–f**) based on the multi-model mean. The colored shading in (**b–f**) was applied to areas that were statistically significant at the 10% level according to Wilcoxon rank sum test; the stippling in (**b–f**) was added to regions with a multi-model mean signal-to-noise ratio (SNR) > 1.

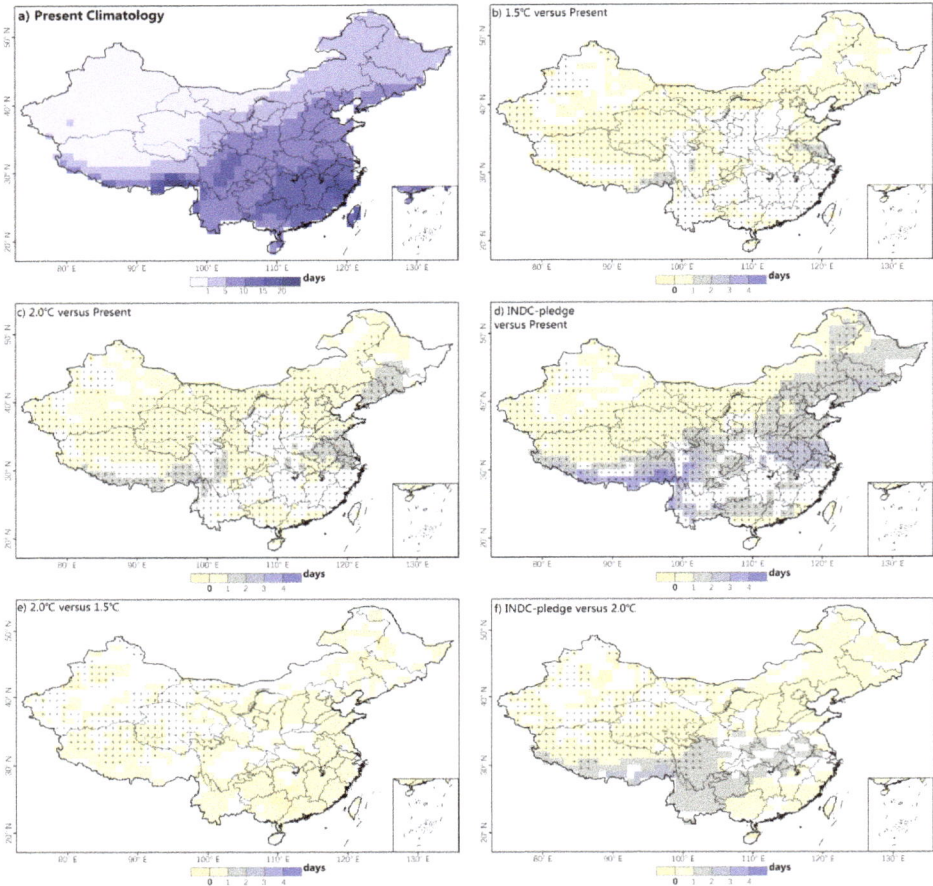

Figure 3. Present climatology of the annual R25 over Asia (**a**) and changes in the annual R25 under different scenarios (**b**–**f**) based on the multi-model mean. The colored shading in (**b**–**f**) was applied to areas that were statistically significant at the 10% level according to Wilcoxon rank sum test; the stippling in (**b**–**f**) was added to regions with a multi-model mean SNR > 1.

The intensity and frequency of extreme precipitation consistently increase in most regions of China, but the magnitudes of the changes are widespread across different subregions and models. Specifically, the Rx5day over China will increase by 4.2% (3.1–7.3%; range of the 25–75% confidence interval) in the 1.5 °C scenario, by 8.3% (6.9–9.8%) in the 2.0 °C scenario, and by 16.0% (13.5–20.1%) in the INDC-pledge scenario compared with the present-day baseline. The spatial patterns of changes of the SDII index are similar, but the magnitudes are smaller. Hotspots, that is, locations in which the increase in the intensity of extreme precipitation is the most prominent, are observed in the Huang–Huai River Valley, Northeast China, and northern and southern periphery of the Tibetan Plateau. In terms of the increase in the frequency of heavy precipitation, the R25 indices in the southeastern periphery of the Tibetan Plateau and Yangtze–Huai River Valley are projected to exhibit the greatest rates (for example, under the INDC, the increase in the R25 in those areas is ~3–4 days compared with the present-day level), while the vast eastern monsoon region of China will experience slight increases in the number of heavy precipitation days. The daily precipitation in arid areas in western China hardly exceeds 25 mm.

The change of the extreme precipitation indices can be shown more intuitively in probability density (frequency distribution) diagrams (Figure 6). The probability density curves of three extreme precipitation indices over China only slightly change, indicating insignificant changes in the mean value of extreme precipitation associated with global warming. However, the shape of the curves becomes wider, suggesting increases in the standard deviations of the three indices. In particular, the probability of record-breaking heavy precipitation occurring in China increases, despite a relatively small rise in the mean value. The RR shows a stronger increase with global warming, which confirms the above-mentioned proposition. Note that the RRs in the East Asian Monsoon Region are significantly larger than the national average (Figures S9–S11).

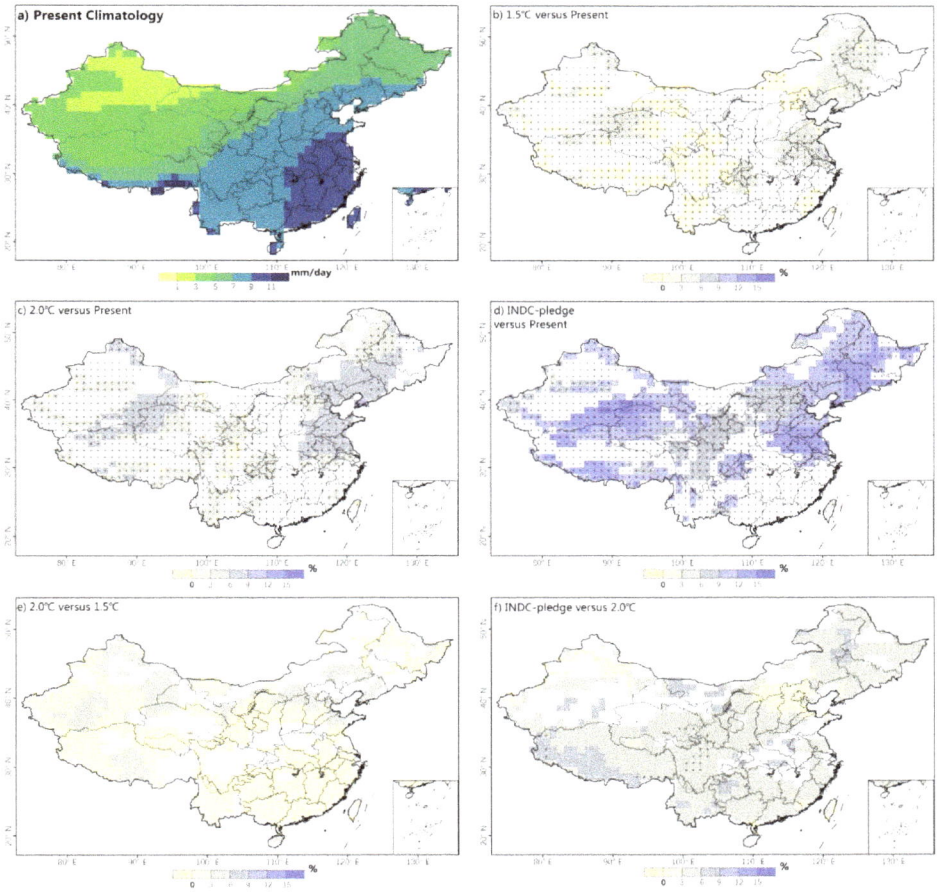

Figure 4. Present climatology of the annual SDII over Asia (**a**) and changes in the annual SDII under different scenarios (**b–f**) based on the multi-model mean. The colored shading in (**b–f**) was applied to areas that were statistically significant at the 10% level according to Wilcoxon rank sum test; the stippling in (**b–f**) was added to regions with a multi-model mean SNR > 1.

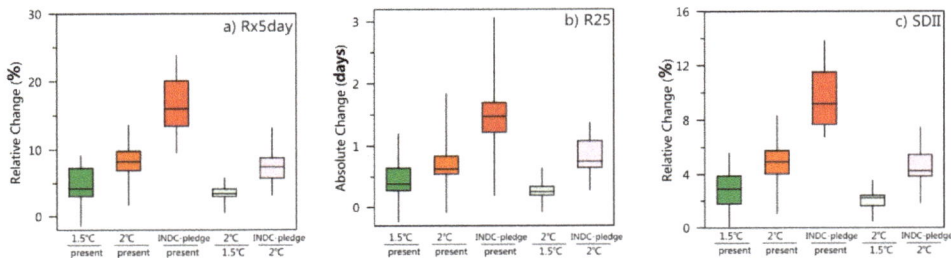

Figure 5. National average differences among different scenarios for the annual Rx5day (**a**), R25 (**b**), and SDII (**c**) changes in China based on the multi-model mean ensemble. Boxes (Whiskers) indicate the 25th and 75th percentiles (maximum and minimum) of 14 climate models, and the horizontal lines represent the multi-model median. The scenarios are the same as those shown in Figures 2–4, with corresponding colors. The differences between different sets of scenarios are labeled on the X-axis. The first three bars show the differences of each scenario (1.5 °C, 2.0 °C, and INDC-pledge) relative to the present-day baseline scenario and the last two bars show the differences between the 1.5 °C and 2.0 °C and INDC scenarios and 2.0 °C scenario, as labeled at the bottom. Detailed information for each subregion is provided in Figure S6 (for Rx5day), Figure S7 (for R25), and Figure S8 (for SDII).

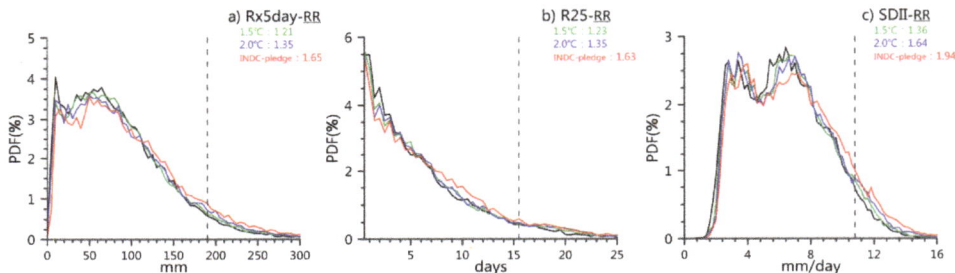

Figure 6. Frequency distributions (probability distribution function, PDF) of extreme precipitation indices: Rx5day (**a**), R25 (**b**), and SDII (**c**). Zero R25 values were omitted. The black lines indicate the results for the period of 1985–2005. The green, blue, and red lines indicate the results during the 1.5 °C, 2.0 °C, and ΔT_{INDC} warming periods, respectively. The dashed lines indicate the 5% extreme values of the baseline period (1985–2005). Detailed information for each subregion is provided in Figure S9 (for Rx5day), Figure S10 (for R25), and Figure S11 (for SDII).

The uncertainties related to the projected precipitation extremes were quantified using the coefficient of variation (C_v) between the models. Figure S12 shows that the spatial pattern of C_v under different scenarios is similar. Compared with other regions, the C_v value is much larger in the South China, Xinjiang Autonomous Region, and the southeastern edge of the Tibetan Plateau. It's noticed that C_v is only a quantification of climate model uncertainty.

3.2. Changes in the Exposure to Precipitation-Related Extremes

To further analyze the possible impact of future precipitation-related extremes, we defined dangerous extreme events as those exceeding the 10 and 20 year RVs compared with the present-day baseline (1961–2005) The 10 and 20 year thresholds represent different levels of danger. Here, we only show the population exposure estimated from the projected population for 2100 under the SSP2 scenario [36]. The results of the population exposures based on projections under other SSP scenarios are qualitatively similar. The evolution of the exposure (especially population-weighted density of the exposure) depending on the warming levels indicates the probability of the human system being impacted by these dangerous extremes.

China and its subregions will be exposed to these events of different RV increases consistently with global warming, and record-breaking events will be more frequent (Figure 7). For example, for heavy precipitation events that exceed the 20 year baseline RV, the population exposure over China increases by 5.7% (5.1–6.0%) to 8.4% (7.5–9.5%) in the 1.5 °C scenario, by 10.6% (8.7–11.2%) in the 2.0 °C scenario, and by 15.9% (14.2–16.4%) in the INDC-pledge scenario compared with the present-day level. An approximately linear increase in the exposure to mean global warming can be observed in all regions, although the degrees of approximation to linear equations varies among the eight subregions (Figure 7). Population exposures in the Tibetan Plateau and Southwest China (SWC1 and 2) are notably higher than in other regions; however, the exposure in those two subregions shows a greater intermodel variability.

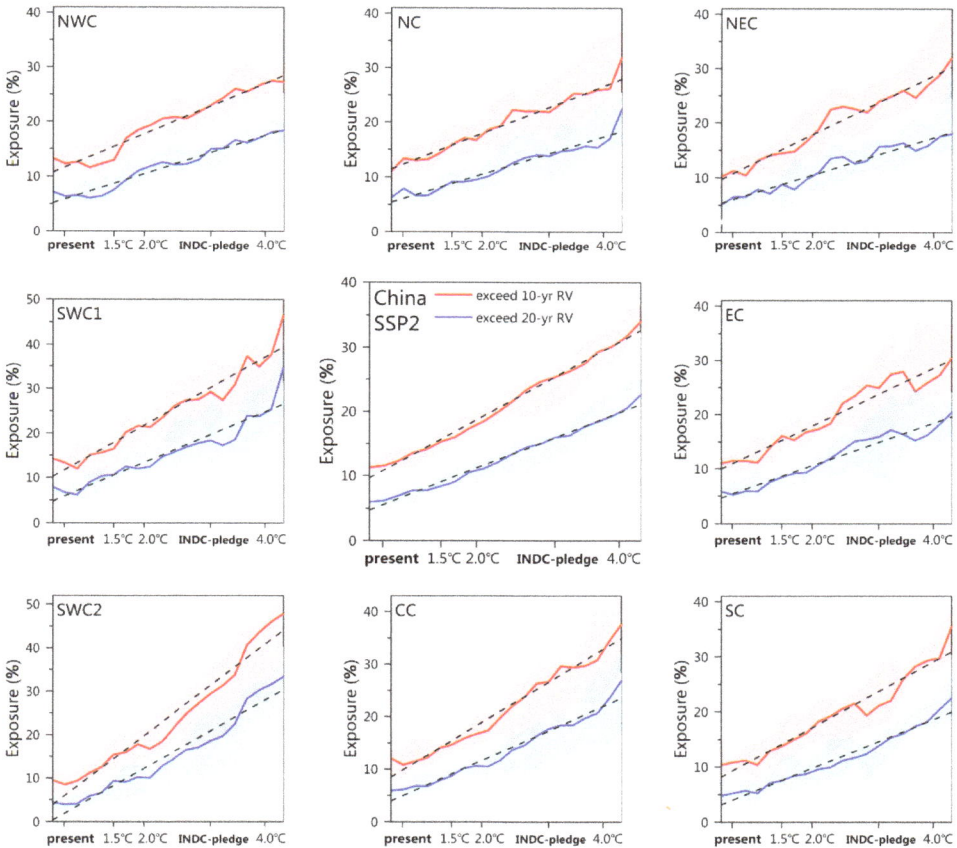

Figure 7. Population exposure to heavy precipitation events with different return values (RVs) at corresponding warming levels over Asia and eight subregions (labeled on the top-left corner). The multi-model means are represented by solid lines and the interquartile ranges are shaded. The dashed black lines denote the linear trend of the exposure depending on the mean global warming. The population-weighted density means are estimated from the 2100 population prediction under the Shared Socioeconomic Pathways 2 (SSP2) scenario (future scenarios). The reliability of the curves decreases at the right edge because of the inconsistent number of ensemble members which corresponds to this figure. The results of the calculations based on the individual models are provided in Figures S15 and S16.

3.3. Impacts Avoided in Low-Warming Scenarios

If the warming is limited to a lower level, China is projected to benefit from robust reductions in the population exposure to dangerous extremes (Figure 8). Across China, the population exposures based on heavy precipitation events that exceed the 10 year baseline RV will increase to 138% (111–160%) in the 1.5 °C scenario, 165% (147–179%) in the 2.0 °C scenario, and 217% (199–238%) in the INDC-pledge scenario compared with the present-day level. Thus, the median values of the avoided impacts (expressed in units of present-day exposure) are estimated to be 27% (14–36%; 1.5 °C versus 2.0 °C) and 55% (41–80%; 2.0 °C versus INDC-pledge), respectively. The avoided impacts of the 2.0 °C versus INDC-pledge scenarios are greater than those of the 1.5 °C versus 2.0 °C scenarios, indicating that further efforts to increase the emission reduction based on the INDC-pledge will lead to more benefits.

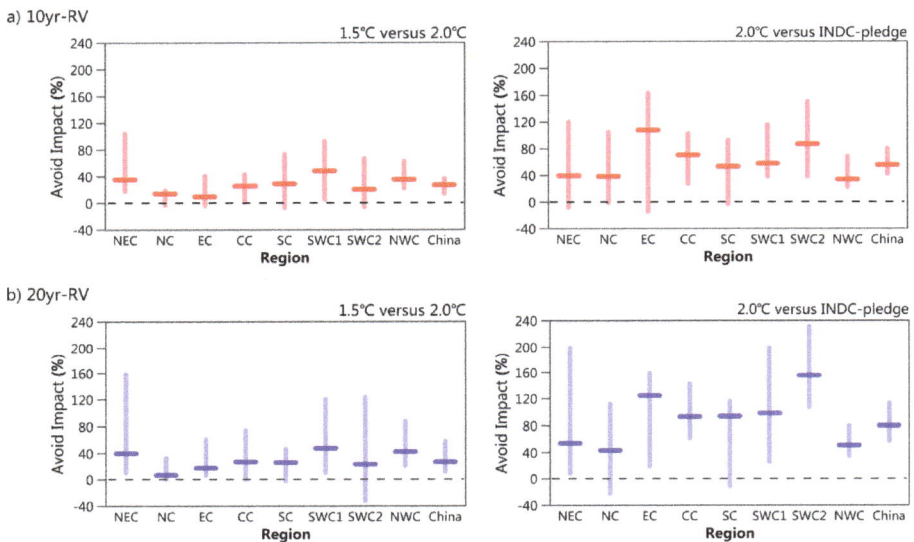

Figure 8. Changes in the avoided extreme precipitation in China and its subregions in low-warming scenarios. The population exposure to heavy precipitation events that exceed the baseline (**a**) 10 and (**b**) 20 year return values is reduced in the low-warming scenarios (1.5 °C compared with 2.0 °C, 2.0 °C compared with Intended Nationally Determined Contributions (INDC). The projected changes are expressed in units of present-day exposure (1985–2005). The central lines and bars denote the multi-model medians and interquartile ranges, respectively.

Note that the avoided impacts are more remarkable for more intense extremes. For example, for heavy rainfall events exceeding the 20 year baseline RV, 27% (12–58%; 1.5 °C versus 2.0 °C) and 80% (57–113%; 2.0 °C versus INDC-pledge) of the population exposure could be reduced by less warming, respectively. More than half of the subregions would experience such a robust avoidance of impacts, although the magnitudes would differ.

Under low-warming scenarios, almost all regions in China will face less risk (exposure) of heavy precipitation. Here, we only considered the fractional population exposure under SSPs. If the absolute population growth is exceeding expectations, the avoided impacts will be larger.

4. Discussion

The response of global warming to actual emission reductions based on the Paris Agreement is of ongoing interest. Most of the existing research is based on RCP or 1.5 °C/2.0° C scenarios, such as Xu et al. (2017) [37], Dosio and Fischer Erich (2017) [9], Schleussner et al. (2016) [10] and Zhang et al.

(2018) [12]. However, these scenarios do not consider current mitigation commitments negotiated by worldwide governments. Our approach is based on self-determined emission reduction commitments established during climate negotiations as the starting point to assess the future climate response.

Our results indicate that climate warming under the INDC scenarios is projected to greatly exceed the long-term Paris Agreement goal of stabilizing the global mean temperature above the 2 °C or 1.5 °C level, which is consistent with previously published studies [15,22,23]. Furthermore, we quantified regional climate change in response to INDC pledges. Based on our results, the regional precipitation extremes in China under the INDC scenarios are projected to show great increases in intensity and frequency compared with the current level. The differences in the exposure to extreme events between different scenarios are higher than that of the mean state. If the global emissions are further reduced to achieve the ambitious temperature target, the benefits on the regional exposure to heavy precipitation may be more pronounced than those based on the mean state of precipitation. We focus not only on the assessment of extreme climate impacts under INDC scenarios, but also on the extreme climate impacts at 1.5 °C. Our results for the 1.5 °C scenario are consistent with those reported in Chapter III of the IPCC 1.5 °C report [13].

We also observed a nearly linear relationship between the regional extreme precipitation and global mean warming, which is consistent with the results of [38]. Linking global warming targets to regional consequences, such as changing climate extremes, would benefit political decision-making with respect to climate negotiations and adaptation. However, linear relationships are only meaningful if associated projection uncertainties are kept within reasonable bounds. Some changes in the climate system may be abrupt due to tipping points [39]. Therefore, the limitations of this approach need to be determined.

It is known that the ability of the model to simulate extreme precipitation indices is lower than the ability to simulate extreme temperature indices [40]. The ability of GCMs to simulate features of extreme precipitation is closely related to the ability of the models to simulate atmospheric circulation such as the Northwest Pacific Subtropical High (NPSH) and East Asian Westerly Jet [41–43]. Moreover, the simulation bias between the GCMs and observation is attributed to the low resolution of GCMs. Regional climate models (RCMs) have finer resolutions and thus can better capture the characteristics of extreme precipitation on a regional scale. For example, the RCM can effectively remove the false strong precipitation center simulated by the GCM along the southeastern edge of the Tibetan Plateau [25]. Thus, an experimental design for INDC emission pledges based on RCMs is needed to better project the changes in extreme precipitation at smaller regional scales in China and provide decision-makers with more precise information.

5. Conclusions

By using fully coupled simulations from 14 CMIP5 models and their ensemble, we analyzed the changes in the extreme precipitation over China under global INDC scenarios and compared the results with those of the 1.5 °C/2.0 °C warming targets. The main findings are as follows:

1. With the enhancement of global warming, the intensity and frequency of extreme precipitation will gradually increase in most regions in China. Changes in the Rx5day are more prominent than those of SDII, while the spatial patterns are similar. The number of days during which the daily precipitation exceeds 25 mm (i.e., R25) increases in the monsoon area in eastern China. The probability distributions of extreme precipitation events are projected to become wider, resulting in the occurrence of more record-breaking heavy precipitation in China in the future despite the relatively small rise in the mean value.
2. The population exposure to dangerous (e.g., exceeding the 20-year RV) extreme precipitation events in China is projected to consistently increase because of climate change, following an approximately linear trend. Less warming would reduce the population exposure to once-in 20 years extreme rainfall events by 27% (12–58%; 1.5 °C versus 2.0 °C) and 57% (80–113%; 2.0 °C versus INDC), respectively. Our results improve the understanding of future risks associated

with extreme precipitation, which is critical for the design of mitigation and adaptation policies in China, which is home to more than 1.4 billion people.

Supplementary Materials: Supplementary information is available in the online version of the paper. The following are available online at http://www.mdpi.com/2073-4441/11/6/1167/s1. Figure S1: Future emission pathways analyzed in this example. The black vertical line represents the range of conditional and unconditional INDC pledges in 2030; thin lines in different colors show the selected emission pathways clustered into the six groups. The range of the 1.5 °C and 2 °C pathways are plotted for reference in grey and orange shaded areas, respectively [4,8,9]. The estimates of current warming above the pre-industrial level (ΔT) for each scenario group are labelled on the right (uncertainty range of 33–66% and median in brackets). Figure S2: Time series of global mean annual temperature changes relative to the baseline climatology of the pre-industrial (1861–1900), as derived from individual GCMs under RCP8.5 (thin line in different colors). Multi-model ensemble with 21 year running mean (black bold line and shade) is also shown on the graph. The horizontal dashed lines indicate a given warming target. Figure S3: Median years projected by the 14 CMIP5 models for three global warming targets (ΔT_{INDC}, 1.5 °C, and 2.0 °C) under the RCP8.5 scenario. Table S2: Details of eight subregions in China [10]. Figure S4: Observed (blue line) and simulated (black line and shade, multi-model median and interquartile range, respectively) time series of three extreme precipitation indices from 1961 to 2005. linear trends of observation (dotted line in red) and multi-model median (dotted line in yellow) are shown in the same graph. Yellow cross denotes the linear trend passed the 95% confidence level. Figure S5: Spatial distribution of extreme precipitation indices for Rx5day (unit: days), SDII (unit: mm day^{-1}), and R25 (unit: days) during the period of present (1985–2005). Column 1: observation; Column 2: MME; Column 3: bias (model ensemble simulation minus observation). Figure S6: Regional average differences among different scenarios in the annual Rx5day in China and its eight subregions, based on the multi-model mean ensemble. Boxes (Whiskers) indicate the 25th and 75th percentiles (maximum and minimum) of 14 climate models, and the horizontal lines represent the multi-model median. Scenarios are the same as Figure 5, with corresponding colors. The differences between different sets of scenarios are labelled on the x-axis. The first three bars show differences for each scenario (1.5 °C, 2.0 °C, and INDC-pledge) relative to the present-day baseline scenario, and the last two bars show differences between the 1.5 °C and 2.0 °C scenarios and INDC scenario relative to the 2.0 °C scenario, as labelled. Eight subregions (labelled on the top left corner) are defined in Text S4. Figure S7: Corresponds to Figure S6, but for R25. The national average of R25 (the central sub-figure), does not include NWC. Figure S8: Corresponds to Figure S6, but for SDII. Figure S9: Frequency distributions of extreme precipitation indices Rx5day over China and its eight subregions. Black lines indicate the results during 1985–2005. Green, blue, and red lines indicate the results during the 1.5 °C, 2.0 °C, and ΔT_{INDC} warming periods, respectively. The dashed lines indicate the 5% extreme values for the baseline period 1985–2005. Risk ratios (RRs) for three scenarios labeled on the top-right (including medians and interquartile ranges). Figure S10: Corresponds to Figure S9, but for R25. Zero values of R25 are omitted. Figure S11: Corresponds to Figure S9, but for SDII. Figure S12: Spatial distribution of the intermodel coefficient of variations of three extreme precipitation indices simulated by the 14 CMIP5 models for scenarios of present, 1.5 °C, 2.0 °C, and INDC-pledge. Figure S13: Three extreme precipitation indices over China (regional average) simulated by 14 individual CMIP5 models (in different color lines), with corresponding global mean warming. Figure S14: Multi-model ensemble areal exposure to heavy rainfall of different return values (RVs). Corresponds to Figure 7 but for areal exposure. Corresponds to this figure; calculation results of individual models refer to Figures S17 and S18. Figure S15: Population exposure to heavy precipitation events of 10 year RV at corresponding warming levels over Asia and eight subregions (labelled in top left corner), based on results of individual models (line in different colors). Figure S16: Corresponds to Figure S15, but for population exposure to heavy precipitation events of 20 year RV. Figure S17: Areal exposure to heavy precipitation events of 10 year RV at corresponding warming levels over Asia and eight subregions (labelled in top left corner), based on results of individual models (line in different colors). Figure S18: Corresponds to Figure S17 but for population precipitation events of 20 year RV. Figure S19: Corresponds to Figure 8 but for areal exposure. Table S1: Details of the 14 CMIP5 global climate models used in this research.

Author Contributions: Writing—original draft preparation, J.Z.; writing—review and editing, F.W.

Funding: This work was supported by the Strategic Leading Science and Technology Program of the Chinese Academy of Sciences (XDA20020202), the National Key Research and Development Program of China (2016YFA0602704) and the National Natural Science Foundation of China (41771050).

Acknowledgments: We would like to thank two anonymous reviewers for their helpful comments.

Conflicts of Interest: The authors declare no conflict of interest.

References

1. Yin, H.; Changan, L.I. Human impact on floods and flood disasters on the Yangtze River. *Geomorphology* **2001**, *41*, 105–109. [CrossRef]

2. Wang, K.; Wei, Y.M.; Ye, M. Beijing storm of July 21, 2012: Observations and reflections. *Nat. Hazards* **2013**, *67*, 969–974. [CrossRef]

3. Trenberth, K.E. Conceptual Framework for Changes of Extremes of the Hydrological Cycle with Climate Change. *Clim. Chang.* **1999**, *42*, 327–339. [CrossRef]
4. Haerter, J.O.; Berg, P. Unexpected rise in extreme precipitation caused by a shift in rain type? *Nat. Geosci.* **2009**, *2*, 372–373. [CrossRef]
5. Collins, M.; Knutti, R.; Arblaster, J.; Dufresne, J.-L.; Fichefet, T.; Friedlingstein, P.; Gao, X.; Gutowski, W.J.; Johns, T.; Krinner, G.; et al. Long-term Climate Change: Projections, Commitments and Irreversibility. In *Climate Change 2013: The Physical Science Basis. Contribution of Working Group I to the Fifth Assessment Report of the Intergovernmental Panel on Climate Change*; Stocker, T.F., Qin, D., Plattner, G.-K., Tignor, M., Allen, S.K., Boschung, J., Nauels, A., Xia, Y., Bex, V., Midgley, P.M., et al., Eds.; Cambridge University Press: Cambridge, UK; New York, NY, USA, 2013.
6. UNFCCC. *Adoption of the Paris Agreement: Proposal by the President*, FCCC/CP/2015/L.9/Rev.1, ed.; UNFCCC: Geneva, Switzerland, 2015.
7. UNFCCC. *Synthesis Report on the Aggregate Effect of the Intended Nationally Determined Contributions*, FCCC/CP/2015/7, ed.; UNFCCC: Paris, France, 2015.
8. Gupta, S.; Tirpak, D.; Burger, N.; Gupta, J.; Höhne, N.; Boncheva, A.; Kanoan, G.; Kolstad, C.; Kruger, J.; Michaelowa, A.; et al. Policies, instruments, and co-operative arrangements. In *Climate Chang. 2007: Mitigation. Contribution of Working Group III to the Fourth Assessment Report of the Intergovernmental Panel on Climate Change*; Metz, O.B., Bosch, P., Dave, R., Meyer, L., Eds.; Cambridge University Press: Cambridge, UK; New York, NY, USA, 2007; pp. 745–807.
9. Dosio, A.; Fischer Erich, M. Will Half a Degree Make a Difference? Robust Projections of Indices of Mean and Extreme Climate in Europe Under 1.5 °C, 2 °C, and 3 °C Global Warming. *Geophys. Res. Lett.* **2017**, *45*, 935–944. [CrossRef]
10. Schleussner, C.-F.; Lissner, T.K.; Fischer, E.M.; Wohland, J.; Perrette, M.; Golly, A.; Rogelj, J.; Childers, K.; Schewe, J.; Frieler, K.; et al. Differential climate impacts for policy-relevant limits to global warming: The case of 1.5 °C and 2 °C. *Earth Syst. Dyn.* **2016**, *7*, 327–351. [CrossRef]
11. Xu, Y.; Zhou, B.-T.; Wu, J.; Han, Z.-Y.; Zhang, Y.-X.; Wu, J. Asian climate change under 1.5–4 °C warming targets. *Adv. Clim. Chang. Res.* **2017**, *8*, 99–107. [CrossRef]
12. Zhang, W.; Zhou, T.; Zou, L.; Zhang, L.; Chen, X. Reduced exposure to extreme precipitation from 0.5 °C less warming in global land monsoon regions. *Nat. Commun.* **2018**, *9*, 3153. [CrossRef]
13. Hoegh-Guldberg, O.; Jacob, D.; Taylor, M.; Bindi, M.; Brown, S.; Camilloni, I.; Diedhiou, A.; Djalante, R.; Ebi, K.; Engelbrecht, F.; et al. Impacts of 1.5°C global warming on natural and human systems. In *Global Warming of 1.5°C. An IPCC Special Report on the Impacts of Global Warming of 1.5°C above Pre-Industrial Levels and Related Global Greenhouse Gas Emission Pathways, in the Context of Strengthening the Global Response to the Threat of Climate Change, Sustainable Development, and Efforts to Eradicate Poverty*; Masson-Delmotte, V., Zhai, P., Pörtner, H.O., Roberts, D., Skea, J., Shukla, P.R., Pirani, A., Moufouma-Okia, W., Péan, C., Pidcock, R., et al., Eds.; Cambridge University Press: Cambridge, UK; New York, NY, USA, 2018; in press.
14. Sanderson, B.M.; O'Neill, B.C.; Tebaldi, C. What would it take to achieve the Paris temperature targets? *Geophys. Res. Lett.* **2016**, *43*, 7133–7142. [CrossRef]
15. Rogelj, J.; den Elzen, M.; Höhne, N.; Fransen, T.; Fekete, H.; Winkler, H.; Schaeffer, R.; Sha, F.; Riahi, K.; Meinshausen, M. Paris Agreement climate proposals need a boost to keep warming well below 2 °C. *Nature* **2016**, *534*, 631. [CrossRef]
16. CAT. Addressing Global Warming. Available online: https://climateactiontracker.org/global/temperatures/ (accessed on 31 December 2017).
17. UNEP. *The Emissions Gap Report*; UNEP: Nairobi, Kenya, 2017.
18. IPCC. *Climate Change 2014: Synthesis Report. Contribution of Working Groups I, II and III to the Fifth Assessment Report of the Intergovernmental Panel on Climate Change*; Cambridge University Press: Cambridge, UK; New York, NY, USA, 2014.
19. IPCC. *Global Warming of 1.5 °C*; Cambridge University Press: Cambridge, UK; New York, NY, USA, 2018.
20. UNFCCC. *National Inventory Submissions*; UNFCCC: Geneva, Switzerland, 2019.
21. Taylor, K.E.; Stouffer, R.J.; Meehl, G.A. An Overview of CMIP5 and the Experiment Design. *Bull. Am. Meteorol. Soc.* **2011**, *93*, 485–498. [CrossRef]
22. Wang, F.; Tokarska, K.B.; Zhang, J.; Ge, Q.; Hao, Z.; Zhang, X.; Wu, M. Climate Warming in Response to Emission Reductions Consistent with the Paris Agreement. *Adv. Meteorol.* **2018**, *2018*, 1–9. [CrossRef]

23. Wang, F.; Ge, Q.; Chen, D.; Luterbacher, J.; Tokarska, K.B.; Hao, Z. Global and regional climate responses to national-committed emission reductions under the Paris agreement. *Geografiska Annaler Ser. A Phys. Geogr.* **2018**, *100*, 240–253. [CrossRef]
24. Jia, W. A gridded daily observation dataset over China region and comparison with the other datasets. *Chin. J. Geophys.* **2013**, *56*, 1102–1111.
25. Gao, X.-J.; Wang, M.-L.; Filippo, G. Climate Change over China in the 21st Century as Simulated by BCC_CSM1.1-RegCM4.0. *Atmos. Ocean. Sci. Lett.* **2013**, *6*, 381–386.
26. Ji, Z.; Kang, S. Evaluation of extreme climate events using a regional climate model for China. *Int. J. Climatol.* **2015**, *35*, 888–902. [CrossRef]
27. Robert, V.; Andreas, G.; Stefan, S.; Erik, K.; Annemiek, S.; Paul, W.; Thomas, M.; Oskar, L.; Grigory, N.; Claas, T.; et al. The European climate under a 2 °C global warming. *Environ. Res. Lett.* **2014**, *9*, 034006.
28. Huang, J.; Yu, H.; Dai, A.; Wei, Y.; Kang, L. Drylands face potential threat under 2 °C global warming target. *Nat. Clim. Chang.* **2017**, *7*, 417. [CrossRef]
29. Zhang, X.; Alexander, L.; Hegerl, G.C.; Jones, P.; Tank, A.K.; Peterson, T.C.; Trewin, B.; Zwiers, F.W. Indices for monitoring changes in extremes based on daily temperature and precipitation data. *Wiley Interdiscip. Rev. Clim. Chang.* **2011**, *2*, 851–870. [CrossRef]
30. Sun, Y.; Solomon, S.; Dai, A.; Portmann, R.W. How Often Does It Rain? *J. Clim.* **2006**, *19*, 916–934. [CrossRef]
31. Huang, J.; Sun, S.; Zhang, J. Detection of trends in precipitation during 1960–2008 in Jiangxi province, southeast China. *Theor. Appl. Climatol.* **2013**, *114*, 237–251. [CrossRef]
32. Zhai, P.; Zhang, X.; Wan, H.; Pan, X. Trends in Total Precipitation and Frequency of Daily Precipitation Extremes over China. *J. Clim.* **2005**, *18*, 1096–1098, 1100–1108. [CrossRef]
33. Lavell, A.; Oppenheimer, M.; Diop, C.; Hess, J.; Lempert, R.; Li, J.; Muir-Wood, R.; Myeong, S. Climate change: New dimensions in disaster risk, exposure, vulnerability, and resilience. In *Managing the Risks of Extreme Events and Disasters to Advance Climate Change Adaptation*; Field, C.B., Barros, V., Stocker, T.F., Qin, D., Dokken, D.J., Ebi, K.L., Mastrandrea, M.D., Mach, K.J., Plattner, G.-K., Allen, S.K., et al., Eds.; A Special Report of Working Groups I and II of the Intergovernmental Panel on Climate Change (IPCC); Cambridge University Press: Cambridge, UK; New York, NY, USA, 2012; pp. 25–64.
34. Cox, D.R.; Hinkley, D.V. *Theoretical Statistics*; Chapman and Hall: London, UK, 1974; p. 511.
35. Kharin, V.V.; Zwiers, F.W. Estimating Extremes in Transient Climate Change Simulations. *J. Clim.* **2005**, *18*, 1156–1173. [CrossRef]
36. Jones, B.; O'Neill, B.C. Spatially explicit global population scenarios consistent with the Shared Socioeconomic Pathways. *Environ. Res. Lett.* **2016**, *11*, 084003. [CrossRef]
37. Ying, X.; Jie, W.; Ying, S.; Bo-Tao, Z.; Rou-Ke, L.; Jia, W. Change in Extreme Climate Events over China Based on CMIP5. *Atmos. Ocean. Sci. Lett.* **2015**, *8*, 185–192. [CrossRef]
38. Seneviratne, S.I.; Donat, M.G.; Pitman, A.J.; Knutti, R.; Wilby, R.L. Allowable CO$_2$ emissions based on regional and impact-related climate targets. *Nature* **2016**, *529*, 477. [CrossRef] [PubMed]
39. Lenton, T.M.; Held, H.; Kriegler, E.; Hall, J.; Lucht, W.; Rahmstorf, S.; Schellnhuber, H. Tipping elements in the Earth's climate system. *Proc. Natl. Acad. Sci. USA* **2008**, *105*, 1786–1793. [CrossRef] [PubMed]
40. Jiang, D.; Sui, Y.; Lang, X. Timing and associated climate change of a 2 °C global warming. *Int. J. Climatol.* **2016**, *36*, 4512–4522. [CrossRef]
41. Jiang, Z.; Wei, L.; Xu, J.; Li, L. Extreme Precipitation Indices over China in CMIP5 Models. Part I: Model Evaluation. *Adv. Sci. Serv. Soc.* **2015**, *28*. [CrossRef]
42. Rajendran, K.; Kitoh, A.; Yukimoto, S. South and East Asian summer monsoon climate and variation in MRI coupled model (MRI-CGCM2). *Adv. Sci. Serv. Soc.* **2004**, *17*, 763–782. [CrossRef]
43. Lin, Z.; Fu, Y.; Lu, R. Intermodel Diversity in the Zonal Location of the Climatological East Asian Westerly Jet Core in Summer and Association with Rainfall over East Asia in CMIP5 Models. *Adv. Atmos. Sci.* **2019**, *36*, 614–622. [CrossRef]

water

MDPI

Article

A Multi-GCM Assessment of the Climate Change Impact on the Hydrology and Hydropower Potential of a Semi-Arid Basin (A Case Study of the Dez Dam Basin, Iran)

Roya Sadat Mousavi, Mojtaba Ahmadizadeh and Safar Marofi *

Department of Water Sciences Engineering, Bu-Ali Sina University, Hamedan 6517838695, Iran;
roya.s.moosavi@gmail.com (R.S.M.); mojtaba_ahmadizade@yahoo.com (M.A.)
* Correspondence: safarmarofi59@gmail.com or marofisafar59@gmail.com; Tel.: +98-918-314-3686

Received: 28 August 2018; Accepted: 8 October 2018; Published: 16 October 2018

Abstract: In this paper, the impact of climate change on the climate and discharge of the Dez Dam Basin and the hydropower potential of two hydropower plants (Bakhtiari and Dez) is investigated based on the downscaled outputs of six GCMs (General Circulation Models) and three SRES (Special Report on Emission Scenarios) scenarios for the early, mid and late 21st century. Projections of all the scenarios and GCMs revealed a significant rise in temperature (up to 4.9 °C) and slight to moderate variation in precipitation (up to 18%). Outputs of the HBV hydrologic model, enforced by projected datasets, show a reduction of the annual flow by 33% under the climate change condition. Further, analyzing the induced changes in the inflow and hydropower generation potential of the Bakhtiari and Dez dams showed that both inflow and hydropower generation is significantly affected by climate change. For the Bakhtiari dam, this indicates a consistent reduction of inflow (up to 27%) and electricity generation (up to 32%). While, in the Dez dam case, the inflow is projected to decrease (up to 22%) and the corresponding hydropower is expected to slightly increase (up to 3%). This contrasting result for the Dez dam is assessed based on its reservoir and hydropower plant capacity, as well as other factors such as the timely releases to meet different demands and flow regime changes under climate change. The results show that the Bakhtiari reservoir and power plant will not meet the design-capacity outputs under the climate change condition as its large capacity cannot be fully utilized; while there is room for the further development of the Dez power plant. Comparing the results of the applied GCMs showed high discrepancies among the outputs of different models.

Keywords: global warming; statistical downscaling; HBV model; flow regime; uncertainty

1. Introduction

Anthropogenic global warming and its consequences, especially in the arid and semi-arid regions, received particular attention in recent years as many scholars documented the occurrence and dominance of droughts, a rise in the temperature, and an increase in the atmospheric water demand, accompanied by a reduction in the precipitation and runoff [1–4]. Additionally, in Iran, many studies support the fact that during recent decades the climate has experienced variations, mostly toward hot and dry conditions [5–14].

Anticipated climatic changes can alter the hydrological regimes, such as the amount of discharge or the timing of the surface flow on both the regional and local (catchment) scales [15–17], with socio-economic and environmental consequences. Projections of the impact of climate change on the hydrological conditions on a global scale show that for low and mid-latitude regions, a reduction of freshwater could exacerbate their water-management problems; while the higher latitudes will experience higher amounts of surface flow [18,19].

Several studies worldwide predicted that the combination of the temperature increase and precipitation variation during the present century will result in a significant change of the runoff. According to the study by Tong et al. on the variability of the discharge, under the B1 scenario and the urbanization in the Las Vegas Wash watershed in the USA, climate change is the main deriving factor of future changes in the watershed, causing a wintertime discharge decrease and a summertime discharge increase [20]. The study on the future water availability in Bangladesh revealed that climate change has a significant impact on runoff and evapotranspiration because the region will face a higher irrigation demand, a decline of groundwater, and a variability of rainfall and runoff (both increasing and decreasing) [21].

The projection of future climatic conditions in Senegal by Tall et al. showed an increase in the temperature, evaporation, and precipitation by the mid-21st century and a decrease of these parameters by the late-21st century. They also reported that dependent on the applied scenarios, the runoff will change (both increasingly and decreasingly). This could lead to an arid climate dominance in the region [22]. He et al. assessed the hydrologic sensitivity to climate change in the upper San Joaquin River basin in California by employing projected temperature and precipitation datasets, the reported temperature rise (between 1.5 and 4.5 °C) and precipitation variations (between 80 and 120%) [23]. They showed that climate change can lead to annual streamflow variations between −41 and 16%. They also detected an earlier shift of most of the streamflow by 15 to 46 days as a result of the temperature increase, which is the cause of the higher seasonal variability of streamflow. Modeling the future climatic and hydrologic response to climate change in Spain revealed a 1.5–3.3 °C temperature increase, a 6–32% precipitation decrease, and a 2–54% runoff decrease [24]. Xu and Luo, by employing seven GCMs (General Circulation Models) and the A1B scenario in two semi-arid and humid regions in China, reported dramatic changes in the temperature (up to +8.6 °C), precipitation (up to 139%), and seasonal discharge (up to 304%), as well as an increase in the extreme flows and seasonal shifts of discharge [25]. Investigating the scenarios of climate change impact on the river flow in Western Kenya using different GCMs and SRES (Special Report on Emission Scenarios) showed that climate change has the potential to significantly alter the river flow [26].

Additionally, assessments of climate change impact on the river flow in Iran show considerable variations of surface water resources across different parts of the country. Simulation of streamflow in the North of Iran, through forcing the hydrologic model with climatic projections based on different SRES scenarios, revealed increases and decreases of the discharge for the wet and dry seasons, respectively, with an overall increase in the annual discharge [27]. According to another study on the future changes of the climatic condition and runoff across Iran, the future temperature and precipitation are projected to vary by ±6 °C and ±60%, respectively [28]. Additionally, these changes will result in higher rates of annual evaporation and a runoff reduction, as well as seasonal variations of runoff (an increase in winter and a decrease in spring). Rafiei Emam et al. showed that the hydrologic response of the Raza-Ghahavand region (a semi-arid region in Iran) to climate change is a decrease in precipitation and an increase in temperature, which leads to less groundwater recharge and a lower soil water content [29]. A study on the impact of climate change on the surface water supply in the Zayandeh-Rud River Basin in Central Iran showed an increase of 0.4–0.76 °C in annual temperature, a decrease of 14–38% in the precipitation and a decrease of 8–43% in the runoff [30]. Modeling the future climate and water resources under the A1B, A2, and B2 scenarios in the Karkheh River Basin (located in the West and Southwest of Iran) revealed a temperature increase and reduction of the water yield in the basin. This reduction is considerable from April to September as a result of the temperature increase and precipitation decrease [31].

It is widely documented that based on the future simulations, the modeled wintertime discharge increase is usually followed by a springtime discharge decrease, which is due to the temperature increase which causes the snow to melt sooner and for there to be less snow [15–17,32,33].

Additionally, a global scale study on the river flow variations caused by the A2 and B1 emission scenarios suggested an increase in the seasonality of the river flow, i.e., an increase in the high flows and a decrease in the low flows, for about one-third of the Earth [34]. Elsner et al., using the A1B and B1 scenarios, showed that climate change causes significant shifts and changes in the amount and timing of the runoff in the Pacific Northwest [35]. Pervez and Henebry reported that a combination of the land use and climatic change under different SRES scenarios can cause some changes and shifts in the amount of discharge and flow timing in the Brahmaputra River Basin [36]. Boyer et al. studied the projected variability of the wintertime and springtime discharges at a regional scale and found that under climate change, the wintertime discharge increases, while the springtime discharge decreases [16].

One of the vulnerable industries to climate change is hydropower generation, which completely relies on the amount of precipitation, snow cover, snowmelt, streamflow, and the timing of the flow, all of which show high inter-annual variability [37–39]. Based on the IPCC report, as a result of climate change, the hydropower generation potential is projected to drop by up to 6% [40]. Therefore, it is essential to adapt the water resources management to the future climatic changes [41]. In Iran, climatic changes and their repercussion on hydropower generation have been studied by Jahandideh et al. [42] and Jamali et al. [43], focusing on the Karun and Karkheh river basins, respectively. They reported a reduction of the hydropower generation potential in both of the studied basins.

Problem Statement and Objectives of the Study

The Dez Dam Basin is located in Iran's Southwest and the discharge from this basin is planned to provide water for different sectors, such as the agriculture, industry, fishery and hydropower generation sectors through existing and currently-in-constructed dams. In the planning and management of water resources, the base period recorded dataset is normally considered. However, the records from recent years and future projections of climate models suggest that global warming and climate change could alter the climate indicators and hydrologic conditions. Therefore, for basins like the Dez Dam Basin, which is not completely developed yet, it is crucial to assess the variations in the amount of discharge on a basin scale with respect to the climate change scenarios; this is essential for climate change adaptation. In this study, we investigated the variations of hydroclimatic conditions induced by climate change in the Dez Dam Basin and its consequences with regards to the hydropower generation potential through two large dams in the basin. For this aim, first, the precipitation and temperature values were projected based on the three SRES scenarios. Then, using the HBV hydrologic model, the discharge was simulated under climate change conditions. Next, the hydropower generation potential was calculated for the two hydropower plants of the Bakhtiari and Dez dams. Since many scholars identified GCMs as one of the significant sources of uncertainty in hydro-climatic studies [25,33,44–49], in this study, an ensemble of six GCMs and three emission scenarios were employed to consider the different climatic conditions within a range, offered by the outputs of different GCMs and scenarios. This assessment provides a useful means to the modify water resources management strategies, considering the repercussions of climate change on the surface water resources and the hydropower plants of the Dez Dam Basin.

2. Materials and Methods

2.1. Study Area and Data

The Dez Dam Basin (hereafter referred to as DDB) is located in the Southwest of Iran between 31°35′51″–34°7′46″ N and 48°9′15″–50°18′37″ E and is the upstream tributary of the great Karun catchment. Figure 1 shows the location of the Dez Dam Basin in Iran, as well as the streams and hydrometric stations in this basin, based on the data layers acquired from the Iran Water Resources Management Company. Figure 1 also illustrates the four sub-basins of the DDB (delineated by the ArcSWAT tool in ArcGIS), namely, Tireh, Marbereh, Sazar, and Bakhtiari. The two main rivers of Sazar and Bakhtiari drain the basin and join at the point known as Tange Panj to form the Dez River.

Figure 1 shows the location of the Dez dam, which is currently operational, and the Bakhtiari dam, which is under construction. As depicted in Figure 1, the Bakhtiari dam is located at the end point of the Bakhtiari sub-basin. Additionally, the inflow to the multi-purpose Dez dam comes from the Sazar and Bakhtiari rivers. The Bakhtiari dam is mainly designated for hydropower generation. While the storage of the Dez reservoir provides water for the domestic, agricultural, and industrial sectors, it is also used to generate electricity. Table 1 presents a description of the studied sub-basins and the hydro-meteorological indicators of the study area. The presented coordinates and elevations in Table 1 belong to the centroid of each sub-basin.

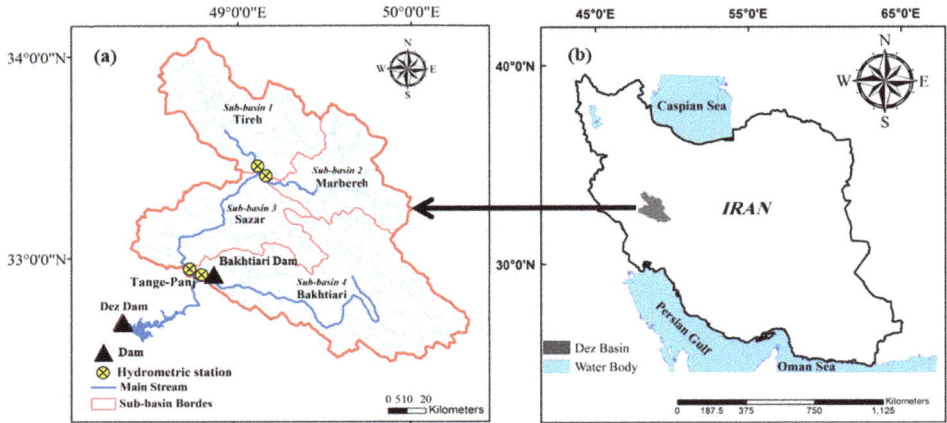

Figure 1. The map of the Dez River Basin; (**a**) the sub-basins, main streams, and location of the hydrometric stations and reservoirs (**b**) its location in Iran.

Table 1. The description of the sub-basins of the Dez River basin.

Sub-Basins	Area (Km^2)	Elevation	Average		
			T (°C)	P (mm/y)	Q (Mm^3)
Tireh (SUB-1)	3477	1551	13.93	603	486
Marbereh (SUB-2)	2553	1943	12.35	472.3	282
Sazar (SUB-3)	3281	1574	14.19	791.2	3231
Bakhtiari (SUB-4)	5973	2460	13.89	673.3	4830

Long-term daily Streamflow (Q), precipitation (P), evaporation (E), and temperature (T) time series were acquired from the Iran Water Resources Management Company and the Islamic Republic of the Iran Meteorological Organization (IRIMO). The observation period for Q is 1989–2009 and, for other parameters, it is 1986–2010.

The catchment is located within a mountainous area known as the Zagros Mountains with limited shallow aquifers in some parts of the sub-basins 1 to 3. The basin's climate is characterized as semi-arid to Mediterranean with warm summers and cold winter and less than 800 mm of precipitation per year.

2.2. GCM-Scenario Ensemble

To project the future T and P, six different GCMs were applied, including CCSM3, ECHAM5-OM, GFDL-CM2.1, HadCM3, INM-CM3.0, and IPSL-CM4. Each of the applied GCMs couple different components of the Earth system on different grid resolutions. Applying a multi-model ensemble of GCMs enables us to consider a wide range of predictions. Table 2 provides a description of the applied GCMs. The three SRES emission scenarios of A1B, A2, and B1 were applied to project the future T and P for the 2011–2030 (the 2020s), 2046–2065 (the 2050s), and 2081–2100 (the 2080s) time horizons. Therefore, in this study, an ensemble of 54 combinations of GCMs, scenarios, and time horizons (hereafter referred to as GSTs) were used to assess the climatic change and its impact on the discharge and hydropower potential of the DDB.

Table 2. The description of the GCMs (General Ciculation Models) and the IPCC-AR4 SRES (Special Report on Emission Scenarios) emission scenarios.

Research Centre	Country	GCM	Acronym	Resolution
National Centre for Atmospheric Research	USA	CCSM3	CCSM	1.4 × 1.4°
Max-Planck Institute for Meteorology	Germany	ECHAM5-OM	ECHAM5	1.9 × 1.9°
Geophysical Fluid Dynamics Lab	USA	GFDL-CM2.1	GFDL	2 × 2.5°
UK Meteorological Office	UK	HadCM3	HadCM3	2.5 × 3.75°
Institute for Numerical Mathematics	Russia	INM-CM3.0	INCM3	4 × 5°
Institute Pierre Simon Laplace	France	IPSL-CM4	IPSL	2.5 × 3.75°

The related assumptions regarding each SRES scenario and the corresponding CO_2 concentrations can be found in the IPCC's fourth assessment technical report (AR4) [50].

To downscale the future T and P in the study area, under three SERS scenarios (A1B, B1, and A2) during the 2020s, the 2050s, and the 2080s, the stochastic weather generator of LARS-WG was applied. This generated the future time series based on the probability distribution of the base period data and the correlations between the observations. The detailed description of LARS-WG is provided by Semenov [51] and Semenov and Stratonovitch [52].

2.3. Hydrological Modeling

The HBV-light (HBV-light-GUI, V. 4.0.0.6) semi-distributed hydrological model is used to simulate the streamflow in the DDB. The model considers snow routine, soil moisture routine, response function, and the flood routing of the basin. This model simulates Q with T, P, and E as the input data. The model provides the option to link sub-basins, therefore, for SUB-1, SUB-2, and SUB-3, the semi-distributed mode was applied. A detailed description of the HBV model structure and routines are provided by Seibert and Vis [53]. Prior to the application of the model, it is necessary to calibrate its parameters by the trial and error procedure, as recommended by Bergström [54]. After model calibration, validation, and after ensuring its efficiency, it was run with a projected climate series to simulate the future runoff in the basin.

2.4. Model Calibration and Validation

There are several statistics to identify the efficiency of a model. In this research, we examined the goodness of fit with different criteria (Equations (1)–(3)) to diagnose the efficiency of the model in the simulation of the streamflow in the calibration and validation periods. In details, the Nash–Sutcliff measure (R_{eff}), the coefficient of determination (R^2), and the mean annual difference (M_{diff}) are utilized to evaluate the model performance regarding efficiency, the timing of the flow, and the average error, respectively.

$$R_{eff} = 1 - \frac{\sum (Q_{Sim} - Q_{Obs})^2}{\sum (Q_{Obs} - \overline{Q}_{Obs})^2} \tag{1}$$

$$R^2 = \frac{\left(\sum (Q_{Obs} - \overline{Q_{Obs}})(Q_{Sim} - \overline{Q_{Sim}})\right)^2}{\sum (Q_{Obs} - \overline{Q_{Obs}})^2 \sum (Q_{Sim} - \overline{Q_{Sim}})^2} \tag{2}$$

$$M_{Diff} = 100 \left(\frac{\sum (Q_{Obs} - Q_{Sim})}{n Q_{Obs}} 365\right) \tag{3}$$

where, Q_{sim} and Q_{obs} are the simulated and observed discharge data and n is the number of data.

2.5. Modeling the Two Reservoirs System

The two reservoirs systems of the Bakhtiari and Dez dams are modeled based on the water balance equation (Equation (4)) by considering all the restrictions on the operation of the reservoirs and the power plant, for power generation.

$$S_{t+1} = S_t + Q_t - RE_t - RD_t - SPL_t - EVAP_t + ADD_t \tag{4}$$

where, S_{t+1} is the reservoir volume at the end of the t period (beginning of the $t + 1$ period); S_t is the reservoir volume at the beginning of the t period; Q_t is the inflow to the reservoir during the t period; RE_t is the outflow from the reservoir to generate energy during the t period; RD_t is the outflow from the reservoir to meet the downstream demand during the t period; SPL_t is the spill from the reservoir during the t period; $EVAP_t$ is the evaporation from the reservoir during the t period; and ADD_t is the volume of the added flow from the upstream reservoir during the t period.

Additionally, based on the minimum operational reservoir volume (S_{min}) and the maximum reservoir volume (S_{max}), the following conditions are considered in the model:

- If $S_{min} < S_t < S_{max}$, the outflow from the reservoir and the hydropower generation is equal to the water and energy demands of that specific month and neither deficit nor spill will occur.
- If $S_{max} < S_t$, considering the upper limit for the reservoir volume ($S_t \leq S_{max}$), the reservoir volume (S_t) is equal to S_{max} and the excessive amount of water ($S_{max} - S_t$) will spill. In this condition, there is no water or energy deficit and the secondary energy could be produced.
- If $S_t < S_{min}$, considering the lower limit for the reservoir operation ($S_{min} \leq S_t$), the reservoir volume will be substituted with the minimum operational reservoir storage, i.e., S_{min}. Therefore, in that month, the deficit is equal to ($S_t - S_{min}$). In this condition, some or all the demands may not be met. If $S_t - S_{min} \geq 0$, the water is released based on the priorities to meet high prioritized demands. Additionally, the energy generation will be affected in accordance with the reduction of the amount of water flowing through a turbine.

The power generation is associated with the installed capacity, efficiency, and plant factor as well as to the inflow and hydraulic head. Equation (5) shows the relationship that is used to calculate the energy generation.

$$P_t = \gamma Q_t H_t e_t \tag{5}$$

where, P_t is the power generated during the t period (W); γ is water specific weight (N/m^3); Q_t is the inflow to the turbine during the t period (m^3/s); H_t is the net hydraulic head on the turbine during the t period (m); and e_t is the power plant efficiency during the t period.

The following constraints are applied for the calculation of the hydropower generation based on the limits of the hydropower plant (Equations (6)–(8)).

$$P_t \leq PCC \tag{6}$$

$$Q_{min} < Q_t < Q_{max} \tag{7}$$

$$H_{min} < H_t < H_{max} \tag{8}$$

where, PCC is the installed capacity (MW); Q_{min} is the minimum turbine inflow (m^3/s); Q_{max} is the maximum turbine inflow (m^3/s); H_{min} is the minimum required head to operate the power plant (m); H_{max} is the maximum head to operate the power plant (m).

3. Results

3.1. Projected Impact of Climate Change on Temperature and Precipitation Rates

Figure 2 illustrates the average of the projected temperatures, obtained by different GSTs and the average temperature of the base period (1986–2010). As depicted in Figure 2, all the GCMs and scenarios suggest a small temperature change during the 2020s. However, for SUB-4, a temperature rate increase up to 1.2 °C is projected for the first time horizon. While, based on all the scenarios, a significant rise in the temperature is projected for the time horizons of the 2050s and the 2080s. Additionally, the average of the projected temperatures in the 2080s is higher than in the 2050s. Generally, the B1 scenario suggests a smaller temperature increase during the 2050s and the 2080s; while, scenarios A2 and A1B revealed a higher temperature during these time horizons. Additionally, regarding the differences between the GCMs, it seems that INCM3 mostly shows the lowest rates of the temperature increase, especially during the 2050s and 2080s; while GFDL, in most cases, revealed the highest temperature increase during the 2050s, based on the scenarios A1B, A2, and B1, respectively. Additionally, during the 2080s, the highest temperatures based on all the scenarios are suggested by ECHAM5. As it is shown in Figure 2, the amplitude of the multi-GCM projections of temperature suggests greater uncertainties in the 2080s, compared to the 2050s and the 2020s. Conversely, the values of the projected temperatures by different GCMs are approximately similar for the first time horizon (the 2020s).

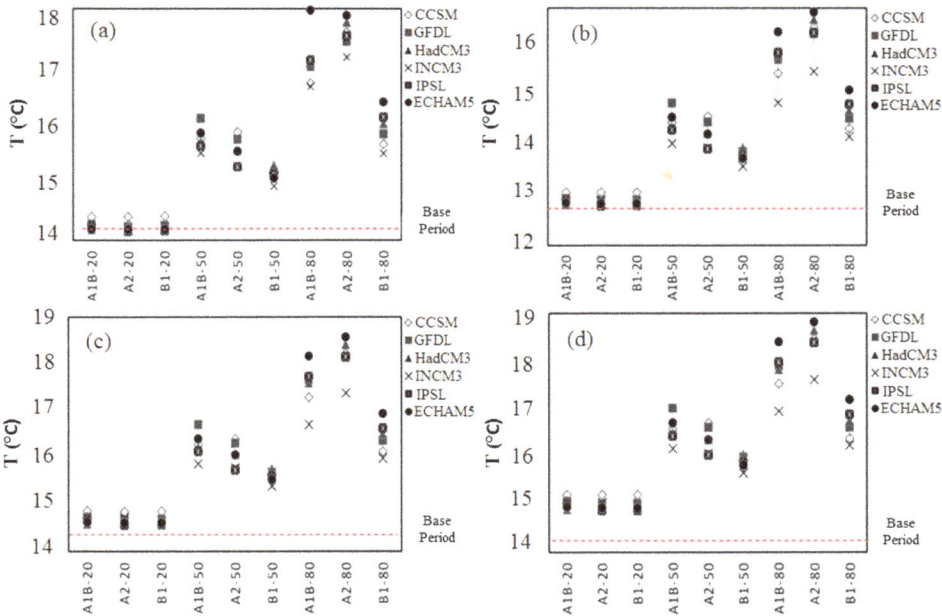

Figure 2. The daily averaged downscaled temperatures based on the A1B, A2, and B1 scenarios for the three time horizons (the 2020s, 2050s, and 2080s) using different GCMs in the sub-basins: SUB-1 (**a**); SUB-2 (**b**); SUB-3 (**c**); and SUB-4 (**d**).

The maximum temperature increase is consistently identified by all models for the 2080s time horizon under the A2 scenario, except for SUB-1. In SUB-1, the greatest rise in temperature is revealed by projections of ECHAM5 under the A1B scenario. The greatest temperature increase in the 2080s time horizon compared to the base period ranges from 4 °C (SUB-1) to 4.9 °C (SUB-4).

In a similar way, Figure 3 illustrates the average of the projected annual precipitation for different GSTs, as well as the average of the precipitation for the base period (1986–2010). As it is obvious, most of the GSTs suggest increasing the precipitation amounts compared to the base period. However, the projections obtained by employing IPSL mostly suggest a reduction of the amount of precipitation in the future time horizons, except for the outputs of the IPSL-A1B-2050s for SUB-1 and SUB-3; these show a slight increase in the amount of precipitation. Additionally, projections for SUB-2 show both a rise and decline of the precipitation, compared to the base period, which is mostly inclining towards a decrease of the precipitation rate in the 2020s. Based on the results, the greatest precipitation changes mostly occur in the 2080s time horizon.

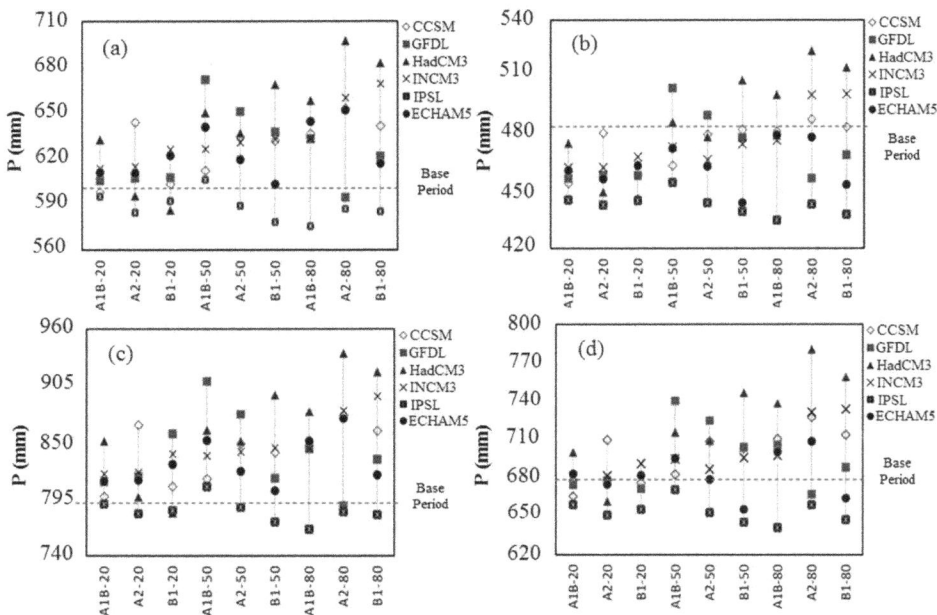

Figure 3. The annual downscaled precipitation based on the A1B, A2, and B1 scenarios for the three time horizons (the 2020s, 2050s, and 2080s) using different GCMs in the sub-basins: SUB-1 (**a**); SUB-2 (**b**); SUB-3 (**c**); and SUB-4 (**d**).

Regarding the differences between GCMs, it is clear that the IPSL performs differently. Additionally, GFDL shows greater deviations in the 2050s from the base period. In addition, some inconsistent results were found among the sub-basins. For example, the projections mostly suggest a precipitation decrease in the first time horizon (the 2020s) in SUB-2, while for other sub-basins the precipitation is mostly projected to be increased.

In total, the projections of most of the GSTs indicate a warmer future and a moderate increase in the precipitation amount. However, the results of the precipitation are more anomalous, showing both an increase and decrease of the amount of precipitation. Additionally, for the late 21st century, the GCM-scenario ensemble suggests a wider range of projections for the temperature and precipitation, implying that the uncertainties of the projections increase in a wider time span. The ensemble revealed similar results for the 2020s.

3.2. Hydrological Modeling of the Dez Dam Basin

To simulate the streamflow in the DDB, the semi-distributed HBV model is applied. First, the model calibration was completed using at least 12 years of observed discharge data. After the calibration, the model performance was evaluated using at least 5 years of observed discharge data, independent of the calibration period. It is taken into consideration that the calibration and validation periods cover both dry and wet conditions to ensure that the model is capable of working with different conditions.

Table 3 represents the information regarding the length of the calibration and validation periods, as well as the model performance at each sub-basin. Overall, the statistics show an acceptable performance of the calibrated model. As it is obvious for the validation period, negative values of *Mdiff* (%) are obtained for SUB-1, 2, and 4, which are mainly due to the reduction of the annual flow during the validation period which coincides with the final years of the observation period. Figure 4a–d shows the observed and simulated streamflow at the outlet of the four sub-basins during the calibration and validation periods.

Table 3. The model performance statistics for the calibration and validation periods.

		Period		R_{eff}	R^2	M_{Diff} (%)
SUB-1	Calibration	1989–2002		0.68	0.70	6
	Validation	2004–2008		0.70	0.79	−19
SUB-2	Calibration	1989–2004		0.62	0.62	8
	Validation	2004–2009		0.43	0.78	−24
SUB-3	Calibration	1989–2004		0.63	0.64	1
	Validation	2004–2009		0.50	0.63	−3
SUB-4	Calibration	1987–2001		0.65	0.68	2
	Validation	2001–2007		0.45	0.60	−18

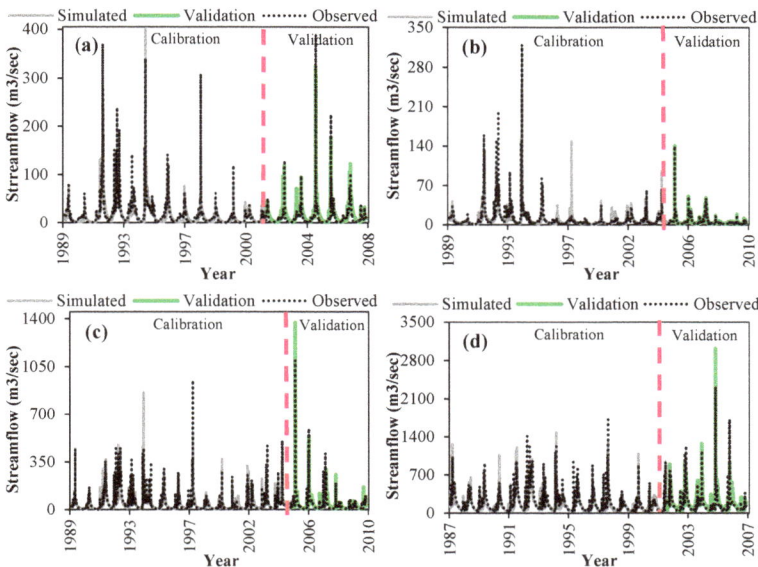

Figure 4. The observed and simulated streamflow during the calibration and validation periods in the sub-basins: SUB-1 (**a**); SUB-2 (**b**); SUB-3 (**c**); and SUB-4 (**d**). The striped line shown on the graphs separates the calibration and validation periods.

3.3. Hydrological Simulation under Climate Change Scenarios

After ensuring the efficiency of the calibrated model, the streamflows were simulated under different climate change scenarios with the projected T and P time series as inputs. Figures 5–8 illustrate the annual pattern of the simulated streamflows under the considered GSTs, as well as the observed discharge during the base period (1989–2009), for SUB-1 to SUB-4, respectively. As it is shown, the amplitude of the future simulations shows a significant variability implying inconsistencies between the simulated streamflows by different GCMs, mainly for the 2050s and 2080s. Obviously, the differences between the simulations are smaller during the 2020s, while the highest variations are associated with the late 21st century. These findings are in agreement with Vidal and Wade [55] who found a highly increasing variance in the late 21st century due to the spread of the outputs of different GCMs. In addition, the greatest inconsistencies between the simulated streamflows correspond to the higher amounts of discharge, which mainly occur during April and March.

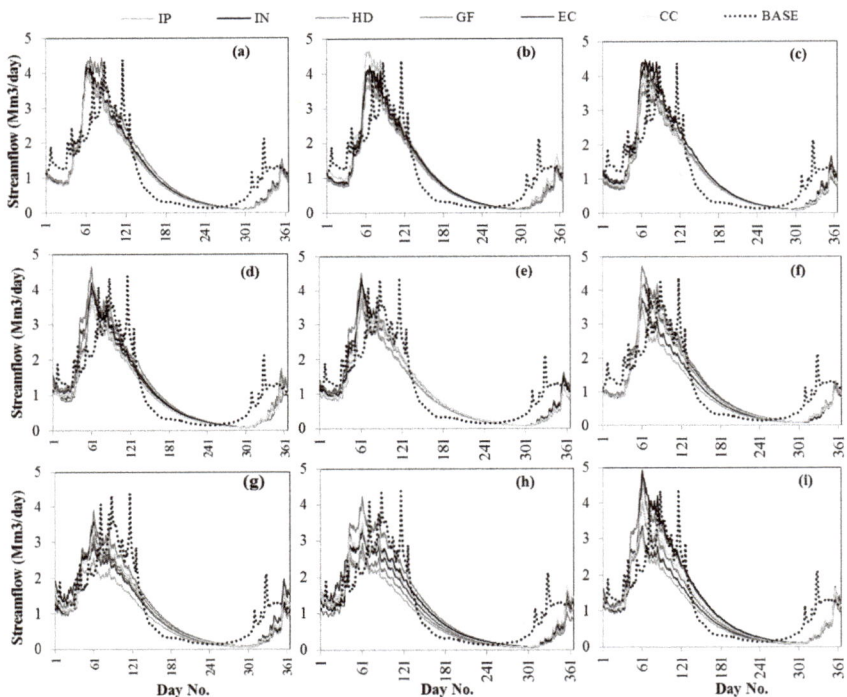

Figure 5. The annual pattern of observed streamflow (dotted graph) and future projected streamflows of SUB-1. (**a**): A1B-2020s; (**b**): A2-2020s; (**c**): B1-2020s; (**d**): A1B-2050s; (**e**): A2-2050s; (**f**): B1-2050s; (**g**): A1B-2080s; (**h**): A2-2080s; (**i**): B1-2080s.

Based on the simulations presented in Figures 5–8, not only the amount of discharge, but also the flow regime experiences some changes under the studied climate change conditions. As for SUB-1 (Figure 5), the advancing shift of the peak flows compared to the base period is obvious for all time horizons, while for SUB-3 (Figure 7), the advancing shift occurs during the 2080s, and for SUB-4, (Figure 8) it was found that the peak of the annual hydrograph occurs earlier during both the 2050s and 2080s time horizons. Similar results indicating the advancing shifts of the peak flows under climate change conditions are widely documented by different researchers. Regarding the timing of the peak flow, Nohara et al. reported that under climate change conditions, the peak flows shift earlier [18].

This is because of the temperature increase which causes the snowmelt to start sooner. Similarly, the hydrologic regime change and advancing shifts of the peak flows is reported by Gan et al. in the Naryn river basin (Central Asia) [17].

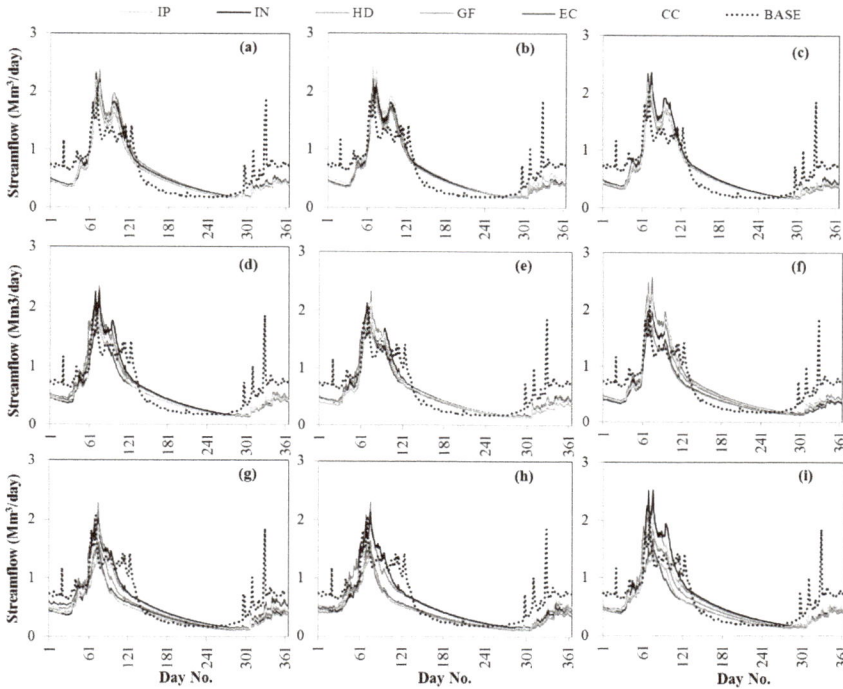

Figure 6. The annual pattern of observed streamflow (dotted graph) and future projected streamflows of SUB- 2. (**a**): A1B-2020s; (**b**): A2-2020s; (**c**): B1-2020s; (**d**): A1B-2050s; (**e**): A2-2050s; (**f**): B1-2050s; (**g**): A1B-2080s; (**h**): A2-2080s; (**i**): B1-2080s.

Table 4 summarizes the percentage of the variations of discharge for each sub-basin based on the applied GSTs.

As it is obvious in Table 4, the sub-basins of the DDB respond differently to climate change. For example, the surface flow of the SUB-3 (in most cases) seems to be less affected by climate change compared to other sub-basins. There are similar reports by other researchers about the different responses to climate change of nearby sub-basins within a region. Nazif and Karamouz, by studying the variability of streamflows in Central Iran under climate change, showed that the streamflows are significantly altered [56]. However, they noticed that the responses of three adjacent basins to climate change were different. Additionally, Ashraf Vaghefi et al. employed the CGCM model, forced by three SRES scenarios, to project the future water resource availability in the Karkheh River Basin, located in West and Southwest of Iran and found that the freshwater availability increases in the northern parts of the basin but it decreases in the southern regions [57]. Additionally, Musau et al. reported that the sensitivity of the four adjacent watersheds, considering their hydrologic response to climate change, were different [26]. This highly spatial variability of the responses of the adjacent sub-basins underlines the importance of studying the impact of climate change on the hydrological conditions in local scales in order to get a better perspective of the behavior of each basin, rather than a holistic view.

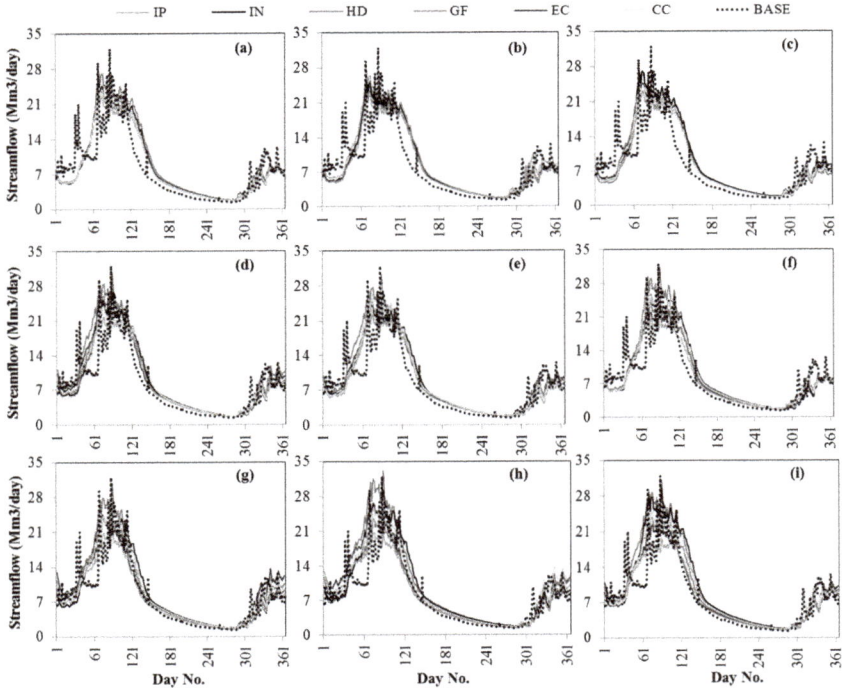

Figure 7. The annual pattern of observed streamflow (dotted graph) and future projected streamflows of SUB- 3. (**a**): A1B-2020s; (**b**): A2-2020s; (**c**): B1-2020s; (**d**): A1B-2050s; (**e**): A2-2050s; (**f**): B1-2050s; (**g**): A1B-2080s; (**h**): A2-2080s; (**i**): B1-2080s.

As shown in Figures 5–8 and Table 4, the results vary based on the GCMs used. For instance, the amount of annual flow of SUB-3 for the 2080s is projected to increase based on INCM3 (A2 and B1). However, IPSL suggests a reduction of flow in SUB-3 for the same time horizon and scenarios. Generally, among the six applied GCMs, IPSL revealed higher reductions in the flow. Therefore, not only were the results for the studied sub-basins different, but the projected response of a certain sub-basin by different GCMs could also be dissimilar. This result is in good agreement with other studies as many researchers have so far identified the GCMs as the main source of uncertainty and reported that the GCM selection can cause significant deviations in the results of the climate change impact assessments [25,44,46]. In this regard, Graham et al., by examining different RCMs, GCMs and hydrological models, concluded that the choice of GCMs is more determinative than other factors [45]. Likewise, Habets et al. applied several climate models, hydrological models, downscaling methods, and emission scenarios to simulate the future water resources and reported that the uncertainties caused by the climate models are 3–4 times greater than other factors [47]. As mentioned by Turco et al., significant differences and deviations between the results of different GCMs/RCMs imply large uncertainties regarding the use of a certain combination of GCMs and RCMs [48]. Additionally, Fang et al. documented that most of the streamflow uncertainties correspond to the structural uncertainty of GCMs [33]. Vidal and Wade mentioned that the highly increasing variance of the late 21st century is mostly due to the spread of the results obtained from different GCMs [55].

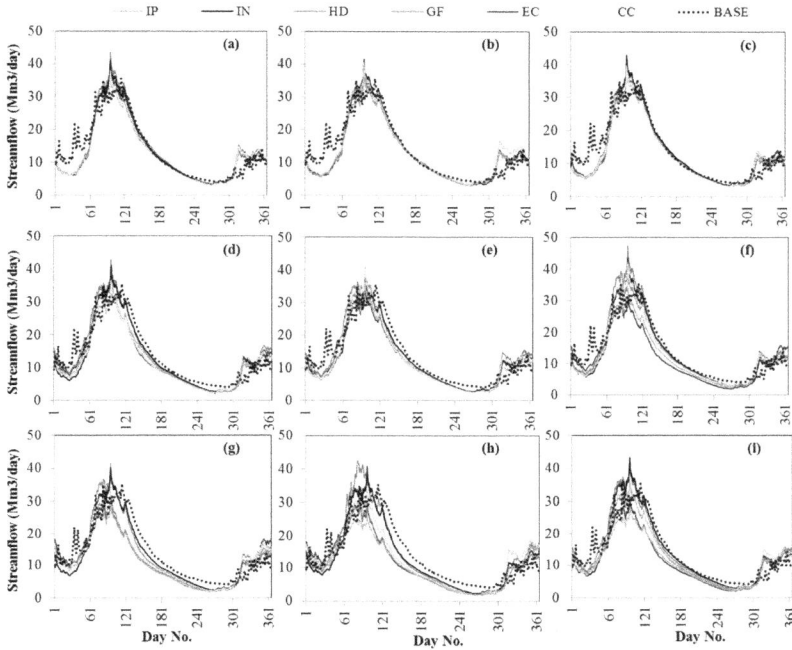

Figure 8. The annual pattern of observed streamflow (dotted graph) and future projected streamflows of SUB- 4. (**a**): A1B-2020s; (**b**): A2-2020s; (**c**): B1-2020s; (**d**): A1B-2050s; (**e**): A2-2050s; (**f**): B1-2050s; (**g**): A1B-2080s; (**h**): A2-2080s; (**i**): B1-2080s.

Table 4. The percentage of the variation of the discharge from sub-basins 1 to 4 under the climate change conditions projected by the six GCMs and three SRES scenarios for the 2020s, 2050s, and 2080s time horizons compared to the base period.

GCM		Scenario-Time Horizon								
		A1B			A2			B1		
		2020	2050	2080	2020	2050	2080	2020	2050	2080
ECHAM5	SUB-1	−7.8	−11.6	−15.3	−7.6	−15.4	−13	−3.6	−19.1	−19.3
	SUB-2	1.7	−12.3	−19.3	−0.6	−10.3	−26.1	2.8	−16.8	−20.4
	SUB-3	−5.3	−1.3	−1.5	−5.9	−6	1.7	−3.1	−10.4	−7.2
	SUB-4	−7	−14.9	−19.8	−7.7	−14.8	−20.4	−6	−15.4	−20.8
HadCM3	SUB-1	−0.8	−5.9	−9.4	−13.1	−10.6	1.7	−16.5	−0.4	1.8
	SUB-2	6.8	−5.5	−8.6	−3.9	−4.7	−4.7	−5.9	6.4	4.1
	SUB-3	0.4	0.1	3.1	−9.2	−1.2	13	−11.8	6.7	10.2
	SUB-4	−13.6	−8.2	−10.4	−10.2	−9.5	−6.4	−10.8	−1.8	−3.1
GFDL	SUB-1	−9.8	−1.4	−17.5	−8.5	−5.9	−28.6	−9	−9.2	−15.2
	SUB-2	−1.3	−0.9	−18.8	0.1	−3.3	−29.6	−1.5	−4.3	−10.7
	SUB-3	−5.9	9.1	−2.6	−5.4	3.6	−12.4	−0.7	−5.8	−4.6
	SUB-4	−9.3	−9	−18.5	−7.1	−9.3	−24.5	−8.6	−9.7	−13.6
CCSM	SUB-1	−12.5	−18.1	−16	3.8	−13.2	−11.1	−11.8	−9.4	−9
	SUB-2	−1.5	−10.9	−13.7	11.2	−7.5	−14.2	−0.9	0	−2.3
	SUB-3	−8.2	−6.9	−1.8	4.2	−2.1	3.3	−6.9	−2	1
	SUB-4	−20.8	−15.4	−15.3	−3.3	−11.5	−14.6	−8.9	−8.9	−9.9
INCM3	SUB-1	−6.5	−12.3	−17.3	−5.3	−9.9	−12.4	−0.2	−8.2	1.7
	SUB-2	2.2	−2.4	−11	1.7	−5.1	−7.3	6.1	−2	7.3
	SUB-3	−4.8	−3.5	−4.1	−4.3	−2.8	2.6	−0.6	−2	7
	SUB-4	−5.9	−8.8	−12	−6.3	−10.8	−10.7	−3.9	−7.6	−3.2
IPSL	SUB-1	−12.2	−18.9	−33.3	−14.5	−22.7	−29.6	−12.6	−25.5	−25.8
	SUB-2	−5.1	−12.9	−33.3	−4.3	−15.9	−30.8	−4.2	−16	−21
	SUB-3	−9.2	−8.7	−16.9	−11	−12.2	−13.4	−10.5	−14.2	−13.3
	SUB-4	−11.8	−15.5	−26.7	−11.6	−17.2	−25	−10.3	−18.1	−21.9

3.4. Variation of the Inflow to Dez and Bakhtiari Reservoirs and Hydropower Generation under Climate Change

The number of annual inflows to the Bakhtiari and Dez reservoirs under climate change conditions and the percentage of changes of the simulated inflows compared to the base period are obtained based on the two reservoir system of the Bakhtiari and Dez dams. The percentage of changes in the inflow to each of these reservoirs compared to the base period are shown for each GST in Table 5. The long-term averages of the discharges at the outlet of Bakhtiari and Sazar are 5106.88 and 2938.21 Mm^3, respectively. Considering that the Bakhtiari Dam is located at the end-point of the Bakhtiari basin and that the Dez dam is located downstream of Tange Panj, the inflow to the Bakhtiari reservoir equals the Bakhtiari basin's discharge and the inflow to the Dez reservoir is the total discharge from the upstream basins, which equals 8045.1 Mm^3. Based on the results, the percentages of changes of the simulated inflow to the Bakhtiari reservoir under climate change varies between −1.7% (HadCM3-B1-2050) and −26.8% (IPSL-A1B-2080). Additionally, the inflow to the Dez reservoir for future time horizons is simulated to decrease by up to 21.8% (IPSL-A1B-2080) and increase by up to 3.6% (HadCM3-B1-2080) compared to the base period.

Table 5. The percentage of variation of the inflows and the hydropower generation of the Bakhtiari and Dez reservoirs under climate change conditions projected by the six GCMs and three SRES scenarios for the 2020s, 2050s, and 2080s time horizons compared to the base period.

GCM		Scenario-Time Horizon								
		A1B			A2			B1		
		2020	2050	2080	2020	2050	2080	2020	2050	2080
ECHAM5	$\Delta I_{(B)}$ [1]	−7	−14.9	−19.9	−7.7	−14	−20.5	−5.9	−15.3	−20.8
	$\Delta E_{(B)}$	−6.8	−16.9	−22.7	−7.6	−16.8	−23.6	−5.3	−17.2	−24.1
	$\Delta I_{(D)}$ [2]	−4.7	−8.3	−11.5	−5.4	−10	−10.7	−3.2	−12	−14.3
	$\Delta E_{(D)}$	2.4	2.5	2.3	2.4	2.4	2.3	2.1	2.2	2
HadCM3	$\Delta I_{(B)}$	−13.6	−8.2	−10.4	−10.1	−9.5	−6.5	−10.7	−1.7	−3.1
	$\Delta E_{(B)}$	−15	−8.4	−11.1	−10.4	−10	−5.8	−11	−0.9	−2.1
	$\Delta I_{(D)}$	−6.8	−3.5	−3.7	−8.2	−4.8	2.5	−9.6	3.2	3.6
	$\Delta E_{(D)}$	1.5	2.4	2.6	2.2	2.5	2.2	2.1	2	1.9
GFDL	$\Delta I_{(B)}$	−9.3	−9.1	−18.6	−7	−9.4	−24.6	−8.5	−9.7	−13.6
	$\Delta E_{(B)}$	−9.4	−9.4	−21.1	−6.7	−9.8	−28.9	−8.6	−9.9	−15
	$\Delta I_{(D)}$	−6.4	−0.6	−11.1	−4.8	−2.9	−18.7	−4	−6.7	−8.7
	$\Delta E_{(D)}$	2.3	2.1	2.3	2.2	2.3	0.5	1.8	2.4	2.4
CCSM	$\Delta I_{(B)}$	−20.8	−15.4	−15.3	−3.3	−11.5	−14.6	−8.8	−8.8	−9.9
	$\Delta E_{(B)}$	−24.1	−17.5	−17.3	−2.2	−12.2	−16.3	−8.9	−9	−10.3
	$\Delta I_{(D)}$	−14.6	−10.7	−8.7	1.3	−6.4	−6.3	−6.5	−4.6	−4.2
	$\Delta E_{(D)}$	1	2.5	2.6	1.7	2.2	2.8	2.4	2.2	2.6
INCM3	$\Delta I_{(B)}$	−5.9	−8.8	−12	−6.3	−10.8	−10.7	−3.8	−7.6	−3.2
	$\Delta E_{(B)}$	−5.5	−8.8	−12.9	−6	−11.2	−11.3	−3	−7.6	−2.4
	$\Delta I_{(D)}$	−3.8	−5.2	−7.5	−3.9	−6.2	−4.1	−0.9	−3.9	2.3
	$\Delta E_{(D)}$	2.2	2.6	2.4	2.1	2.3	2.4	1.9	2	1.7
IPSL	$\Delta I_{(B)}$	−11.7	−15.5	−26.8	−11.5	−17.2	−25.1	−10.2	−18.1	−22
	$\Delta E_{(B)}$	−12.4	−17.4	−31.8	−12	−19.3	−29.7	−10.5	−20.6	−25.6
	$\Delta I_{(D)}$	−9.2	−11.5	−21.8	−9.8	−13.9	−19.4	−8.8	−15.2	−17.4
	$\Delta E_{(D)}$	2	2.3	−1.3	2.1	1.9	0	2.3	1.6	1.5

[1] B: Bakhtiari dam; [2] D: Dez dam.

The hydropower generation potential for the base period and for the future time horizons were also calculated using the inflows obtained from the previous step. Studying the monthly average of hydropower generation for the Bakhtiari and Dez reservoirs under climate change conditions versus the base period (Figures 9 and 10) shows that the ranges of simulations by the use of different GCMs and scenarios are greater in the late 21st century, while the amount of produced energy for different GCMs in the 2020s time horizon indicates a higher agreement of the results.

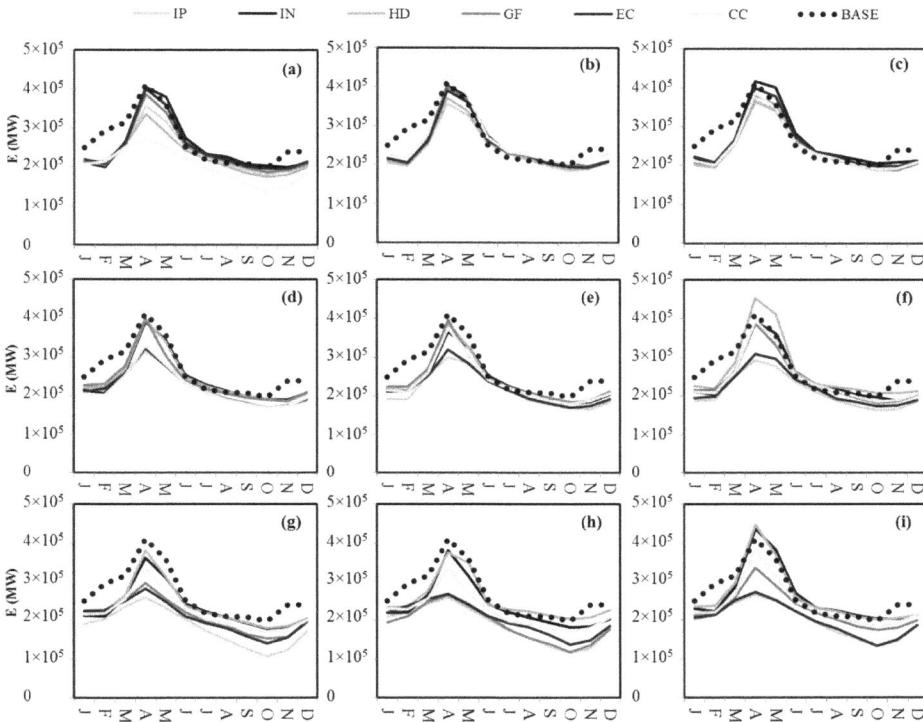

Figure 9. The monthly hydropower generation of the Bakhtiari power plant for the base period and under climate change conditions projected by the six GCMs and three SRES scenarios for the 2020s, 2050s, and 2080s time horizons; (**a**): A1B-2020s; (**b**): A2-2020s; (**c**): B1-2020s; (**d**): A1B-2050s; (**e**): A2-2050s; (**f**): B1-2050s; (**g**): A1B-2080s; (**h**): A2-2080s; (**i**): B1-2080s.

The percentages of the deviations of the projected hydropower generation potential from the base period for both reservoirs are presented in Table 5 for all the GSTs. Based on the results of the different GSTs, the potential of the hydropower generation of the Bakhtiari power plant was simulated to decrease between 0.9% (HadCM3-B1-2050) and 31.8% (IPSL-A1B-2080) compared to the base period. This is in agreement with the reduction of discharges to the Bakhtiari reservoir. On the contrary, the hydropower generation potential of the Dez power plant is simulated to be slightly higher than the base period, for all the GSTs because the highest increase reaches up to 2.8% (CCSM-A2-2080). Meanwhile, its inflow is mostly projected to decrease for future time horizons.

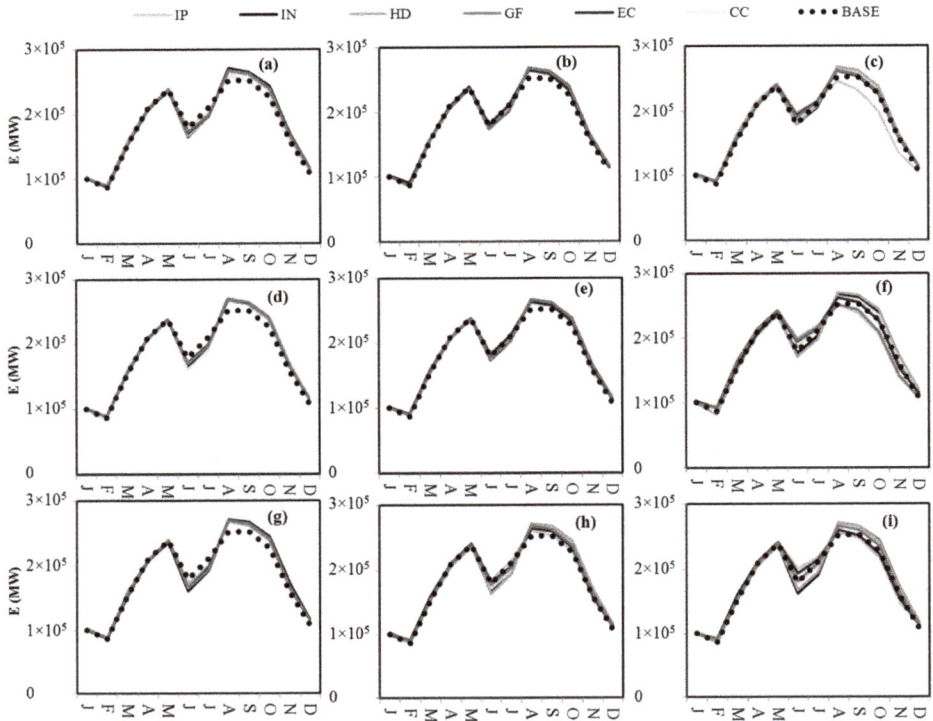

Figure 10. The monthly hydropower generation of the Dez power plant for the base period and under climate change conditions projected by the six GCMs and three SRES scenarios for the 2020s, 2050s, and 2080s time horizons; (**a**): A1B-2020s; (**b**): A2-2020s; (**c**): B1-2020s; (**d**): A1B-2050s; (**e**): A2-2050s; (**f**): B1-2050s; (**g**): A1B-2080s; (**h**): A2-2080s; (**i**): B1-2080s.

As mentioned above, in the Dez reservoir case, there are some inconsistencies between the changes in its inflow and hydropower generation caused by climate change. While the results for the Bakhtiari reservoir shows more consistency. The difference between the responses of the two reservoirs in terms of their hydropower generation potential could be attributed to the different operational rules due to different purposes, as well as due to other factors such as the size of the reservoirs and the installed capacity of their hydropower plants. To study the possible reasons, the time series of inflow, outflow, spill, and the water level in the reservoirs for the base period and for the future time horizons were assessed. The time series pertaining to the base period and the future projections under climate change (for one of the GSTs as an example) are presented in Figures 11 and 12.

Regarding the different responses of the two studied reservoirs, the following facts should be taken into consideration:

- It should be noted that the Dez dam is a multi-purpose dam which provides water for different purposes during specific times to meet specified demands. Therefore, the releases from its reservoir are planned and only the part of the release or spill, which is not greater than the penstock or turbine capacity, contributes to the power generation. However, the release from Bakhtiari is only for hydropower generation purposes.
- Additionally, the small capacity of Dez reservoir (2.7 Bm^3), compared to the discharge from its draining catchment (8 Bm^3), causes considerable spills (Figures 11b and 12b).

- Having considered that the whole capacity of the Dez hydropower plant is relatively small, compared to its inflow and releases, a significant proportion of the releases (or spills) does not contribute to power generation. Meanwhile, the Bakhtiari reservoir, with a capacity of 5.16 Bm³ and an average inflow of 5.11 Bm³, can save most of the inflows with negligible spills, both in the base period and in the future time horizons (Figures 11a and 12a).
- Additionally, the large capacity of the hydropower plant does not pose any limitation on the energy production. Therefore, in the case of Bakhtiari, there is a direct relationship between the changes in the rates of inflow and energy generation.
- A comparison between the simulated inflow of the Dez reservoir during the future time horizons and the base period suggests that under the climate change conditions, a fewer number of floods and fewer inflows and peak flows would lead to fewer losses through spill (Figures 11b and 12b), which means that more water could be saved in the reservoir to be used to generate electricity.

Considering the reasons explained above, the future changes induced by climate change in the hydropower generation potential of the Dez power plant are not consistent with the changes in the inflow of the Dez reservoir. Additionally, the slight increase in the hydropower generation potential could be attributed to the changing of the regime of discharges with smaller peaks, leading to fewer spills and more water remaining to produce electricity.

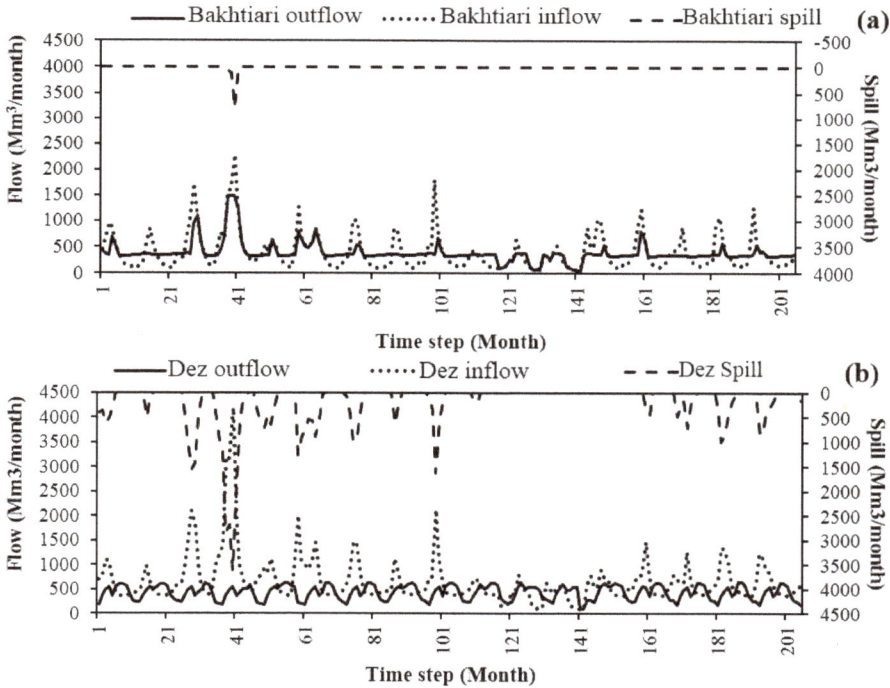

Figure 11. The monthly inflows, outflows, spills, and reservoir storages for the base period; (**a**): inflow, outflow, and spill of the Bakhtiari reservoir; (**b**): inflow, outflow, and spill of the Dez reservoir.

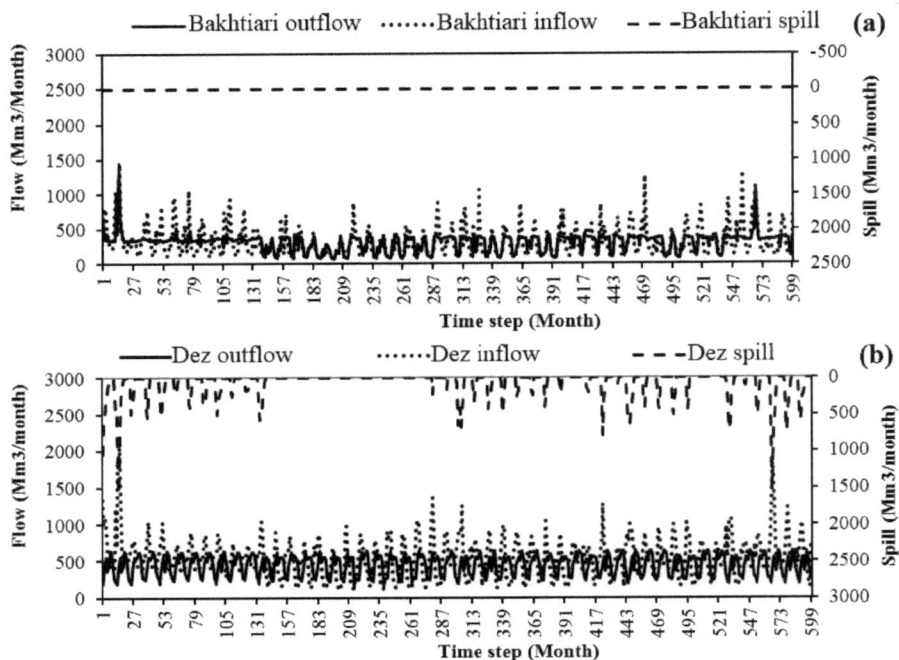

Figure 12. The monthly inflows, outflows, spills, and reservoir storages for the 2080s future time horizon based on the projections of the IPSL under the B1 scenario; (**a**): inflow, outflow, and spill of the Bakhtiari reservoir; (**b**): inflow, outflow, and spill of the Dez reservoir.

4. Conclusions

The impact of climate change on the climate, discharge, and the hydropower generation potential in the Dez Dam Basin was studied based on the downscaled outputs from six GCMs and three SRES scenarios for the three time horizons. The study revealed that the basin experiences a significant temperature rise in the mid and late 21st century (up to 4 °C). These changes are accompanied by variations in precipitation, which mostly lean towards a slight increase in the amount of precipitation. In total, the projections of all the scenarios and GCMs indicate a warmer future and small-to-moderate increase in the amount of precipitation. However, the obtained results for precipitation are more anomalous, showing both an increase and a decrease in the amount of precipitation. To simulate the future discharge at the outlet of each sub-basin under climate change conditions, the calibrated HBV hydrologic model was enforced with the projected temperature and precipitation time series. The results mostly suggest a reduction of the annual discharge in the study area. The most significant reduction of the annual flow, compared to the base period, reaches up to 33% in the two sub-basins located upstream of the basin, namely Tireh and Marbereh. Meanwhile, the greatest reductions of the annual flow of the Sazar and Bakhtiari sub-basins reach up to 17 and 27%, respectively. Considering the climate change impacts on the temperature, precipitation, and discharge in different sub-basins of the DDB, it was found that the responses of the four sub-basins are different in many cases, highlighting the noteworthiness of analyzing the impacts of climate change on local scales.

Moreover, in this study, the impact of climate change on the hydropower generation potential of the two hydropower plants in the DDB was investigated. Based on the results, climate change has the potential to significantly alter the hydropower generation potential in this basin. The results showed a reduction in the inflow and electricity generation for the Bakhtiari reservoir. Meanwhile, for the Dez reservoir, the reduction in the inflow was accompanied by a slight increase in the generated electricity.

This contrasting result obtained for the Dez reservoir was assessed and has been attributed to the small size of the reservoir (2.7 Bm^3) compared to the basin's discharge (8 Bm^3), the low capacity of hydropower plant, the different purposes and timely releases from its reservoir, as well as the flow regime changes in the future which cause less spills due to the lower peaks of floods. Overall, the results showed a reduction of the electricity generation at the Bakhtiari power plant. Therefore, its capacity seems to be high considering the future climatic conditions, while, based on the findings, there is room for the further development of the Dez power plant in order to increase its capacity for the production of more electricity.

This study used multiple GCMs for the projection of temperature, precipitation, discharge, and hydropower generation based on the three future time horizons scenarios. The results showed considerable discrepancies in the projected variables when obtained from different GCMs, which indicate the important role of the GCMs in future climatic impact assessments. Therefore, it is highly recommended that for future projections, different GCMs be employed to cover a range of likely projections.

Author Contributions: The authors of this research paper, R.S.M., M.A. and S.M. contributed significantly in the design of the study and data analysis and validation. Hydrologic modeling and modeling of the reservoirs system carried out by R.S.M. and M.A. All the authors contributed in interpretation of the results, writing the original draft of the paper and revision process. This work was supervised by S.M.

Funding: This research received no external funding.

Acknowledgments: Authors wish to thank the Islamic Republic of Iran Meteorological Organization (IRIMO) and Iran Water Resources Management Co. for providing the necessary data for this research.

Conflicts of Interest: The authors declare no conflict of interest.

References

1. Kallis, G. Droughts Annual Reviews. *Environ. Resour.* **2008**, *33*, 85–118. [CrossRef]
2. Souvignet, M.; Gaese, H.; Ribbe, L.; Kretschmer, N.; Oyarzún, R. Statistical downscaling of precipitation and temperature in north-central Chile: An assessment of possible climate change impacts in an arid Andean watershed. *Hydrol. Sci. J.* **2010**, *55*, 41–57. [CrossRef]
3. Guo, Y.; Shen, Y. Quantifying water and energy budgets and the impacts of climatic and human factors in the Haihe River Basin, China: 2. Trends and implications to water resources. *J. Hydrol.* **2015**, *527*, 251–261. [CrossRef]
4. Lalika, M.C.; Meire, P.; Ngaga, Y.M.; Chang'a, L. Understanding watershed dynamics and impacts of climate change and variability in the Pangani River Basin, Tanzania. *Ecohydrol. Hydrobiol.* **2015**, *15*, 26–38. [CrossRef]
5. Modarres, R.; Silva, V.P.R. Rainfall trends in arid and semi-arid regions of Iran. *J. Arid Environ.* **2007**, *70*, 344–355. [CrossRef]
6. Dinpashoh, Y.; Jhajharia, D.; Fakheri-Fard, A.; Singh, V.P.; Kahya, E. Trends in reference crop evapotranspiration over Iran. *J. Hydrol.* **2011**, *399*, 422–433. [CrossRef]
7. Tabari, H.; Somee, B.S.; Zadeh, M.R. Testing for long-term trends in climatic variables in Iran. *Atmos. Res.* **2011**, *100*, 132–140. [CrossRef]
8. Masih, I.; Uhlenbrook, S.; Maskey, S.; Smakhtin, V. Streamflow trends and climate linkages in the Zagros Mountains, Iran. *Clim. Chang.* **2011**, *104*, 317–338. [CrossRef]
9. Abghari, H.; Tabari, H.; Talaee, P.H. River flow trends in the west of Iran during the past 40 years: Impact of precipitation variability. *Glob. Planet. Chang.* **2013**, *101*, 52–60. [CrossRef]
10. Ahani, H.; Kherad, M.; Kousari, M.R.; Roosmalen, L.V.; Aryanfar, R.; Hosseini, S.M. Non-parametric trend analysis of the aridity index for three large arid and semi-arid basins in Iran. *Theor. Appl. Climatol.* **2013**, *112*, 553–564. [CrossRef]
11. Golian, S.; Mazdiyasni, O.; AghaKouchak, A. Trends in meteorological and agricultural droughts in Iran. *Theor. Appl. Climatol.* **2015**, *119*, 679–688. [CrossRef]
12. Kousari, M.R.; Ahani, H.; Hendi-zadeh, R. Temporal and spatial trend detection of maximum air temperature in Iran during 1960–2005. *Glob. Planet. Chang.* **2013**, *111*, 97–110. [CrossRef]

13. Marofi, S.; Soleymani, S.; Salarijazi, M.; Marofi, H. Watershed-wide trend analysis of temperature characteristics in Karun-Dez watershed, southwestern Iran. *Theor. Appl. Climatol.* **2012**, *110*, 311–320. [CrossRef]

14. Soltani, M.; Laux, P.; Kunstmann, H.; Stan, K.; Sohrabi, M.M.; Molanejad, M.; Sabziparvar, A.A.; Ranjbar SaadatAbadi, A.; Ranjbar, F.; Rousta, I.; et al. Assessment of climate variations in temperature and precipitation extreme events over Iran. *Theor. Appl. Climatol.* **2015**, *126*, 775–795. [CrossRef]

15. Adam, J.C.; Hamlet, A.F.; Lettenmaier, D.P. Implications of global climate change for snowmelt hydrology in the twenty-first century. *Hydrol. Process.* **2009**, *23*, 962–972. [CrossRef]

16. Boyer, C.; Chaumont, D.; Chartier, I.; Roy, A.G. Impact of climate change on the hydrology of St. Lawrence tributaries. *J. Hydrol.* **2010**, *384*, 65–83. [CrossRef]

17. Gan, R.; Luo, Y.; Zuo, Q.; Sun, L. Effects of projected climate change on the glacier and runoff generation in the Naryn River Basin, Central Asia. *J. Hydrol.* **2015**, *523*, 240–251. [CrossRef]

18. Nohara, D.; Kitoh, A.; Hosaka, M.; Oki, T. Impact of Climate Change on River Discharge Projected by Multimodel Ensemble. *J. Hydrometeorol.* **2006**, *7*, 1076–1089. [CrossRef]

19. Arnell, N.W.; Gosling, S.N. The impacts of climate change on river flow regimes at the global scale. *J. Hydrol.* **2013**, *486*, 351–364. [CrossRef]

20. Tong, S.T.Y.; Yang, H.; Chen, H.; Yang, J.Y. Hydrologic impacts of climate change and urbanization in the Las Vegas Wash Watershed, Nevada. *J. Water Clim. Chang.* **2016**, *7*, 598–620. [CrossRef]

21. Kirby, J.M.; Mainuddin, M.; Mpelasoka, F.; Ahmad, M.D.; Palash, W.; Quadir, M.E.; Shah-Newaz, S.M.; Hossain, M.M. The impact of climate change on regional water balances in Bangladesh. *Clim. Chang.* **2016**, *135*, 481–491. [CrossRef]

22. Tall, M.; Sylla, M.B.; Diallo, I.; Pal, J.S.; Faye, A.; Mbaye, M.L.; Gaye, A.T. Projected impact of climate change in the hydroclimatology of Senegal with a focus over the Lake of Guiers for the twenty-first century. *Theor. Appl. Climatol.* **2017**, *129*, 655–665. [CrossRef]

23. He, Z.; Wang, Z.; Suen, C.J.; Ma, X. Hydrologic sensitivity of the Upper San Joaquin River Watershed in California to climate change scenarios. *Hydrol. Res.* **2013**, *44*, 723–736. [CrossRef]

24. Senent-Aparicio, J.; Pérez-Sánchez, J.; Carrillo-García, J.; Soto, J. Using SWAT and Fuzzy TOPSIS to Assess the Impact of Climate Change in the Headwaters of the Segura River Basin (SE Spain). *Water* **2017**, *9*, 149. [CrossRef]

25. Xu, H.; Luo, Y. Climate change and its impacts on river discharge in two climate regions in China. *Hydrol. Earth Syst. Sci.* **2015**, *19*, 4609–4618. [CrossRef]

26. Musau, J.; Sang, J.; Gathenya, J.; Luedeling, E. Hydrological responses to climate change in Mt. Elgon watersheds. *J. Hydrol. Reg. Stud.* **2015**, *3*, 233–246. [CrossRef]

27. Azari, M.; Moradi, H.R.; Saghafian, B.; Faramarzi, M. Climate change impacts on streamflow and sediment yield in the North of Iran. *Hydrol. Sci. J.* **2016**, *16*, 123–133. [CrossRef]

28. Shahni Danesh, A.; Ahadi, M.S.; Fahmi, H.; Habibi Nokhandan, M.; Eshraghi, H. Climate change impact assessment on water resources in Iran: Applying dynamic and statistical downscaling methods. *J. Water Clim. Chang.* **2016**, *7*, 551–577. [CrossRef]

29. Rafiei Emam, A.; Kappas, M.; Hosseini, S.Z. Assessing the impact of climate change on water resources, crop production and land degradation in a semi-arid river basin. *Hydrol. Res.* **2015**, *46*, 854–870. [CrossRef]

30. Gohari, A.; Bozorgi, A.; Madani, K.; Elledge, J.; Berndtsson, R. Adaptation of surface water supply to climate change in central Iran. *J. Water Clim. Chang.* **2014**, *5*, 391–407. [CrossRef]

31. Solaymani, H.R.; Gosain, A.K. Assessment of climate change impacts in a semi-arid watershed in Iran using regional climate models. *J. Water Clim. Chang.* **2015**, *6*, 161–180. [CrossRef]

32. Zhou, Y.; Xu, Y.J.; Xiao, W.; Wang, J.; Huang, Y.; Yang, H. Climate Change Impacts on Flow and Suspended Sediment Yield in Headwaters of High-Latitude Regions—A Case Study in China's Far Northeast. *Water* **2017**, *9*, 966. [CrossRef]

33. Fang, G.; Yang, J.; Chen, Y.; Li, Z.; De Maeyer, P. Impact of GCM structure uncertainty on hydrological processes in an arid area of China. *Hydrol. Res.* **2018**, *49*, 893–907. [CrossRef]

34. Van Vliet, M.T.H.; Franssen, W.H.P.; Yearsley, J.R.; Ludwig, F.; Haddeland, I. Global river discharge and water temperature under climate change. *Glob. Environ. Chang.* **2013**, *23*, 450–464. [CrossRef]

35. Elsner, M.M.; Cuo, L.; Voisin, N.; Deems, J.S.; Hamlet, A.F.; Vano, J.A.; Mickelson, K.E.B.; Lee, S.-Y.; Lettenmaier, D.P. Implications of 21st century climate change for the hydrology of Washington State. *Clim. Chang.* **2010**, *102*, 225–260. [CrossRef]

36. Pervez, M.S.; Henebry, G.M. Assessing the impacts of climate and land use and land cover change on the freshwater availability in the Brahmaputra River basin. *J. Hydrol. Reg. Stud.* **2015**, *3*, 285–311. [CrossRef]
37. Bartolini, E.; Claps, P.; D'Odorico, P. Interannual variability of winter precipitation in the European Alps: Relations with the North Atlantic oscillation. *Hydrol. Earth Syst. Sci.* **2009**, *13*, 17–25. [CrossRef]
38. Lambrecht, A.; Mayer, C. Temporal variability of the nonsteady contribution from glaciers to water discharge in western Austria. *J. Hydrol.* **2009**, *376*, 353–361. [CrossRef]
39. Abrishamchi, A.; Jamali, S.; Madani, K.; Hadian, S. Climate Change and Hydropower in Iran's Karkheh River Basin. In Proceedings of the World Environmental and Water Resources Congress: Crossing Boundaries, Albuquerque, NM, USA, 20–24 May 2012; Loucks, E.D., Ed.; ASCE Library: Reston, VA, USA, 2012.
40. Bates, B.C.; Kundzewicz, Z.W.; Wu, S.; Palutikof, J.P. Climate Change and Water. Technical Paper of the Intergovernmental Panel on Climate Change, IPCC secretariat, Geneva. *Clim. Chang. Policy Renewed Environ. Ethic* **2008**, *21*, 85–101.
41. Xu, C.-Y.; Singh, V.P. Review on regional water resources assessment models under stationary and changing climate. *Water Resour. Manag.* **2004**, *18*, 591–612. [CrossRef]
42. Jahandideh-Tehrani, M.; Haddad, O.B.; Loáiciga, H.A. Hydropower reservoir management under climate change: The Karoon reservoir system. *Water Resour. Manag.* **2015**, *29*, 749–770. [CrossRef]
43. Jamali, S.; Abrishamchi, A.; Marino, M. Climate Change Impact Assessment on Hydrology of Karkheh Basin. *Proc. Inst. Civ. Eng.-Water Manag.* **2013**, *166*, 93–104. [CrossRef]
44. Wang, S.; McGrath, R.; Semmler, T.; Sweeney, C.; Nolan, P. The impact of the climate change on discharge of Suir River Catchment (Ireland) under different climate scenarios. *Nat. Hazard. Earth Syst.* **2006**, *6*, 387–395. [CrossRef]
45. Graham, L.P.; Hagemann, S.; Jaun, S.; Beniston, M. On interpreting hydrological change from regional climate models. *Clim. Chang.* **2007**, *81*, 97–122. [CrossRef]
46. Prudhomme, C.; Davies, H. Assessing uncertainties in climate change impact analyses on the river flow regimes in the UK. Part 1: Baseline climate. *Clim. Chang.* **2009**, *93*, 177–195. [CrossRef]
47. Habets, F.; Boé, J.; Déqué, M.; Ducharne, A.; Gascoin, S.; Hachour, A.; Martin, E.; Pagé, C.; Sauquet, E.; Terray, L.; et al. Impact of climate change on the hydrogeology of two basins in northern France. *Clim. Chang.* **2013**, *121*, 771–785. [CrossRef]
48. Turco, M.; Sanna, A.; Herrera, S.; Llasat, M.C.; Gutiérrez, J.M. Large biases and inconsistent climate change signals in ENSEMBLES regional projections. *Clim. Chang.* **2013**, *120*, 859–869. [CrossRef]
49. Hosseinzadehtalaei, P.; Tabari, H.; Willems, P. Uncertainty assessment for climate change impact on intense precipitation: How many model runs do we need? *Int. J. Climatol.* **2017**, *37*, 1105–1117. [CrossRef]
50. Solomon, S.; Qin, D.; Manning, M.; Chen, Z.; Marquis, M.; Averyt, K.B.; Tignor, M.; Miller, H.L. *IPCC, 2007: Climate Change 2007: The Physical Science Basis*; Contribution of Working Group I to the Fourth Assessment Report of the Intergovernmental Panel on Climate Change; Cambridge University Press: Cambridge, UK; New York, NY, USA, 2007; p. 966. ISBN 978-0-521-70596-7.
51. Semenov, M.A. Development of high-resolution UKCIP02-based climate change scenarios in the UK. *Agric. For. Meteorol.* **2007**, *144*, 127–138. [CrossRef]
52. Semenov, M.A.; Stratonovitch, P. Use of multi-model ensembles from global climate models for assessment of climate change impacts. *Clim. Res.* **2010**, *41*, 1–14. [CrossRef]
53. Seibert, J.; Vis, M.J.P. Teaching hydrological modeling with a user-friendly catchment-runoff-model software package. *Hydrol. Earth Syst. Sci.* **2012**, *16*, 3315–3325. [CrossRef]
54. Bergström, S. *The HBV Model-Its Structure and Applications*; SMHI Report RH No. 4; SMHI: Northkoping, Sweden, 1992.
55. Vidal, J.P.; Wade, S.D. Multimodel projections of catchment-scale precipitation regime. *J. Hydrol.* **2008**, *353*, 143–158. [CrossRef]

56. Nazif, S.; Karamouz, M. Evaluation of climate change impacts on streamflow to a multiple reservoir system using a data-based mechanistic model. *J. Water Clim. Chang.* **2014**, *5*, 610–624. [CrossRef]
57. Ashraf Vaghefi, S.; Mousavi, S.J.; Abbaspour, K.C.; Srinivasan, R.; Yang, H. Analyses of the impact of climate change on water resources components, drought and wheat yield in semiarid regions: Karkheh River Basin in Iran. *Hydrol. Process.* **2014**, *28*, 2018–2032. [CrossRef]

Article

Combing Random Forest and Least Square Support Vector Regression for Improving Extreme Rainfall Downscaling

Quoc Bao Pham, Tao-Chang Yang, Chen-Min Kuo, Hung-Wei Tseng and Pao-Shan Yu *

Department of Hydraulic and Ocean Engineering, National Cheng-Kung University, Tainan 701, Taiwan; pbquoc92@gmail.com (Q.B.P.); tcyang58@hotmail.com (T.-C.Y.); jemkuo@mail.ncku.edu.tw (C.-M.K.); bigwei1618@gmail.com (H.-W.T.)
* Correspondence: yups@mail.ncku.edu.tw; Tel.: +886-6-2757575 (ext. 63248)

Received: 2 January 2019; Accepted: 26 February 2019; Published: 3 March 2019

Abstract: A statistical downscaling approach for improving extreme rainfall simulation was proposed to predict the daily rainfalls at Shih-Men Reservoir catchment in northern Taiwan. The structure of the proposed downscaling approach is composed of two parts: the rainfall-state classification and the regression for rainfall-amount prediction. Predictors of classification and regression methods were selected from the large-scale climate variables of the NCEP reanalysis data based on statistical tests. The data during 1964–1999 and 2000–2013 were used for calibration and validation, respectively. Three classification methods, including linear discriminant analysis (LDA), random forest (RF), and support vector classification (SVC), were adopted for rainfall-state classification and their performances were compared. After rainfall-state classification, the least square support vector regression (LS-SVR) was used for rainfall-amount prediction for different rainfall states. Two rainfall states (i.e., dry day and wet day) and three rainfall states (dry day, non-extreme-rainfall day, and extreme-rainfall day) were defined and compared for judging their downscaling performances. The results show that RF outperforms LDA and SVC for rainfall-state classification. Using RF for three-rainfall-states classification and LS-SVR for rainfall-amount prediction can improve the extreme rainfall downscaling.

Keywords: statistical downscaling; random forest; least square support vector regression; extreme rainfall

1. Introduction

Statistical precipitation downscaling is the process of making a link between a set of large-scale atmospheric variables (i.e., mean sea level pressure, vorticity, and geopotential height) and predictand (i.e., local precipitation). The large-scale predictors are essential for climate change research, but they do not actually provide a truthful presentation of the climate in a small basin. Generally, they have a spatial resolution coarser than 2 by 2 degrees in latitude and longitude, whereas hydrologists are more concerned with the catchment scale which is usually up to a few hundred square kilometers. This leads to a need for downscaling large-scale predictors to local precipitation. The NCEP reanalysis data set is a continually updated globally gridded data set that represents the state of the Earth's atmosphere, incorporating observations and numerical weather prediction model output from 1948 to present. The NCEP reanalysis data is commonly used to develop a statistical relationship between large-scale climate factors with local rainfall for building (or training) downscaling models. The GCM outputs under climate change scenarios are then used as the inputs of downscaling models to project future precipitations for studying climate-change impacts [1]. The current study used the NCEP reanalysis data for building the proposed downscaling approach.

To date, there are many methods proposed for statistical downscaling using different techniques such as stochastic weather generators [2–5], weather typing method [6–8], resampling methods [9–11], and regression methods. The regression methods are attracting more attention and preferred to apply due to their flexibility and straightforwardness. There are numerous variant approaches of regression-based downscaling techniques such as logistic regression model [12], local polynomial regression [13], linear and non-linear regression [14], canonical correlation analysis [15], principal components [16], artificial neural network [17,18], support vector machine (SVM) [19–22], and beta regression [23].

Among these statistical downscaling methods, SVM shows its elegant and remarkable advantages comparing to the other methods. There are several studies which proved that SVM and variants of SVM are superior to ANN [19,24], multivariate analysis, and the Statistical DownScaling Model (SDSM) [24]. For instance, SVM performed better than ANN in predicting groundwater levels [25], runoff and sediment yield simulation [26], flood stage prediction [27], rainfall–runoff modeling [28], river flow forecasting [29,30], long-term discharge prediction [31], and modeling discharge-suspended sediment relationship [32]. SVM is also superior to multiple linear regression (MLR) in streamflow forecasting [33], autoregressive moving average (ARMA) in discharge prediction [31,34], autoregressive integrated moving average (ARIMA) in streamflow prediction [35], neural networks (NN), and MLR in daily water demand and inflow forecasting [36] and prediction of reservoir inflows [37]. In addition, SVM performed better than NN and empirical models in modeling daily reference evapotranspiration [38], neuro fuzzy inference system (ANFIS) in river flow forecasting [29] and daily forecasting of dam water levels [39], and genetic programming (GP) in forecasting monthly discharge time series [34].

However, many researches for downscaling precipitation at the catchment scale using SVM [19,24,40–43] conclude that the downscaling methods based on SVM performed well for normal rainfall but unsatisfactorily for extreme rainfall (i.e., underestimated extreme-rainfall amount). Tripathi et al. [19] detected that monthly precipitation downscaling by SVM could not reproduce the high rainfall observed in the historical records since the regression-based statistical downscaling models regularly cannot explain entire variance of the downscaled variable. They suggested that investigation of more large-scale predictor variables and a much longer validation period might likely provide more insight into this problem. A similar finding about the inability of SVM to mimic high rainfall has also been reported by Anandhi et al. [40].

In Taiwan, the downscaling methods based on SVM have been proposed by Chen et al. [24] and Yang et al. [41] for Shih-Men Reservoir catchment in northern Taiwan. The main structure of their proposed downscaling approach comprises the rainfall-state classification and the regression for rainfall amount. Chen et al. [24] used support vector classification (SVC) and linear discriminant analysis (LDA) for rainfall-state classification, while Yang et al. [41] only used LDA. Both the studies use the support vector regression (SVR) for the rainfall-amount prediction for wet days. Chen et al. [24] compared the performance of SVM to linear multiple regression and SDSM. The downscaled results showed that the SVM produced more accurate daily precipitation than SDSM and linear multiple regression. Yang et al. [41] found that the proposed downscaling model performed well in capturing the magnitude and variation of daily precipitations below 50 mm/day but underestimated the extreme rainfalls.

The aforementioned weakness of SVM in downscaling extreme rainfall inspires the current study to propose a modified statistical downscaling approach based on the methods developed by Chen et al. [24] and Yang et al. [41] for improving the extreme rainfall downscaling. The main structure of the proposed downscaling approach comprises the rainfall-state classification and the regression for rainfall-amount prediction. Three classification methods, including LDA, random forest (RF) and SVC, were adopted for rainfall-state classification and their performances were compared. The least square support vector regression (LS-SVR) was used for the rainfall-amount prediction for different rainfall states. Two rainfall states (i.e., dry day and wet day) and three rainfall states (dry day, non-extreme-rainfall day, and extreme-rainfall day) were defined and compared for judging their downscaling performances. Through the above comparisons, the optimal classification method with proper rainfall-state delineation can be found and linked with the rainfall-amount prediction method for improving the extreme rainfall downscaling.

The remaining part of this paper is organized as follows. Section 2 "Study Area and Data Set" provides a summary description of the study area and the data set including local rainfall and large-scale predictors. Section 3 "Methodology" describes three types of the proposed approach (i.e., Approach Type-I, Approach Type-II and Approach Type-III) and briefly introduces LDA, RF, and LS-SVR. Section 4 "Results and Discussion" describes the analysis results of rainfall-states classification and regression for rainfall-amount prediction by different classification methods and types of approach. Comparison of different classification methods (i.e., LDA, RF, and SVC) and different types of approach were made. Finally, Section 5 "Conclusions and Future Work" concludes the paper.

2. Study Area and Data Set

Shih-Men Reservoir, located in the Danshuei River basin in northern Taiwan, was completed in 1964 as a multifunction reservoir for water supply, agriculture, hydropower generation, and flood control. The Shih-Men Reservoir is a major reservoir with a storage capacity of around 3×10^8 m^3. Its upstream catchment (Figure 1) has an area of 763 km^2, and the basin ground elevation varies from 209 to 2609 meters. The average annual rainfall of the catchment is around 2250 mm.

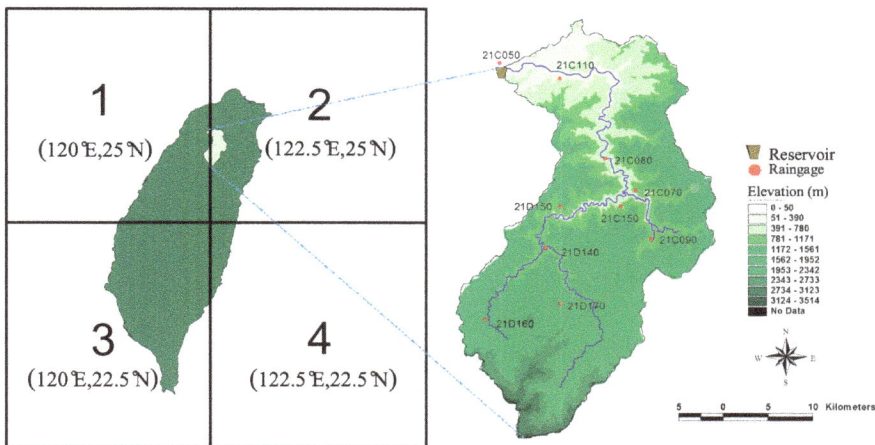

Figure 1. Shih-Men Reservoir basin (Source: [41]).

Taiwan's climate is governed by the East Asian Monsoon, which is divided into the summer and winter monsoons. Therefore, the Water Resources Bureau in Taiwan divided a year into the wet season (May–October) and the dry season (November–April) based on the summer and winter monsoons, respectively. The proportion of rainfall during the wet and dry seasons is about 7:3. The long-term daily rainfall from 1964 to 2013 at 10 rain gauges in the study area were collected to serve as the dataset (Table 1). The daily areal rainfalls in Shih-Men Reservoir catchment were calculated by using the Thiessen polygons method which determined the weights of all the stations listed in Table 1.

The daily data of 28 climate variables at the nearest grid point (i.e., Grid #2 at 122.5° E, 25° N in Figure 1) from 1964 to 2013 are obtained from the re-analysis data of National Centre for Environmental Prediction (NCEP)/National Centre for Atmospheric Research (NCAR) as listed in Table 2. These climate variables were used as the candidates of model predictors. The areal rainfalls and the NCEP reanalysis data during 1964–1999 (calibration period) and 2000–2013 (validation period) were used for building statistical downscaling models and for examining and comparing downscaling results, respectively.

Table 1. Information on rain gauges in Shih-Men reservoir catchment.

Station Name	Station Code	Location		Elevation (m)	Areal Weight
		Longitude (°E)	Latitude (°N)		
Shih-Men	21C050	121.23	24.81	255	0.018
Ba-Ling	21C070	121.39	24.69	1220	0.075
Kao-Yi	21C080	121.35	24.71	620	0.127
Ka-La-Ho	21C090	121.39	24.64	1260	0.123
Chang-Hsing	21C110	121.30	24.80	350	0.151
San-Kuang	21C150	121.36	24.67	630	0.038
Hsiu-Luan	21D140	121.28	24.62	840	0.045
Yu-Feng	21D150	121.29	24.66	780	0.049
Hsin-Pai-Shih	21D160	121.25	24.59	1620	0.115
Chen-His-Pao	21D170	121.30	24.58	630	0.259

Table 2. Large-scale climate factor (from NCEP).

No.	Acronym	Predictor
1	Mslp	Mean sea level pressure
2	p5_z	Vorticity at 500 hPa height
3	p8_z	Vorticity at 850 hPa height
4	p300	300 hPa geopotential height
5	p500	500 hPa geopotential height
6	p850	850 hPa geopotential height
7	p_f	Near surface geostrophic airflow velocity
8	p_z	Near surface vorticity
9	r500	Relative humidity at 500 hPa height
10	r850	Relative humidity at 850 hPa height
11	rhum	Near surface relative humidity
12	shum500	500 hPa specific humidity
13	Temp	Near surface air temperature
14	uas	Zonal surface wind speed
15	ua_700	700 hPa zonal wind speed
16	ua_850	850 hPa zonal wind speed
17	pr_wtr	Precipitable water
18	lftx	Surface lifted index
19	prec	Precipitation total
20	dswrf	Surface downwelling shortwave flux in air
21	dlwrf	Surface downwelling long flux in air
22	vas	Meridional surface wind speed
23	ta_700	700 hPa temperature
24	ta_850	850 hPa temperature
25	ta_925	925 hPa temperature
26	va_925	925 hPa meridional wind speed
27	uswrf	Surface upwelling shortwave flux in air
28	ulwrf	Surface upwelling longwave flux in air

3. Methods

3.1. Proposed Approach

The main structure of the proposed downscaling approach comprises rainfall-state classification and regression for rainfall-amount prediction. Three classification methods, including LDA, RF, and SVC, were adopted for rainfall-state classification and their performances were compared. The LS-SVR was used for the rainfall-amount prediction for different rainfall states. Two rainfall states (i.e., dry day and wet day) and three rainfall states (i.e., dry day, non-extreme-rainfall day, and extreme-rainfall day) were defined and compared for judging their downscaling performances. Three types of approach were constructed and described as follows.

3.1.1. Approach Type-I

Two rainfall states (i.e., dry day and wet day) are defined for rainfall-state classification by using LDA, RF and SVC. The classification performances of LDA, RF, and SVC are compared to decide the best classification method for linking to the rainfall-amount prediction method. The LS-SVR is used for rainfall-amount prediction for the rainfall state of "wet day". The flowchart of Approach Type-I is shown in Figure 2. Dry day and wet day are defined as rainfall = 0 mm/day and rainfall > 0 mm/day, respectively. Previous researches used the SVC and LDA [24] and only LDA [41] for rainfall-state classification and the SVR for rainfall-amount prediction for the rainfall state "wet day".

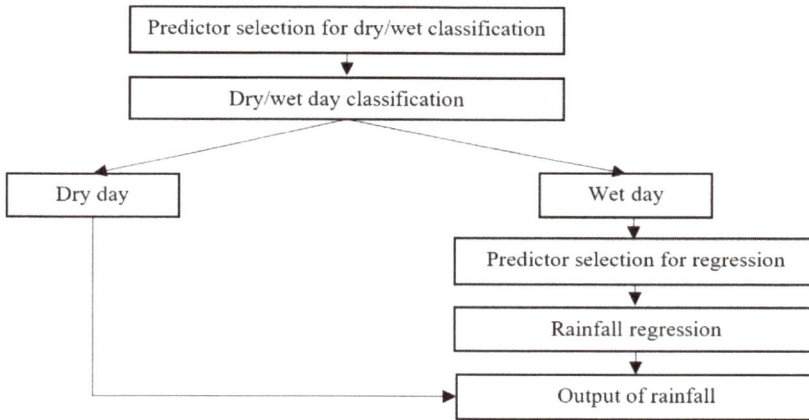

Figure 2. Flowchart of Approach Type-I.

3.1.2. Approach Type-II

Two-steps classification is used for this type. The first step defines two rainfall states (dry day and wet day) and uses LDA, RF, and SVC for rainfall-state classification. For the second step, the rainfall state of "wet day" is further divided into two states "non-extreme-rainfall day" and "extreme-rainfall day" and the LDA, RF and SVC are also used for rainfall-state classification and compared to judge their performances. Non-extreme-rainfall day and extreme-rainfall day are defined as rainfall < 50 mm/day and rainfall ≥ 50 mm/day, respectively. The threshold of 50 mm/day is defined by the Central Weather Bureau of Taiwan, which is based on the historical cases for catchments where occurred torrents, landside, or rockfall with a rainfall greater than the threshold. After rainfall-state classification by the best classification method, the LS-SVR is used for rainfall-amount prediction for the rainfall states of "non-extreme-rainfall day" and "extreme-rainfall day". The flowchart of Approach Type-II is shown in Figure 3.

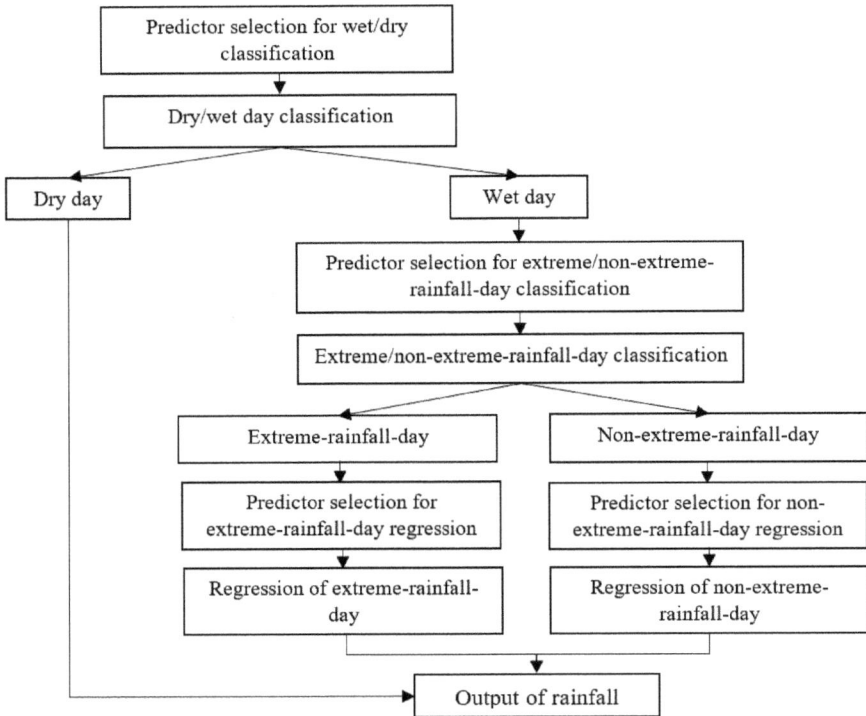

Figure 3. Flowchart of Approach Type-II.

3.1.3. Approach Type-III

One-step classification for three rainfall states (dry day, non-extreme-rainfall day, and extreme-rainfall day) is used for this type, which means a day is directly classified into one of the three rainfall states. Three rainfall states are defined as Approach Type II and LDA, RF, and SVC are used for rainfall-state classification and compared to judge their performances. Coupled with the best classification method, the LS-SVR is used for rainfall-amount prediction for the rainfall states of "non-extreme-rainfall day" and "extreme-rainfall day". The flowchart of Approach Type-III is shown in Figure 4.

The above three types of approach (i.e., Approach Type-I, Approach Type-II, and Approach Type-III) are used for daily rainfall downscaling and their performances are compared. Through the comparisons, the optimal classification method with proper rainfall-state delineation can be found and linked with the rainfall-amount prediction method for improving the extreme rainfall downscaling.

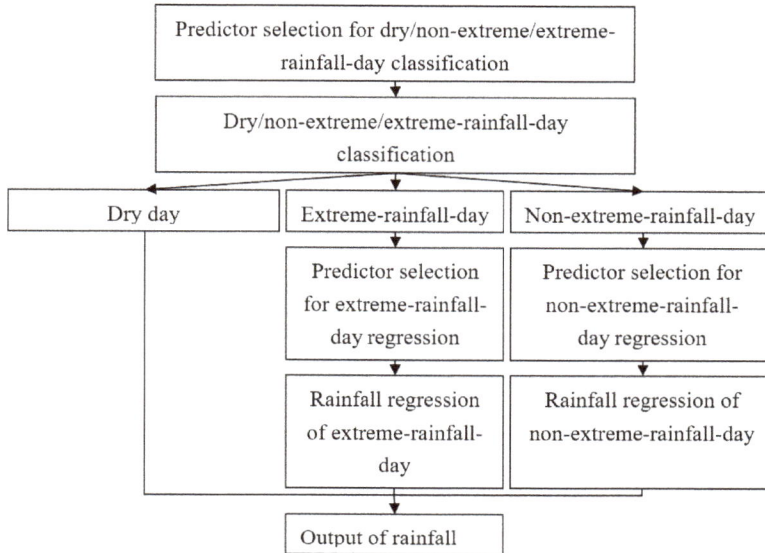

Figure 4. Flowchart of Approach Type-III.

3.2. Linear Discriminant Analysis

LDA, originally developed by Fisher (1936) [44], finds a linear discriminant function L to determine the class of a predictand based on a set of n predictors $(x_1, x_2, \ldots, x_n)$.

$$L = a_0 + a_1 x_1 + a_2 x_2 + \ldots + a_n x_n \tag{1}$$

The parameters $(a_0, a_1, a_2, \ldots, a_n)$ are calibrated from the training data of predictors and a predefined class label (for example, +1 and −1) of the predictand. The linear discriminant function L is then used to predict the class of a new predictand according to the estimated class label. In the current study, LDA was performed by the "fitcdiscr" function provided by MathWorks.

3.3. Random Forest

Random forests (RFs) are very flexible and powerful ensemble classifiers based on decision trees which were firstly developed by Breiman (2001) [45–47]. Very recently, there has been increasing interest in RF and it was applied in different areas to solve classification problems [48–51]. However, there are few applications of RFs to classify rainfall states. The only such application of RFs was recently proposed to predict rainfall occurrence in Besut station, on the east coast of Peninsular Malaysia [52]. RFs have two calibration parameters which consist of the number of variables (mtry) and the number of trees (ntree). In the present study, the value of mtry which equal the square of number of features were implemented for each classification model. Such value can generally give near optimum results for classification tasks [53]. The value of ntree ranging from 0 to 2000 was used for searching the optimal value (ntree = 500) adopted in this work. The randomForest package [54] is used in this study.

3.4. Least Square-Support Vector Machine

The least squares support vector machine (LS-SVM) algorithm is an improved algorithm of standard SVM, which provides a computational advantage (reduces the computational burden) over standard SVM by converting quadratic optimization problem into a system of linear equations [55]. In the LS-SVM algorithm, a solution is obtained by solving a linear set of equations instead of solving a quadratic programming problem involving standard SVM. The LS-SVM can be used for both classification and regression problems. In the current study, the LS-SVR is used for constructing rainfall state classification and the daily rainfall downscaling models. The description of SVC for rainfall states classification can be found in more detail in Chen et al. [24]. The brief description on the LS-SVR is as follows.

By considering inputs x_i (predictors: climate variables) and output y_i (predictand: local rainfall). According to the LS-SVR method, the nonlinear LS-SVR function can be expressed as

$$f(x) = w^T \varphi(x) + b \tag{2}$$

where f indicates the relationship between the climate variables (predictors) and local rainfall (predictand), w, φ and b are the m-dimensional weight vector, mapping function and bias term, respectively [56].

Using the function estimation error, the regression problem can be expressed regarding structural minimization principle as

$$\min J(w,e) = \frac{1}{2} w^T w + \frac{\gamma}{2} \sum_{i=1}^{m} e_i^2 \tag{3}$$

which is subjected to the following constraints:

$$y_i = w^T \varphi(x_i) + b + e_i (i = 1, 2, \ldots, m) \tag{4}$$

where γ refers the penalty term and e_i is the training error for x_i.

To find the solutions of w and e, the Lagrange multiplier optimal programming method is employed to solve Equation (3). The objective function can be determined by altering the constraint problem into an unconstraint problem. The Lagrange function L can be expressed as

$$L(w,b,e,\alpha) = J(w,e) - \sum_{i=1}^{m} \alpha_i \{ w^T \varphi(x_i) + b + e_i - y_i \} \tag{5}$$

where α_i are the Lagrange multipliers.

Taking into account the Karush–Kuhn–Tucker (KKT) conditions [56], the optimal conditions can be obtained by taking the partial derivatives of Equation (5) with respect to w, b, e and α, respectively as

$$\begin{cases} w = \sum\limits_{i=1}^{m} \alpha_i \varphi(x_i) \\ \sum\limits_{i=1}^{m} \alpha_i = 0 \\ \alpha_i = \gamma e_i \\ w^T \varphi(x_i) + b + e_i - y_i = 0 \end{cases} \tag{6}$$

Thus, the linear equations can be derived after elimination of e_i and w as

$$\begin{bmatrix} 0 & -Y^T \\ Y & ZZ^T + I/\gamma \end{bmatrix} \begin{bmatrix} b \\ \alpha \end{bmatrix} = \begin{bmatrix} 0 \\ 1 \end{bmatrix} \tag{7}$$

where $Y = (y_1, \ldots, y_m)$, $Z = (\varphi(x_1)^T y_1, \ldots, \varphi(x_m)^T y_m)$, $I = (1, \ldots, 1)$, $\alpha = (\alpha_1, \ldots, \alpha_l)$

By defining kernel function $K(x, x_i) = \varphi(x)^T \varphi(x_i)$, $i = 1, \ldots, m$, which is satisfied with Mercer's condition (the readers could refer to the paper of Suykens et al. [57] to get more explanation of Mercer's condition). As a result, the LS-SVR can be represented as

$$f(x) = \sum_{i=1}^{m} \alpha_i K(x, x_i) + b \tag{8}$$

In this study, the commonly used RBF kernel function given in Equation (9) was used.

$$K(x, x_i) = \exp\left(-\left\|x - x_i\right\|_2 / 2\sigma^2\right) \tag{9}$$

Before calibrating the LS-SVR, the values of local rainfall and predictor variables were normalized by their respective means and standard deviations. The normalized values of local rainfall and predictor variables were then utilized for calibrating the LS-SVR. The LS-SVR needs the calibration of two parameters: the penalty term (γ) and the kernel width (σ). In the training period of LS-SVR, the grid-search method [58] is used to estimate optimal parameters. The grid search method can yield an optimal parameter set and employing a cross-validation procedure can prevent the downscaling model from over-fitting. In the current study, the LS-SVR was performed by the package provided by MATLAB toolbox (http://www.esat.kuleuven.ac.be/sista/lssvmlab).

4. Results and Discussion

4.1. Rainfall-State Classification

For Approach Type-I, the calibration data (including the NCEP reanalysis data and local rainfalls) were separated into two groups (i.e., wet-day group and dry-day groups) according to daily local rainfalls in both dry and wet seasons. The two-sample Kolmogorov–Smirnov test was then performed to choose suitable predictors of the NCEP reanalysis data. This study used the two-sample Kolmogorov–Smirnov test to select predictors of the NCEP reanalysis data that are distinguishable between the dry-day group and the wet-day group. The predictors which showed a significant difference between two groups (with a significance level of 0.01) were considered as the suitable predictors for classification models. In the current study, the test was performed by the "kstest2" function provided by MathWorks. The selected predictors after testing are mean sea level pressure (mslp), vorticity (p_z, p5_z, and p8_z), geopotential height (p300, p500, and p850), relative humidity (r500, r850, and rhum), zonal wind speed (ua_700 and ua_850), meridional wind speed (vas and va_925), and temperature (ta_700, ta_850, and ta_925). The above selected predictors for Approach Type-I were also used for Approach Type-II and Approach Type-III.

For Approach Type-II, after conducting the same aforementioned process of Approach Type-I, the given wet days were further classified into non-extreme-rainfall-day group and extreme-rainfall-day group. For Approach Type-III, the calibration data (including the NCEP reanalysis data and local rainfalls) were separated into dry-day, non-extreme-rainfall-day, and extreme-rainfall-day groups according to the daily local rainfalls. Because there are only few extreme rainfalls during the dry season, the classification of non-extreme-day and extreme-day was only conducted during the wet season.

The accuracies of (1) the dry-day/wet-day classification for Approach Type-I, (2) the non-extreme-rainfall-day/extreme-rainfall-day classification for Approach Type-II and (3) the dry-day/non-extreme-rainfall-day/extreme-rainfall-day classification for Approach Type-III can be estimated respectively as

$$Accuracy(wet/dry) = \frac{D|D + W|W}{D + W} \tag{10}$$

$$Accuracy(non-extreme/extreme) = \frac{N|N + E|E}{N + E} \tag{11}$$

$$Accuracy(dry/non-extreme/extreme) = \frac{D|D + N|N + E|E}{D + N + E} \tag{12}$$

where D is the number of dry days, W is the number of wet days, D I D indicates the number of days that a dry day is correctly classified as a dry day, W I W indicates the number of wet days that a wet day correctly classified as a wet day, N is the number of non-extreme-rainfall days, E is the number of extreme-rainfall days, N I N indicates the number of days that a non-extreme-rainfall day is correctly classified as a non-extreme-rainfall day, and E I E indicates the number of extreme-rainfall days that an extreme-rainfall day correctly classified as an extreme-rainfall day.

Since the formulas for calculating the classification accuracies for Approach Type-I (Equation (10)), Approach Type-II (Equations (10) and (11)) and Approach Type-III (Equation (12)) are different, the results in Table 3 were only used for comparing the classification performances by different methods (LDA, SVC, and RF) in each approach, not for judging which type of approach is the best in the classification step.

Table 3. Classification accuracy (%) of dry/wet day and extreme-rainfall/non-extreme-rainfall day.

Type of Approach	LDA		RF		SVC	
	Wet Season	Dry Season	Wet Season	Dry Season	Wet Season	Dry Season
Type-I	75.38	75.39	79.35	75.64	74.00	72.42
Type-II Step 1 [1]	75.38	75.39	79.35	75.64	74.00	72.42
Type-II Step 2	95.26	97.62	95.31	98.33	93.06	96.83
Type-III	66.72	68.85	74.46	69.71	69.44	68.63

[1] Note: Step 1 in Approach Type-II is similar to Approach Type-I. LDA: linear discriminant analysis; RF: random forest; SVC: support vector classification.

The performances of (1) the dry-day/wet-day classification for Approach Type-I, (2) the non-extreme-rainfall-day/extreme-rainfall-day classification for Approach Type-II and (3) the dry-day/non-extreme-rainfall-day/extreme-rainfall-day classification for Approach Type-III are shown in Table 3. There are three methods (i.e., LDA, RF, and SVC) which were used for classifying rainfall states in both wet and dry season. The accuracies of dry-day/wet-day classification are generally higher than 72%. The performance of the dry-day/wet-day classification models in the wet season are better than those in the dry season for all three methods. The accuracies of dry-day/non-extreme-rainfall-day/extreme-rainfall-day classification are generally higher than 66%. The accuracies of non-extreme-rainfall-day/extreme-rainfall-day classification in Step 2 of Approach Type-II are generally higher than 93%.

The proportions of individual states (dry day, non-extreme-rainfall day, and extreme-rainfall day) during the wet season in the calibration period are 33.83%, 62.72%, and 3.45%, respectively. In the validation period, the proportions of individual states (dry day, non-extreme-rainfall day, and extreme-rainfall day) during the wet season are 34.07%, 61.54%, and 4.39%, respectively. Improvement of extreme rainfall downscaling is the main concern of the current study. For emphasizing the classification accuracy for extreme-rainfall-day state, the classification accuracies (%) of extreme-rainfall day during wet season were presented in Table 4. The dry season was not taken into account because most of extreme-rainfall-day occurred during wet season.

Table 4. Classification accuracy (%) of extreme-rainfall-day state during the wet season.

Type of Approach	LDA	RF	SVC
Type-II Step 2	49.52	56.19	30.48
Type-III	47.36	47.57	15.53

By comparing the performances of the three classification methods, it is found that RF outperforms LDA and SVC by the largest classification accuracy (%) of dry/wet day and extreme-rainfall/non-extreme-rainfall day in Table 3, and the largest classification accuracy of extreme-rainfall-day state during the wet season in Table 4. Therefore, the outputs of RF classification models were selected as inputs for the regression models to simulate rainfall amounts.

4.2. Regression for Rainfall-Amount

Before establishing the regression models, the principal component analysis (PCA) was used to transform the predictors (i.e., the 28 climate variables of the NCEP reanalysis data) to new matrices as the input matrices for LS-SVR models. The purposes of PCA are to eliminate the multicollinearity and reduce the dimension of a large data set. In the current study, PCA was carried out with the NCEP reanalysis data obtained from the nearest grid point of the study area. Nine principal components were selected based on the eigen-values which are greater than 1.0, which can explain more than 85% of the variance of the data set (i.e., the NCEP reanalysis data). The transformed variables by the nine principal components were used as the predictors of the LS-SVR models for different types of approach. Based on the transformed variables by PCA, the LS-SVR models were developed for the wet and dry seasons separately. PCA reduced the dimension of the large data set from a sample size of 143,080 corresponding to 28 predictors to a smaller sample size of 45,990 corresponding to nine principle components, which considerably reduces the computational consumption. The local rainfall and the NCEP reanalysis data during the calibration period were used to tune the two hyper-parameters of each LS-SVR model. Table 5 lists the tuned parameters of the LS-SVR models. Since most of extreme rainfalls occur during the wet season, the observed data were separated into two groups (i.e., non-extreme-rainfall group and extreme-rainfall group) for Approach Type-II and Approach Type-III. As there are too few extreme rainfalls during the dry season, only Approach Type-I approach was used for this season. The rainfalls calculated by the LS-SVR models are normalized values which should be converted to their original scale.

Table 5. The tuned parameters of least square support vector regression (LS-SVR) models.

Season	Model	Penalty Term	Kernel Width
Wet	Approach Type-I for wet day	4.62	6.27
Wet	Approach Type-II for non-extreme-rainfall day	1.64	5.95
Wet	Approach Type-II for extreme-rainfall day	78.50	1.12
Wet	Approach Type-III for non-extreme-rainfall day	2.32	5.40
Wet	Approach Type-III for extreme-rainfall day	73.65	1.05
Dry	Approach Type-I for wet day	10.89	32.60
Dry	Approach Type-II for wet day	11.14	31.23
Dry	Approach Type-III for wet day	24.34	52.81

The data of 1964–1999 were used to train the classification and regression models. During the validation period (2000–2013), the 2990 wet days were extracted for construction and evaluation of the LS-SVR models. In order to demonstrate the accuracy of the proposed approach objectively and evidently, three statistical measures (i.e., Mean, standard deviation (SD) and Skewness) are employed for examining whether the downscaling rainfalls by the proposed approach conserves the statistical characteristics of the observed rainfalls. Tables 6 and 7 list these above measures for comparing the performances of the three types of approach. From the tables, the output of Approach Type-II is slightly better than that of Approach Type-III. The simulated values of Mean, SD, and skewness in Approach Type-II are closer to the observed values than those in Approach Type-III except for SD during the calibration period (Table 6). In general, the Mean and SD of simulated rainfalls from the three types of approach tend to underestimate the observed rainfalls. However, Approach Type-II and Approach Type-III conserve the Mean and SD of observed rainfalls significantly more than Approach Type-I.

Table 6. Statistics of regression results on wet days in the calibration period.

Statistics	Approach Type-I	Approach Type-II	Approach Type-III	Observation
Mean (mm)	10.36	10.33	10.33	10.29
SD (mm)	21.49	24.15	24.12	26.87
Skewness (mm)	10.52	10.22	10.32	9.52

Table 7. Statistics of regression results on wet days during the validation period.

Statistics	Approach Type-I	Approach Type-II	Approach Type-III	Observation
Mean (mm)	10.52	11.51	10.57	12.29
SD (mm)	22.55	30.21	28.50	34.94
Skewness (mm)	9.12	8.05	9.09	8.08

To compare the simulated performances for each type of approach, Figure 5 shows the RMSE of individual months for three types of approach during the wet season in the validation period. Since most of extreme rainfalls occur during the wet season and the efficiency of Approach Type-II and Approach Type-III strongly represents during this season, only the RMSE of individual months during the wet season is presented in the figure. In Figure 5, Approach Type-III and Approach Type-II have the RMSE smaller than that of Approach Type-I in most of months except for the month of July. This is because that the classification models of Approach Type-II and Approach Type-III only have the accuracy around 50% (correctly classified 9 extreme rainfalls in a total of 18 extreme rainfalls) in July. While the accuracy in August and September are 64.29% and 79.16%, respectively, for Approach Type-III, which is much better than Approach Type-I. This implies that the accuracy of extreme rainfall classification has a significant impact on the efficiency of the proposed approach. The classification of the non-extreme-day/extreme-day showed that the performance in August and September are better when compared to July, which might be attributed to the number of heavy rainfalls in August and September.

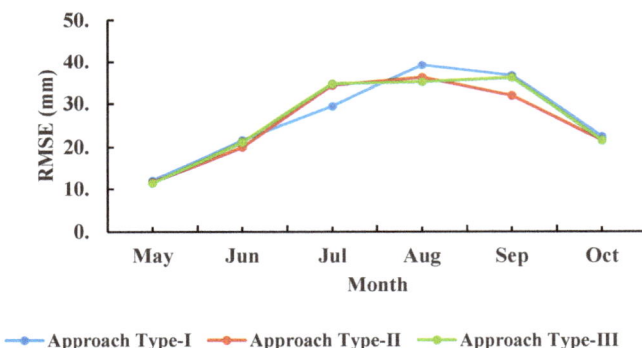

Figure 5. The RMSE of individual months for three types of approach during the wet.

In general, Approach Type-II and Approach Type-III show that their performances in terms of Mean, SD, and skewness are better than the performance of Approach Type-I. Approach Type-II shows its SD significantly better than that of Approach Type-I and Approach Type-III. Approach Type-II is slightly better than Approach Type-III in terms of Mean and skewness. It is apparent that both Approach Type-II and Approach Type-III outperform Approach Type-I in term of generation of extreme rainfalls during both calibration and validation periods (Figures 6 and 7). Approach Type-II and Approach Type-III are quite similar in reproducing extreme rainfalls.

Figures 6 and 7 shows the daily downscaling rainfalls for the three types of approach in the form of quantile–quantile (Q–Q) plots. It reveals that Approach Type-II and Approach Type-III significantly outperform the Approach Type-I when rainfalls are larger than around 50 mm/day. This results are consistent with the comparison results of statistical characteristics for both Approach Type-II and Approach Type-III with a better skewness estimate than that of Approach Type-I. Overall, both Approach Type-II and Approach Type-III models perform better than Approach Type-I in downscaling extreme rainfall amounts. It is worth noting that there are three very extreme rainfalls greater than 450 mm/day during the validation period (Figure 7) and the three very extreme rainfalls were still significantly underestimated. This is because there are too few data of very extreme rainfalls for training the models.

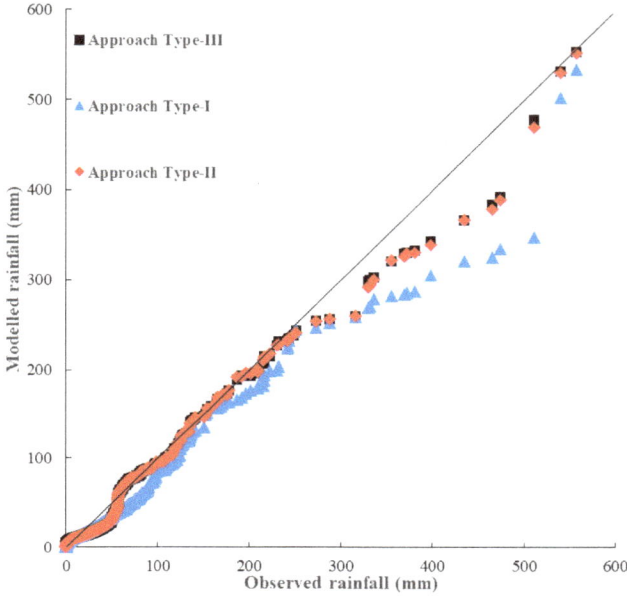

Figure 6. Quantile–quantile (Q–Q) plot of downscaling daily rainfalls during the calibration period (1964–1999).

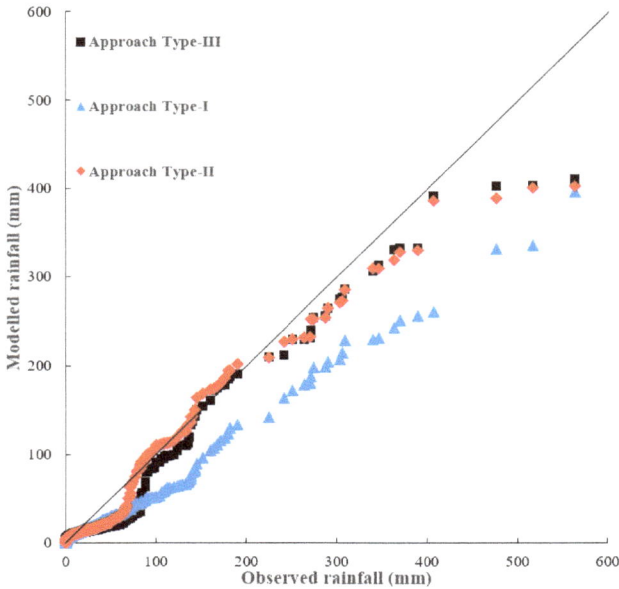

Figure 7. Q–Q plot of downscaling daily rainfalls during the validation period (2000–2013).

4.3. Discussion

Negative output values from the LS-SVR models were set to zero in the current study. The proportion of negative values among the total number of wet days are 3.59% in Approach Type-I, 1.45% in Approach Type-II, and 1.96% in Approach Type-III during the calibration period. Those are 3.89% in Approach Type-I, 2.68% in Approach Type-II, and 3.65 in Approach Type-III during the validation period. It is obvious that Approach Type-II and Approach Type-III have less negative output values than Approach Type-I. Separation of wet days into non-extreme-rainfall-day and extreme-rainfall-day data groups can get the benefit in terms of gaining less negative output values from LS-SVR models. The reason might be that the separation supports the LS-SVR models in Approach Type-II and Approach Type-III to gain more suitable parameters for each data groups (i.e., non-extreme-rainfall day and extreme-rainfall day), while the LS-SVR model in Approach Type-I only tunes one set of parameters for only a wet-day data group.

The poor skill of downscaling in capturing extreme events is attributed to two reasons. First, the standardization may reduce the bias in the mean and variance of the predictor variable, but it is much harder to accommodate the bias in large-scale patterns of atmospheric circulation or unrealistic intervariable relationships between predictor variables. The other reason may be that the NCEP reanalysis data are not able to reproduce the extreme value as many extreme events occur at a much smaller scale.

Even though the poor skill of GCM downscaling in capturing extreme events, it is found that the proposed downscaling approach with three rainfall states classification (i.e., Approach Type-II and Approach Type-III) can improve the extreme-rainfall downscaling by Approach Type-I. These two types of approach (i.e., Approach Type-II and Approach Type-III) can conserve the statistical characteristics (e.g., standard deviation and skewness) of observation data, which is a big challenge of many downscaling models. It is noted that Approach Type-II and Approach Type-III performed the extreme-rainfall downscaling better than Approach Type-I during the wet season.

5. Conclusions and Future Work

The current study proposes a statistical downscaling approach for improving daily extreme rainfall simulation at Shih-Men Reservoir catchment in northern Taiwan, which comprises rainfall-state classification and regression for rainfall-amount prediction. Three classification methods (i.e., LDA, RF, and SVC) were adopted for rainfall-state classification and the LS-SVR was used for the rainfall-amount prediction for different rainfall states. Two rainfall states (i.e., dry day and wet day) and three rainfall states (dry day, non-extreme-rainfall day, and extreme-rainfall day) were defined and compared for judging their downscaling performances.

Three types of approach (i.e., Approach Type-I, Approach Type-II and Approach Type-III) have been developed and tested for rainfall downscaling in the study area. Approach Type-I adopts two rainfall states for rainfall-state classification. Approach Type-II and Approach Type-III adopt three rainfall states for two-steps and one-step rainfall-state classification, respectively. The results reveal that RF outperforms LDA and SVC for the rainfall-state classification for all three types of approach. Approach Type-II and Approach Type-III, which use RF for three-rainfall-states classification and LS-SVR for rainfall-amount prediction, have better extreme rainfall simulation than Approach Type I. Future work can apply the two types of approach for the areas with more extreme-rainfall data to validate the performances for extreme-rainfall downscaling.

Adopting a proper threshold of daily extreme rainfall is essential for extreme/non-extreme-rainfall-day classification. The threshold of extreme rainfall strongly influences the rainfall-state classification performance. The current study adopted 50 mm/day as the threshold of extreme rainfall which is defined by the Central Weather Bureau of Taiwan. Using the thresholds less than 50 mm/day (i.e., 30 mm/day and 10 mm/day) for getting more extreme events (i.e., larger sample size) was also tested and had no improvement for rainfall-state classification in the study case. Therefore, using an inappropriate threshold of extreme rainfall may result in a failure of extreme rainfall classification. Selection of a proper threshold of extreme rainfall should be further investigated scientifically and carefully. The future work may apply the detrended fluctuation analysis (DFA) to choose an appropriate threshold of extreme rainfall for a catchment [59].

The choice of a certain reanalysis dataset is often motivated by either ease of access (availability of the dataset at the institution), ease of use (availability of code to read it), or by the preference for the local provider [60]. In the current study, the NCEP reanalysis data were used for ease of access and ease of use. The other available reanalysis data (e.g., European Centre for Medium-Range Weather Forecasts, ECMWF) with a much better spatial resolution data can be the alternatives for future work.

Author Contributions: P.S.Y. and Q.B.P. conceived and designed the modeling; Q.B.P performed the modeling; T.C.Y. revised the modeling; T.C.Y., P.S.Y., C.M.K. and H.W.T. contributed materials and analysis tools and supervised the work; Q.B.P. and T.C.Y. wrote the paper; T.C.Y., P.S.Y., C.M.K. and H.W.T. revised and edited the manuscript.

Conflicts of Interest: The authors declare no conflict of interest.

References

1. Arora, H.; Ojha, C.S.P.; Kashyap, D. Effect of Spatial Extent of Atmospheric Variables on Development of Statistical Downscaling Model for Monthly Precipitation in Yamuna-Hindon Interbasin, India. *J. Hydrol. Eng.* **2016**, *21*, 05016019. [CrossRef]
2. Khalili, M.; Leconte, R.; Brissette, F. Stochastic multisite generation of daily precipitation data using spatial autocorrelation. *J. Hydrometeorol.* **2007**, *8*, 396–412. [CrossRef]
3. Tung, C.P.; Haith, D.A. Global-warming effects on New York streamflows. *J. Water Resour. Plan. Manag.* **1995**, *121*, 216–225. [CrossRef]
4. Hughes, J.P.; Guttorp, P. A class of stochastic models for relating synoptic atmospheric patterns to regional hydrologic phenomena. *Water Resour. Res.* **1994**, *30*, 1535–1546. [CrossRef]
5. Yu, P.S.; Yang, T.C.; Wu, C.K. Impact of climate change on water resources in southern Taiwan. *J. Hydrol.* **2002**, *260*, 161–175. [CrossRef]
6. Bardossy, A.; Plate, E.J. Space-time model for daily rainfall using atmospheric circulation patterns. *Water Resour. Res.* **1992**, *28*, 1247–1259. [CrossRef]
7. Von Storch, H.; Zorita, E.; Cubasch, U. Downscaling of global climate change estimates to regional scales: An application to Iberian rainfall in wintertime. *J. Clim.* **1993**, *6*, 1161–1171. [CrossRef]
8. Bárdossy, A. Disaggregation of daily precipitation. In Proceedings of the Workshop on Ribamod–River Basin Modelling, Management and Flood Mitigation Concerted Action, Padua, Italy, 25–26 September 1997.
9. Buishand, T.A.; Brandsma, T. Multisite simulation of daily precipitation and temperature in the Rhine basin by nearest-neighbor resampling. *Water Resour. Res.* **2001**, *37*, 2761–2776. [CrossRef]
10. Murphy, J. Predictions of climate change over Europe using statistical and dynamical downscaling techniques. *Int. J. Climatol.* **2000**, *20*, 489–501. [CrossRef]
11. Palutikof, J.P.; Goodess, C.M.; Watkins, S.J.; Holt, T. Generating rainfall and temperature scenarios at multiple sites: Examples from the Mediterranean. *J. Clim.* **2002**, *15*, 3529–3548. [CrossRef]
12. Abaurrea, J.; Asín, J. Forecasting local daily precipitation patterns in a climate change scenario. *Clim. Res.* **2005**, *28*, 183–197. [CrossRef]
13. George, J.; Janaki, L.; Gomathy, J.P. Statistical downscaling using local polynomial regression for rainfall predictions–a case study. *Water Resour. Manag.* **2016**, *30*, 183–193. [CrossRef]
14. Wilby, R.L.; Dawson, C.W.; Barrow, E.M. SDSM—A decision support tool for the assessment of regional climate change impacts. *Environ. Model. Softw.* **2002**, *17*, 145–157. [CrossRef]
15. Landman, W.A.; Mason, S.J. Forecasts of near-global sea surface temperatures using canonical correlation analysis. *J. Clim.* **2001**, *14*, 3819–3833. [CrossRef]
16. Chu, J.L.; Kang, H.; Tam, C.Y.; Park, C.K.; Chen, C.T. Seasonal forecast for local precipitation over northern Taiwan using statistical downscaling. *J. Geophys. Res. Atmos.* **2008**, *113*, D12118. [CrossRef]
17. Hewitson, B.C.; Crane, R.G. Climate downscaling: Techniques and application. *Clim. Res.* **1996**, *7*, 85–95. [CrossRef]

18. Dibike, Y.B.; Coulibaly, P. Temporal neural networks for downscaling climate variability and extremes. *Neural Netw.* **2006**, *19*, 135–144. [CrossRef] [PubMed]

19. Tripathi, S.; Srinivas, V.V.; Nanjundiah, R.S. Downscaling of precipitation for climate change scenarios: A support vector machine approach. *J. Hydrol.* **2006**, *330*, 621–640. [CrossRef]

20. Ghosh, S. SVM-PGSL coupled approach for statistical downscaling to predict rainfall from GCM output. *J. Geophys. Res. Atmos.* **2010**, *115*. [CrossRef]

21. Raje, D.; Mujumdar, P.P. A comparison of three methods for downscaling daily precipitation in the Punjab region. *Hydrol. Process.* **2011**, *25*, 3575–3589. [CrossRef]

22. Kundu, S.; Khare, D.; Mondal, A. Future changes in rainfall, temperature and reference evapotranspiration in the central India by least square support vector machine. *Geosci. Front.* **2017**, *8*, 583–596. [CrossRef]

23. Mandal, S.; Srivastav, R.K.; Simonovic, S.P. Use of beta regression for statistical downscaling of precipitation in the Campbell River basin, British Columbia, Canada. *J. Hydrol.* **2016**, *538*, 49–62. [CrossRef]

24. Chen, S.T.; Yu, P.S.; Tang, Y.H. Statistical downscaling of daily precipitation using support vector machines and multivariate analysis. *J. Hydrol.* **2010**, *385*, 13–22. [CrossRef]

25. Yoon, H.; Jun, S.C.; Hyun, Y.; Bae, G.O.; Lee, K.K. A comparative study of artificial neural networks and support vector machines for predicting groundwater levels in a coastal aquifer. *J. Hydrol.* **2011**, *396*, 128–138. [CrossRef]

26. Misra, D.; Oommen, T.; Agarwal, A.; Mishra, S.K.; Thompson, A.M. Application and analysis of support vector machine based simulation for runoff and sediment yield. *Biosyst. Eng.* **2009**, *103*, 527–535. [CrossRef]

27. Liong, S.Y.; Sivapragasam, C. Flood stage forecasting with support vector machines. *JAWRA J. Am. Water Resour. Assoc.* **2002**, *38*, 173–186. [CrossRef]

28. Dibike, Y.B.; Velickov, S.; Solomatine, D.; Abbott, M.B. Model induction with support vector machines: Introduction and applications. *J. Comput. Civ. Eng.* **2001**, *15*, 208–216. [CrossRef]

29. He, Z.; Wen, X.; Liu, H.; Du, J. A comparative study of artificial neural network, adaptive neuro fuzzy inference system and support vector machine for forecasting river flow in the semiarid mountain region. *J. Hydrol.* **2014**, *509*, 379–386. [CrossRef]

30. Bhagwat, P.P.; Maity, R. Multistep-ahead river flow prediction using LS-SVR at daily scale. *J. Water Resour. Prot.* **2012**, *4*, 528–539. [CrossRef]

31. Lin, J.Y.; Cheng, C.T.; Chau, K.W. Using support vector machines for long-term discharge prediction. *Hydrol. Sci. J.* **2006**, *51*, 599–612. [CrossRef]

32. Kisi, O. Modeling discharge-suspended sediment relationship using least square support vector machine. *J. Hydrol.* **2012**, *456*, 110–120. [CrossRef]

33. Zakaria, Z.A.; Shabri, A. Streamflow forecasting at ungaged sites using support vector machines. *Appl. Math. Sci.* **2012**, *6*, 3003–3014.

34. Wang, W.C.; Chau, K.W.; Cheng, C.T.; Qiu, L. A comparison of performance of several artificial intelligence methods for forecasting monthly discharge time series. *J. Hydrol.* **2009**, *374*, 294–306. [CrossRef]

35. Maity, R.; Bhagwat, P.P.; Bhatnagar, A. Potential of support vector regression for prediction of monthly streamflow using endogenous property. *Hydrol. Process.* **2010**, *24*, 917–923. [CrossRef]

36. Hwang, S.H.; Ham, D.H.; Kim, J.H. A new measure for assessing the efficiency of hydrological data-driven forecasting models. *Hydrol. Sci. J.* **2012**, *57*, 1257–1274. [CrossRef]

37. Okkan, U. Performance of least squares support vector machine for monthly reservoir inflow prediction. *Fresenius Environ. Bull.* **2012**, *21*, 611–620.

38. Kisi, O. Least squares support vector machine for modeling daily reference evapotranspiration. *Irrig. Sci.* **2013**, *31*, 611–619. [CrossRef]

39. Hipni, A.; El-shafie, A.; Najah, A.; Karim, O.A.; Hussain, A.; Mukhlisin, M. Daily forecasting of dam water levels: Comparing a support vector machine (SVM) model with adaptive neuro fuzzy inference system (ANFIS). *Water Resour. Manag.* **2013**, *27*, 3803–3823. [CrossRef]

40. Anandhi, A.; Srinivas, V.V.; Nanjundiah, R.S.; Nagesh Kumar, D. Downscaling precipitation to river basin in India for IPCC SRES scenarios using support vector machine. *Int. J. Climatol.* **2008**, *28*, 401–420. [CrossRef]

41. Yang, T.C.; Yu, P.S.; Wei, C.M.; Chen, S.T. Projection of climate change for daily precipitation: A case study in Shih-Men reservoir catchment in Taiwan. *Hydrol. Process.* **2011**, *25*, 1342–1354. [CrossRef]

42. Devak, M.; Dhanya, C.T.; Gosain, A.K. Dynamic coupling of support vector machine and K-nearest neighbour for downscaling daily rainfall. *J. Hydrol.* **2015**, *525*, 286–301. [CrossRef]

43. Okkan, U.; Kirdemir, U. Downscaling of monthly precipitation using CMIP5 climate models operated under RCPs. *Meteorol. Appl.* **2016**, *23*, 514–528. [CrossRef]

44. Fisher, R.A. The use of multiple measurements in taxonomic problems. *Ann. Eugen.* **1936**, *7*, 179–188. [CrossRef]

45. Breiman, L. Random forests. *Mach. Learn.* **2001**, *45*, 5–32. [CrossRef]

46. Catani, F.; Lagomarsino, D.; Segoni, S.; Tofani, V. Landslide susceptibility estimation by random forests technique: Sensitivity and scaling issues. *Nat. Hazards Earth Syst. Sci.* **2013**, *13*, 2815–2831. [CrossRef]

47. Micheletti, N.; Foresti, L.; Robert, S.; Leuenberger, M.; Pedrazzini, A.; Jaboyedoff, M.; Kanevski, M. Machine learning feature selection methods for landslide susceptibility mapping. *Math. Geosci.* **2014**, *46*, 33–57. [CrossRef]

48. Chan, J.C.W.; Paelinckx, D. Evaluation of Random Forest and Adaboost tree-based ensemble classification and spectral band selection for ecotope mapping using airborne hyperspectral imagery. *Remote Sens. Environ.* **2008**, *112*, 2999–3011. [CrossRef]

49. Ibarra-Berastegi, G.; Saénz, J.; Ezcurra, A.; Elías, A.; Diaz Argandoña, J.; Errasti, I. Downscaling of surface moisture flux and precipitation in the Ebro Valley (Spain) using analogues and analogues followed by random forests and multiple linear regression. *Hydrol. Earth Syst. Sci.* **2011**, *15*, 1895–1907. [CrossRef]

50. Stumpf, A.; Kerle, N. Object-oriented mapping of landslides using Random Forests. *Remote Sens. Environ.* **2011**, *115*, 2564–2577. [CrossRef]

51. Vincenzi, S.; Zucchetta, M.; Franzoi, P.; Pellizzato, M.; Pranovi, F.; De Leo, G.A.; Torricelli, P. Application of a Random Forest algorithm to predict spatial distribution of the potential yield of Ruditapes philippinarum in the Venice lagoon, Italy. *Ecol. Model.* **2011**, *222*, 1471–1478. [CrossRef]

52. Pour, S.H.; Shahid, S.; Chung, E.S. A hybrid model for statistical downscaling of daily rainfall. *Procedia Eng.* **2016**, *154*, 1424–1430. [CrossRef]

53. Yu, P.S.; Yang, T.C.; Chen, S.Y.; Kuo, C.M.; Tseng, H.W. Comparison of random forests and support vector machine for real-time radar-derived rainfall forecasting. *J. Hydrol.* **2017**, *552*, 92–104. [CrossRef]

54. Liaw, A.; Wiener, M. Classification and regression by randomForest. *R News* **2002**, *2*, 18–22.

55. Okkan, U.; Serbes, Z.A. Rainfall–runoff modeling using least squares support vector machines. *Environmetrics* **2012**, *23*, 549–564. [CrossRef]

56. Kisi, O. Pan evaporation modeling using least square support vector machine, multivariate adaptive regression splines and M5 model tree. *J. Hydrol.* **2015**, *528*, 312–320. [CrossRef]

57. Suykens, J.A.; De Brabanter, J.; Lukas, L.; Vandewalle, J. Weighted least squares support vector machines: Robustness and sparse approximation. *Neurocomputing* **2002**, *48*, 85–105. [CrossRef]

58. Van Gestel, T.; Suykens, J.A.; Baesens, B.; Viaene, S.; Vanthienen, J.; Dedene, G.; Vandewalle, J. Benchmarking least squares support vector machine classifiers. *Mach. Learn.* **2004**, *54*, 5–32. [CrossRef]

59. Lin, G.F.; Chang, M.J.; Wu, J.T. A hybrid statistical downscaling method based on the classification of rainfall patterns. *Water Resour. Manag.* **2017**, *31*, 377–401. [CrossRef]

60. Horton, P.; Brönnimann, S. Impact of global atmospheric reanalyses on statistical precipitation downscaling. *Clim. Dyn.* **2018**, 1–23. [CrossRef]

Article

Drought Prediction for Areas with Sparse Monitoring Networks: A Case Study for Fiji

Jinyoung Rhee * and Hongwei Yang

Climate Services and Research Department, APEC Climate Center, Busan 48058, Korea; hwyang@apcc21.org
* Correspondence: jyrhee@apcc21.org; Tel.: +82-51-745-3959; Fax: +82-51-745-3999

Received: 18 May 2018; Accepted: 11 June 2018; Published: 14 June 2018

Abstract: Hybrid drought prediction models were developed for areas with limited monitoring gauges using the APEC Climate Center Multi-Model Ensemble seasonal climate forecast and machine learning models of Extra-Trees and Adaboost. The models provide spatially distributed detailed drought prediction data of the 6-month Standardized Precipitation Index for the case study area, Fiji. In order to overcome the limitation of a sparse monitoring network, both in-situ data and bias-corrected dynamic downscaling of historical climate data from the Weather Research Forecasting (WRF) model were used as reference data. Performance measures of the mean absolute error as well as classification accuracy were used. The WRF outputs reflect the topography of the area. Hybrid models showed better performance than simply bias corrected forecasts in most cases. Especially, the model based on Extra-Trees trained using the WRF model outputs performed the best in most cases.

Keywords: drought prediction; APCC Multi-Model Ensemble; seasonal climate forecast; machine learning; sparse monitoring network; Fiji

1. Introduction

Islands in the South Pacific are vulnerable to climate change [1]. The climate in the South Pacific has become drier by 15% and warmer by 0.8 °C, compared to the earlier 20th century [2]. Fiji, one of the key Pacific Island countries, experiences easterly trade winds on most calendar days. The easterly trade winds or the northeasterly monsoon, when lifted by high mountains, causes moisture condensation and produces heavy rainfall on the windward eastern side of Fiji. The subsidence of the relatively dry air produces less rainfall on the leeward western side.

From a large-scale viewpoint, the El Nino Southern Oscillation (ENSO) is the main cause of climate variability over this region at interannual timescales. La Nina events dominated the interannual sea surface temperature (SST) anomaly (SSTA) over the central Equatorial Pacific during 1950 and 1975; after that time, El Nino events became more frequent [3]. The Pacific Decadal Oscillation (PDO) dominates the climate variability at decadal timescales [4]. PDO was mostly positive prior to 1998 and then shifted to a strong negative phase [5]. Positive PDO is characterized by the similar SSTA of El Nino over the Equatorial Pacific, and thus shifts the weather systems northeastward, but on a decadal timescale. The South Pacific Convergence Zone (SPCZ) is a reverse-oriented monsoon trough with strong low-level convergence and a rainfall band that extends from the Warm Pool southeastward to French Polynesia [6,7]. The interferential impact of ENSO and PDO on the SPCZ is complex [8,9]. El Nino events weaken the strength of the Walker Circulation and shift the dominant weather systems over the Equatorial Pacific toward areas in the northeast such as the SPCZ. When El Nino takes place during the positive PDO, the SPCZ moves northeast towards the equator, and its intensity becomes stronger [8]. The large-scale convection departure decreases precipitation over Fiji and leads to droughts [10].

Fiji has observed more frequent dry conditions since the 1950's compared to previous decades in the western and northern areas based on analysis performed using the Standardized Precipitation Index (SPI). Analysis of observed monthly rainfall for Fiji over the period 1949–2008 showed downward trends at a 99% confidence level with decreases in rainfall of approximately 13–47 mm per year [11]. Although no significant long-term trends were observed in annual rainfall [12], there were more frequent dry seasons during the last 50 years compared to the first 50 years when the nearly 100 years of data since 1900 were examined [13]. The local temperature also increased due to the effects of climate change [14]. The most impacted stations were located in western and northern Fiji, where deficiency in rainfall from 1969–1988 caused an increase in moderate and severe droughts [11]. Risbey et al. [15] projected an increase in rainfall of approximately 3.3% by 2025 and 9.7% by 2100 using a global climate model (GCM). Feresi et al. [16] and Agrawala et al. [17] did not project a definitive change in rainfall. IPCC [18] projected that Fiji will experience an intensified seasonal cycle, i.e., a rainfall decrease in the dry season and a rainfall increase in the wet season. The shift towards extended periods of dry spells causes loss of soil fertility, which could impact negatively on agriculture [1].

Since 1940, severe droughts have occurred in 1942, 1958, 1969, 1978, 1983, 1987, 1992, 1997–1998, 2003, and 2010 [16]. Severe droughts can cause serious socio-economic loss as well as physical damages as drought conditions persist. The ENSO event of 1997–1998 caused a severe drought with damages of up to Fiji $100 million. Rainfall failure occurred across two successive dry seasons, and more significantly during the intervening wet season when precipitation is normally reliable [16]. Since many rural communities are reliant on rainwater, streams, and shallow wells for domestic use, watering crop gardens, and livestock, these communities are especially vulnerable to periods of drought when surface water resources are at a minimum [19]. Schools and businesses were forced to close and caused disruption to residential areas. Such impacts made extreme difficulties for Fiji since the resources of an island country are limited. External aid and governmental assistance were required to ensure supply of sustenance and facilitate recovery in the worst-hit parts of Fiji, which included the western and northern divisions and outer islands.

Drought conditions in Fiji are currently monitored using the 3-, 6-, and 12-month SPI calculated for weather stations with long historical data [20]. The monitoring network over Fiji with long data is quite sparse though, resulting in considerable uncertainty in the estimates of extreme wet and dry events. Evidence shows that estimation of the historical trends has a large noise-to-signal ratio over regions with sparse data networks [21]. Furthermore, most Fiji weather stations with long data are located along the coastline, so the sparse network cannot capture small-scale convective precipitation over land and precipitation from orographic lifting at mountains. Rainfall variability in the high mountains is greater than the variability in cities.

The limited variables and inconsistency in duration of satellite observation introduces difficulties and uncertainties in methods and analysis. For example, the Climate Prediction Center Morphing Technique (CMORPH) data is only available from 1998 onward. Due to the limited number or variables being observed, it is difficult to prepare for droughts because the response of rainfall distribution to large-scale dynamics is unclear. In addition, unlike other types of disasters, the onset and termination of droughts is not always clear. The increase in uncertainty of climate variability makes the reduction of drought impacts even more difficult.

Drought outlook of Fiji is also provided based on SPI: SPI predictions for weather stations are based on the statistically downscaled seasonal forecast data from the Seasonal Climate Outlooks for Pacific Island Countries developed by the Bureau of Meteorology of Australia. If spatially distributed drought prediction is available, possibly reflecting the orographic effect of the main island, it would be helpful to prevent and minimize the adverse impacts of droughts in Fiji. Drought prediction data only available for weather stations or obtained based on low-resolution bias-corrected seasonal forecast data are not sufficient for effective decision making.

This study aims to develop a drought prediction model that can be used for areas with sparse monitoring networks. Fiji is a case study area. By providing spatially detailed drought prediction data, vulnerability to droughts may be reduced while resiliency may be increased. Multi-Model Ensemble seasonal climate forecast data from APEC Climate Center (APCC MME) are used to provide up to 6 months-lead climate forecasting. Machine learning models are used to provide spatially distributed drought information for ungauged areas. In order to overcome the limitation of sparse monitoring networks, dynamically downscaled historical climate data from the Weather Research and Forecasting (WRF) model are used to train machine learning models instead of in-situ data as reference data.

This study ultimately targets national, provincial, and regional officials whose main duties include water resources and agricultural management. The final beneficiaries of the output are residents of the area; water users and farmers for whom decision-making can be helped by drought prediction information with finer spatial resolution.

2. Study Area

Fiji has a total area of about 194,000 km^2 of which approximately 10% is land. Fiji consists of 332 islands. The two largest islands are Viti Levu and Vanua Levu, which account for about three-quarters of the total land area of Fiji [22]. Figure 1 shows the topography of Fiji's main islands. The largest island, Viti Levu, which has an area of 10,388 km^2, is covered with thick tropical forest. The island has a considerable area higher than 500 m in elevation with the peak of Mount Tomanivi at 1324 m above sea level. Viti Levu hosts the capital city of Suva, which contains about three-quarters of the population. Other important towns include Nadi, where the international airport is located, and Lautoka.

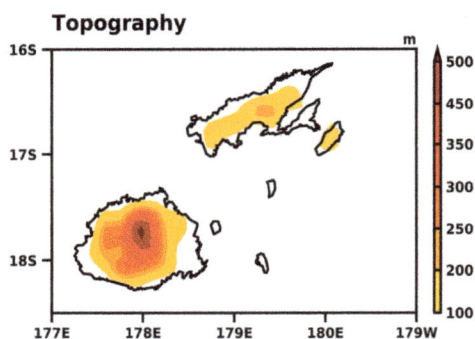

Figure 1. Topography of Fiji's main islands (color shades are in units of meters).

Fiji has a tropical marine climate and is warm year-round with minimal extremes. The warm season lasts from November to April and the cool season lasts from May to October. Temperatures in the cool season average 22 °C. Winds are moderate, though cyclones occur about once a year (10–12 times per decade). Viti Levu is a mountainous volcanic island with a wet-dry tropical climate. The southeast side of the island faces the predominant trade winds and therefore receives more precipitation than the northwest side, which is rain-shadowed by interior highlands. The volcanic mountains force orographic lifting of the saturated air, which can produce extremely heavy rainfall on the windward side of the mountain. Rainfall on the leeward side is much lighter due to the subsidence of the dry air, which largely influences agriculture in those areas. In the dry season, the uneven distribution of rainfall can cause a prolonged lack of moisture on the leeward side. The leeward side only receives 20% of the annual total rainfall in the dry season, compared to 33% received on the windward side [23].

Sugar export is an important source of foreign exchange for Fiji, as sugar cane processing makes up one-third of industrial activity. Coconut, ginger, and copra are also significant industries. These agricultural products are highly influenced by climate extremes; the sugar industry was damaged by drought in 1998.

3. Materials

3.1. In-Situ Data

Figure 2a shows the location of rainfall gauges of the two main islands used in this study (Table 1). In-situ rain-gauge hourly precipitation data for 1981–2010 were obtained and daily data for the period were used for the bias-correction of the WRF model. Monthly data were also used for calculating drought index values for the training of machine learning models. Some data were missing during a short period of time from gauges at Udu Point and Nabouwalu.

Table 1. Fiji rainfall gauges used in the analysis.

Observation Sites	Latitude	Longitude
Udu Point (91652)	16.13° S	180.02° E
Nabouwalu (91659)	16.98° S	178.70° E
Nadi (91680)	17.75° S	177.45° E
Suva (91690)	18.15° S	178.45° E

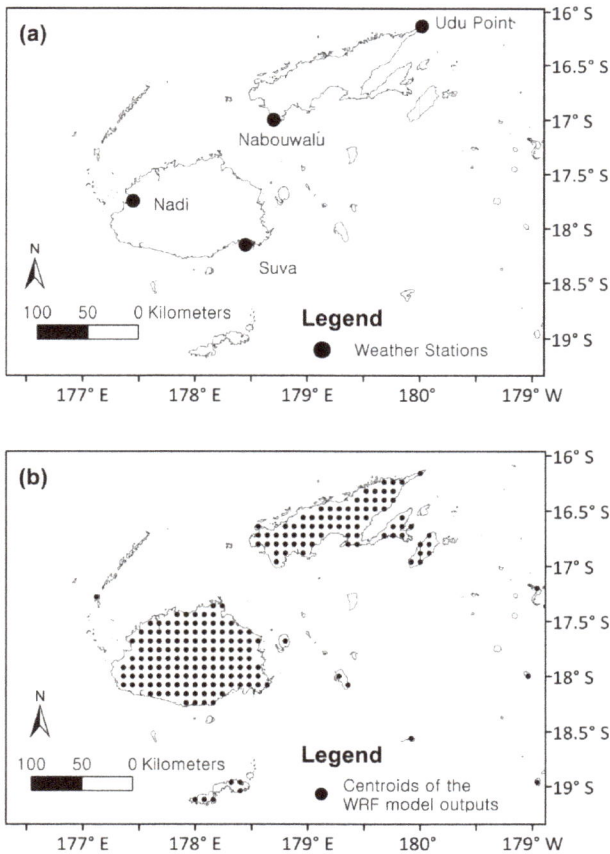

Figure 2. Location of (**a**) the rainfall gauges; and (**b**) the centroids of the Weather Research and Forecasting (WRF) model outputs.

3.2. WRF Model Outputs

Dynamic downscaling of historical climate through the WRF model forced by the European Centre for Medium-Range Weather Forecasts Reanalysis (ERA)-Interim reanalysis dataset in a double nested framework with spectral nudging in the parent domain was used in this study [24]. Many validations show that the WRF outputs are pretty reliable. Precipitation data with 8 km spatial resolution for 1981–2010 were used in this study. Centroids of the 227 grid cells are shown in Figure 2b.

3.3. SPI

The SPI is widely used to characterize meteorological drought on a range of timescales [25,26] (Table 2). It quantifies observed precipitation as a standardized departure from a selected probability distribution function that models the raw precipitation data. The raw precipitation data are fitted to a gamma distribution, for example, and then transformed to a normal distribution. The SPI values can be interpreted as the number of standard deviations by which the observed anomaly deviates from the long-term mean. The SPI can be created for differing periods of 1 to 36 months, using monthly input data. The SPI can be compared across regions with markedly different climates. In this study, 6-month SPI (SPI6) was used to examine the performance of the drought prediction model developed, which is based on APCC MME up to 6 months-lead forecast data. SPI6 is also used by the Fiji Meteorological Service (FMS) to examine agricultural (soil moisture) and hydrological droughts because the 6-month droughts affect deeper rooted plants and medium-sized water bodies [27].

Table 2. Drought categories based on Standardized Precipitation Index (SPI) [26].

Classification	Index Value
Extremely wet (EW)	≥2.00
Very wet (VW)	1.50 to 1.99
Moderately wet (MW)	1.00 to 1.49
Near Normal (NN)	0.99 to −0.99
Moderate drought (MD)	−1.00 to −1.49
Severe drought (SD)	−1.50 to −1.99
Extreme drought (ED)	≤−2.00

3.4. APCC MME Seasonal Climate Forecast

APCC produces the future 6-month global climate forecast using the MME technique, by collecting, standardizing, and utilizing climate prediction data from 17 different climate prediction organizations from all round the world. The MME technique collates data from different high quality climate models resulting in a better forecast than each climate model's independent forecast. For this study, 6-month MME data produced by the Simple Composite Method (SCM) based on six individual models were obtained from the APEC Climate Data Service System [28]. The six individual climate models were APCC model, the Centro Euro-Mediterraneo sui Cambiamenti Climatici model, the Meteorological Service of Canada (MSC) model, the National Aeronautics and Space Administration (NASA) model, the National Centers for Environmental Prediction (NCEP) model, Pusan National University (PNU) model, and the Predictive Ocean Atmosphere Model for Australia.

3.5. Remote Sensing Data

3.5.1. PERSIANN-CDR

The drought prediction model developed in this study relies on remote sensing based precipitation data in order to compensate for the low spatial coverage of weather stations. To secure precipitation data covering a large enough area, the Precipitation Estimation from Remotely Sensed Information using Artificial Neural Networks (PERSIANN)-Climate Data Record (CDR) was used [29]. PERSIANN-CDR data were created based on infrared sensor data for the period with no microwave sensor data. The data cover 60° S–60° N, 180° W–180° E, with a spatial resolution of 0.25° × 0.25°. Daily data were obtained and converted to monthly total precipitation data.

3.5.2. TRMM

The tropical rainfall measuring mission (TRMM) was developed jointly by the United States (US) NASA and the Japan Aerospace Exploration Agency. The TRMM 3B42 product with 3-h data collection intervals was obtained from the NASA Goddard Earth Sciences Data and Information Service Center and converted to monthly total precipitation data. The TRMM data cover 50° S–50° N, 180° W–180° E, and have a spatial resolution of 0.25° × 0.25°. The data are in equirectangular (or geographic) projection with WGS84 datum.

3.5.3. GPM

The Integrated Multi-Satellite Retrievals for the Global Precipitation Measurement Mission (GPM) data were used as remote sensing based precipitation data from April 2014 onward. The data were obtained from the Precipitation Measurement Missions of NASA, and cover 90° S–90° N, 180° W–180° E, and have a spatial resolution of 0.1° × 0.1°. The data are also in equirectangular (or geographic) projection with WGS84 datum. The data were converted to monthly total precipitation data.

3.5.4. MODIS Land Surface Temperature

Daytime and nighttime land surface temperature (LST) data from the Level-3 standard product of the Moderate Resolution Imaging Spectroradiometer (MODIS) onboard the Aqua satellite, MYD11A2 LST and Emissivity 8-day L3 Global 1 km, were obtained from the Earth Observing System Data and Information System EARTHDATA of NASA from July 2002 to December 2016. MYD11A2 data are the average of daily MYD11A1 data of cloud-free days. Temporal and spatial resolutions of the data are 8-day and approximately 1 km × 1 km, respectively. The data are projected in Sinusoidal projection.

Since the time scale of the developed drought prediction model is monthly, the 8-day data were converted into monthly data using the number of days of the 8-day period for each month as weights. Mean LST (LST_MEAN) was also calculated from daytime LST (LST_DAY) and nighttime LST (LST_NIGHT).

3.5.5. MODIS Vegetation Indices

Vegetation indices of the Normalized Difference Vegetation Index (NDVI) and the Enhanced Vegetation Index (EVI) data were obtained from the Level-3 data of MODIS onboard Aqua, MYD13A3 Vegetation Indices Monthly L3 Global 1 km, from EARTHDATA of NASA from July 2002 to December 2016. Temporal and spatial resolutions are monthly and approximately 1 km × 1 km, respectively. The data are also projected in Sinusoidal projection.

The NDVI can be calculated using the changes in reflectance in red and near infrared (NIR) channels (Equation (1)) and has been widely used as an indicator of vegetation vigor [30]. The EVI uses the blue band in addition to red and NIR bands, minimizing the influence of the background effect of soil, snow, and water (Equation (2)). The EVI retains sensitivity to vegetation vitality, which is often shown saturated in the NDVI. The blue band helps to remove the atmospheric effect caused by air and clouds.

$$NDVI = \frac{NIR - RED}{NIR + RED} \tag{1}$$

$$EVI = 2 \times \frac{NIR - RED}{L + NIR + C1 \times RED + C2 \times BLUE} \tag{2}$$

where NIR, RED, and BLUE are reflectance values of NIR, RED, and BLUE channels, respectively; L is a parameter for reducing the background effect of canopy; C1 and C2 are weighting parameters to correct the influence of the aerosol effect of the red band when the blue and red bands are used together [31].

3.5.6. Elevation Data

Global 30 Arc-Second Elevation (GTOPO30) data with 1 km × 1 km spatial resolution were obtained from the US Geological Survey and used for the study area.

3.6. Large-Scale Climate Index

3.6.1. SPCZ

The SPCZ, a reverse-oriented monsoon trough, is a band of low-level convergence, cloudiness, and precipitation extending from the Western Pacific Warm Pool at the maritime continent southeastward toward French Polynesia and as far as the Cook Islands (160° W, 20° S). The SPCZ occurs where the southeast trade winds from transitory anticyclones to the south meet with the semi-permanent easterly flow from the eastern South Pacific anticyclone.

To study the SPCZ and its impacts on weather and climate over the South Pacific islands, previous studies suggested several SPCZ indices [8,32–35]. Here, we adopted the SPCZ strength index from Kidwell et al. [34] to quantify the impact of the SPCZ on rainfall over Fiji. The SPCZ region was encompassed in 0°–30° S, 130° E–110° W. The strength of the SPCZ is defined by the surface wind convergence in this region derived from the ERA-Interim. Divergence was calculated with Equation (3):

$$D(x,y) = \frac{\partial u}{\partial x} + \frac{\partial v}{\partial y} \tag{3}$$

where u and v are the zonal and meridional components of the surface winds. Positive D corresponds to surface divergence, and a negative value corresponds to surface convergence. The SPCZ strength is defined by the monthly mean area-weighted average of convergence within the SPCZ region:

$$s = \sum D(x,y)a(x,y) / \sum a(x,y) \tag{4}$$

where $a(x,y)$ is the area of a grid cell centered at location (x,y), and the spatial summation $\sum$ is performed over grid cells with $D(x,y) < 0$ within the SPCZ region. The anomaly of the SPCZ strength is defined as SPCZ index.

3.6.2. MEI

The ENSO is an irregularly periodic variation in winds and SST over the tropical eastern Pacific Ocean, affecting much of the tropics and subtropics. The warming phase is known as El Nino and the cooling phase as La Nina. Southern Oscillation is the accompanying atmospheric component, coupled with the sea temperature change; El Nino is accompanied with high air surface pressure while La Nina with low in the tropical western Pacific. The two periods last several months each (typically occurring every few years) and their effects vary in intensity. The Multivariate ENSO Index (MEI) from the National Oceanic and Atmospheric Administration (NOAA) were used as a measure of ENSO.

4. Methods

4.1. Drought Modeling

Mishra and Singh [36] reviewed a variety of drought modeling methods and described the components of drought modeling as hydro-meteorological variables, drought indices, climate indices, methodologies, and outputs. Among hydro-meteorological variables, rainfall is the most important variable for meteorological drought forecasting, soil moisture and crop yield are the key variables for agricultural drought forecasting, and stream flow and reservoir level are the most important variables for hydrological drought forecasting. Sometimes many variables are combined to obtain drought characteristics such as drought severity, duration, and spatial extent. Large-scale climate indices such as ENSO or the Arctic Oscillation (AO) index are used to forecast longer droughts. There can be many methods used, including regression models, time-series models, probability models, neural networks models, and statistical-dynamic models [36–41].

Recently, drought prediction methods using machine learning have been developed [42,43]. Rules required by expert systems can be developed either by human experts or derived by machines based on data provided by human beings; this training process is called machine learning [44]. Tadesse et al. [42] developed a rule-based regression tree model forecasting drought conditions and crop yield based on remotely sensed vegetation conditions, SPI, land use, available water capacity of soil, and irrigation areas. Rhee and Im [43] tested decision tree models, random forest models, and extra-trees models to forecast drought indices of the SPI and the Standardized Precipitation-Evapotranspiration Index in South Korea.

4.2. Machine Learning Model Design

As an indicator representing true drought conditions, the target variable was set as SPI6_OBS, which is reference SPI6 calculated either using in-situ precipitation data from four rainfall gauges or using the WRF model outputs from 227 pixel locations (Figure 3).

If we were to monitor current drought conditions, we may rely on SPI6_RS, which is SPI6 calculated from remote sensing based rainfall, since reference SPI6 is only available for the past or for some limited locations. However, there are usually gaps between SPI6_RS and SPI6_OBS. In order to explain or reduce the discrepancy, drought-affected input variables of LST_DAY, LST_NIGHT, LST_MEAN, NDVI, and EVI can be included to the model (Figure 3). Elevation (ELEV) can also be included to consider the topographical effect on rainfall, complementing the coarse spatial resolution of remotely sensed rainfall data (Figure 3).

Since the purpose of the model is drought prediction, long-range climate forecasting can be used to estimate the effect of synoptic and large-scale atmospheric circulation. While SPI6_RS was used for training machine learning models assuming perfect climate forecast, SPI6_FCST was used for test; SPI6_FCST is SPI6 calculated from bias-corrected precipitation data combining the percent increment of the rainfall anomaly of APCC MME and the climatology of remote sensing based rainfall [45] (Figure 3). A 6-month period of accumulated rainfall was divided into two periods according to the lead-time of the forecast; months with observed rainfall and months with forecasted rainfall. Remote sensing-based precipitation data were used as the observed rainfall, and bias-corrected precipitation forecast data were used as the forecasted rainfall. Parameters for the gamma probability distribution functions were pre-fitted based on remote sensing-based precipitation data and used for SPI6_FCST calculations.

Month of the data (MONTH) was also included for temporal information, and large-scale circulation indices of MEI and SPCZ strength (SPCZ) were also included (Figure 3).

Time points of data vary for 1 to 6-month lead drought prediction; initial points of data were used for remote sensing data and large-scale indices (for example, January 2017 values were used for 3-month lead predictions for April 2017), while target points of data were used for MONTH, SPI6_RS (training), and SPI6_FCST (test).

As the machine learning models, the Extra-Trees (ERT hereafter) [46] and the Adaboost [47] models were used in this study. The implementation was done using the Python library scikit-learn 0.18.1. ERT is known to produce stable results against outliers and noise in training data, and had excellent performance in drought forecasting [43]. Adaboost is a weak learner; it enables the model to simulate minor characteristics of training data by assigning higher weights to the subsets that are less reflected during its iteration processes.

The training of the models can be done either using in-situ data or using the WRF model outputs for SPI6_OBS. The models trained using SPI6_OBS based on in-situ data may not be appropriate to be used for other areas because data from only four weather stations are used and the models are trained specific to the locations. Two cases were compared; in one case, the models were trained using 80% of the WRF model outputs and evaluated using 20% of the data. In the other case, the models were trained using all in-situ data and evaluated using the same test dataset of the previous case. Numbers of data samples are shown in Table 3.

Table 3. Numbers of data samples used for training and testing.

Source	Type	Lead Time (Month)	Number of Samples	
			All Categories	Three Drier Categories
WRF model output	Train (80%)	1	16,693	1767
		2	16,545	1787
		3	16,379	1776
		4	16,211	1762
		5	16,043	1761
		6	15,875	1792
	Test (20%)	1	4169	470
		2	4132	445
		3	4091	445
		4	4049	447
		5	4006	456
		6	3964	424
In-situ data	All	1	266	37
		2	264	37
		3	262	37
		4	260	37
		5	258	37
		6	256	36

Although a three-tier approach of training, validation, and testing is often used to optimize parameters for some artificial intelligence models, we used a two-tier approach of training and testing with the fixed number of trees for ERT and Adaboost of 100 and the maximum depth of tree growth of 15 levels. Various numbers of trees and levels of maximum depth of tree growth had been tested using cross-validation of training data; the number of trees larger than 100 did not produce much difference. Although larger levels of maximum depth of tree growth tend to produce better results, the retrieval of the trained model with larger than 15 levels of maximum depth of tree growth including full development was very demanding of computational resources.

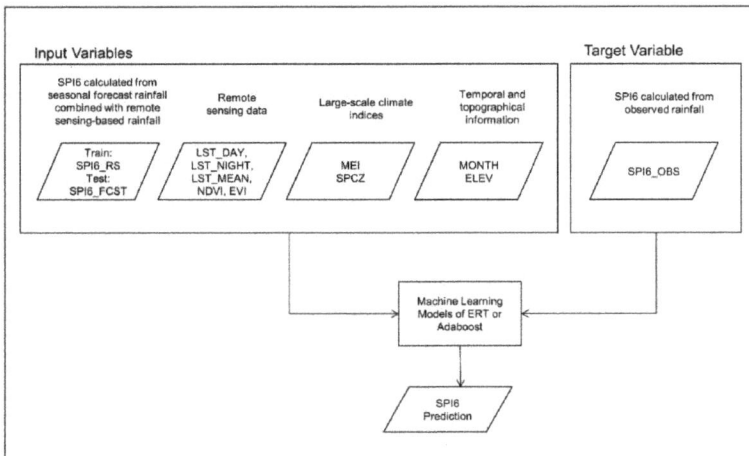

Figure 3. Flow diagram of the drought prediction model.

4.3. Data Pre-Processing

Remote sensing-based variables of LST_DAY, LST_NIGHT, LST_MEAN, NDVI, EVI, and ELEV were all subset to the extent of 176.5° E–178° W, 21.5° S–12.0° S and then resampled to have 0.01° × 0.01° spatial resolution. Since many machine learning models tend to be sensitive to the magnitudes of input variables, these data were scaled using maximum and minimum values of each month for each pixel [48].

Since SPI is inherently Gaussian, the numbers of input data for each drought category of Table 2 are not even. Because some machine learning models are known to be sensitive to the distribution of samples, the following process was performed when preparing input data: additional input data were created with added noise by multiplying the standard deviation of the variable for the location and month with a random number between 0 and 1, so that all drought categories have the same sample numbers during training.

The thirty-year period from 1981 to 2010 was used for calculating SPI. Due to the short history of MODIS, the input data from July 2002 to 2016 were used for the machine learning models.

4.4. Performance Measures

Information on drought index values or corresponding drought categories indicating the severity of drought can be more useful to users than just having binary information of drought or non-drought. Performance measures used in this study include: Total Accuracy, which is the producer's accuracy, and mean absolute error (MAE) for all drought categories in Table 2 (total MAE hereafter). Although there may not be enough serious drought events during the short study period from July 2002 to 2016, performance measures only for the three drier categories of Extreme Drought, Severe Drought, and Moderate Drought were also used: Drought Accuracy, which is a modified producer's accuracy in Rhee and Im [43] focusing on the three drier categories, and MAE for the three drier categories (Drought MAE hereafter).

$$\text{Total or Drought Accuracy} = \frac{\sum C}{\sum N} \tag{5}$$

$$\text{Total or Drought MAE} = \frac{\sum \left| \text{SPI6}_{obs} - \text{SPI6}_{pred} \right|}{\text{Total Number of Samples}} \tag{6}$$

where N is the number of samples for each category, and C is the number of correctly categorized samples for each category. All categories are considered for Total Accuracy and Total MAE, while the three drier categories are considered for Drought Accuracy and Drought MAE.

5. Results and Discussion

5.1. Training of the Models

The machine learning models of ERT and Adaboost were trained using 80% of the WRF model outputs (ERT_WRF and Adaboost_WRF hereafter) or using 100% of the in-situ data (ERT_INSITU and Adaboost_INSITU hereafter). The performance of SPI6 predictions from simply bias-corrected precipitation forecast (FCST_ONLY hereafter) based on the same training dataset of the WRF model outputs was compared to the performance of ERT and Adaboost (Figure 4). Differences in MAE between methods were also statistically tested using two-sided or one-sided Welch's *t*-test for both Total MAE and Drought MAE.

Both ERT_WRF and Adaboost_WRF outperformed FCST_ONLY in most cases, and Total MAE and Drought MAE values of ERT_WRF were especially small (Figure 4a,b). The differences were all statistically significant based on two-tailed *p*-values with a confidence level of 0.01 (data not shown). Only Adaboost_WRF with 1-month lead predictions showed larger Drought MAE than FCST_ONLY based on one-sided *t*-test (Figure 4b). ERT_WRF outperformed FCST_ONLY and Adaboost_WRF based on one-sided *t*-test (data not shown).

In terms of Total Accuracy and Drought Accuracy, ERT_WRF was much higher compared to FCST_ONLY for all lead times (Figure 4c,d). However, Adaboost_WRF could not perform better than FCST_ONLY in terms of Total Accuracy of 2-month lead predictions and Drought Accuracy of 1- and 2-month lead predictions (Figure 4c,d).

We could see that ERT_INSITU is overly fitted based on zero or near-zero Total MAE and Drought MAE values and perfect Total Accuracy and Drought Accuracy, despite the large number of trees (Figure 4). It is not very surprising since the numbers of samples for all categories and the three drier categories are not large; smaller than 270 and 40, respectively (Table 3). Both ERT_INSITU and Adaboost_INSITU outperformed FCST_ONLY in all cases (Figure 4). Total MAE and Drought MAE of ERT_WRF were larger than ERT_INSITU because of possible overfitting based on one-sided *t*-test, no difference in MAE was found between ERT_WRF and Adaboost_INSITU.

Scatter plots of reference SPI6 vs. 1-month lead SPI6 predictions for training are shown in Figure 5.

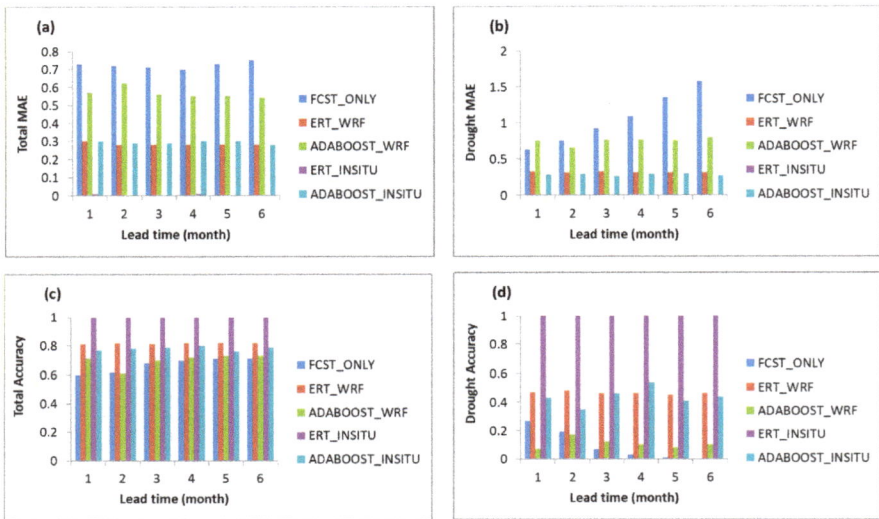

Figure 4. Training performance (**a**) Total MAE; (**b**) Drought MAE; (**c**) Total Accuracy; and (**d**) Drought Accuracy of SPI6 predictions from simply bias-corrected precipitation forecast (FCST_ONLY), Extra-Trees (ERT) and Adaboost trained using 80% of the WRF model outputs (ERT_WRF and Adaboost_WRF), and ERT and Adaboost trained using 100% of in-situ data (ERT_INSITU and Adaboost_INSITU).

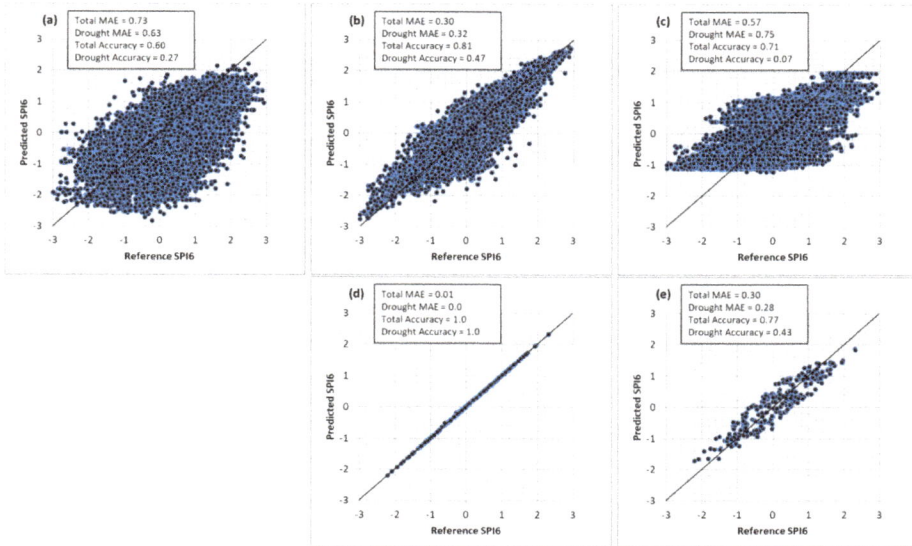

Figure 5. Scatter plots of reference SPI6 vs. 1-month lead SPI6 predictions for training based on (a) FCST_ONLY; (b) ERT_WRF; (c) Adaboost_WRF; (d) ERT_INSITU; and (e) Adaboost_INSITU. Reference SPI6 are based on 80% of the WRF model outputs from (a) to (c) and 100% of in-situ data for (d,e).

5.2. Test of the Models

The performance of SPI6 predictions of the machine learning models (ERT_WRF, Adaboost_WRF, ERT_INSITU and Adaboost_INSITU) as well as FCST_ONLY was evaluated based on the remaining 20% of the WRF model outputs (Figure 6). Differences in MAE between methods were also statistically tested.

ERT_WRF showed the smallest Total MAE, and the differences between ERT_WRF and all other methods were statistically significant based on one-sided *t*-test with the confidence interval of 0.01 (Figure 6a; *p*-values are not shown). Adaboost_WRF also produced smaller Total MAE compared to FCST_ONLY for 1- to 4-month lead predictions, while the differences were not statistically significant for 5- and 6-month lead predictions (two-tailed *p*-values are 0.031 and 0.026, respectively). Even ERT_INSITU and Adaboost_INSITU produced significantly smaller Total MAE than FCST_ONLY for 1- to 3-month lead predictions (Figure 6a). Cases that failed to reject the null hypothesis of equal mean error with FCST_ONLY are shaded (Figure 6a).

In contrast to training where Drought MAE of FCST_ONLY was mostly the largest (Figure 4c), Drought MAE of FCST_ONLY was mostly the smallest for all lead times with the test dataset (Figure 6c). Cases that failed to reject the null hypothesis of equal or larger mean error with FCST_ONLY are shaded based on two-tailed and one-tailed *p*-values, meaning only these cases produce comparable Drought MAE to FCST_ONLY (Figure 6c; data not shown). The one-sided *t*-test with the null hypothesis of larger error of FCST_ONLY in all other cases was rejected, meaning that they produced larger Drought MAE in most cases (Figure 6c).

There were no obvious differences observed in Total Accuracy between the methods; Total Accuracy of ERT_WRF was the highest for all lead times (Figure 6b). FCST_ONLY produced higher Drought Accuracy for 1-month lead SPI6 predictions, while ERT_WRF performed the best for longer-term predictions (Figure 6d). The selection of training data (WRF model outputs versus in-situ data), the selection of a prediction model (FCST_ONLY versus machine learning models of ERT and Adaboost), and the lead time had the greatest effect on Drought Accuracy (Figure 6d).

Scatter plots of reference SPI6 vs. 1-month as well as 3-month lead SPI6 predictions for testing are shown in Figures 7 and 8, respectively.

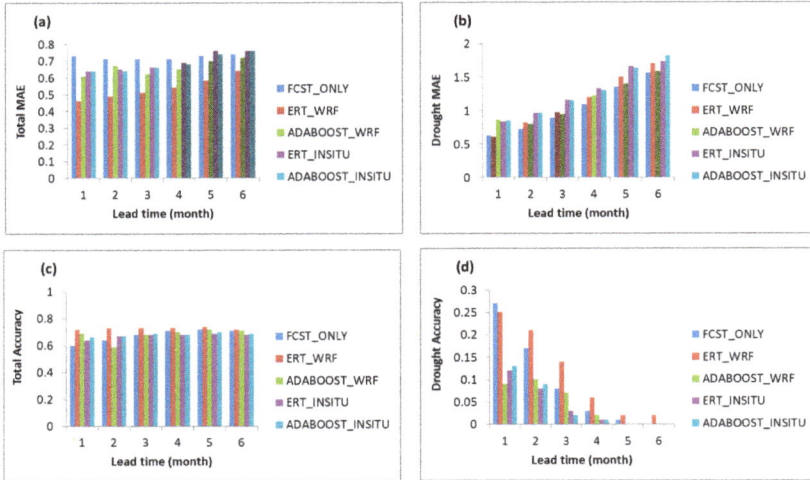

Figure 6. Test performance (**a**) Total MAE; (**b**) Drought MAE; (**c**) Total Accuracy; and (**d**) Drought Accuracy of SPI6 predictions from simply bias-corrected precipitation forecast (FCST_ONLY), ERT and Adaboost trained using 80% of the WRF model outputs (ERT_WRF and Adaboost_WRF), and ERT and Adaboost trained using 100% of in-situ data (ERT_INSITU and Adaboost_INSITU). Test was performed using the 20% remaining WRF model outputs.

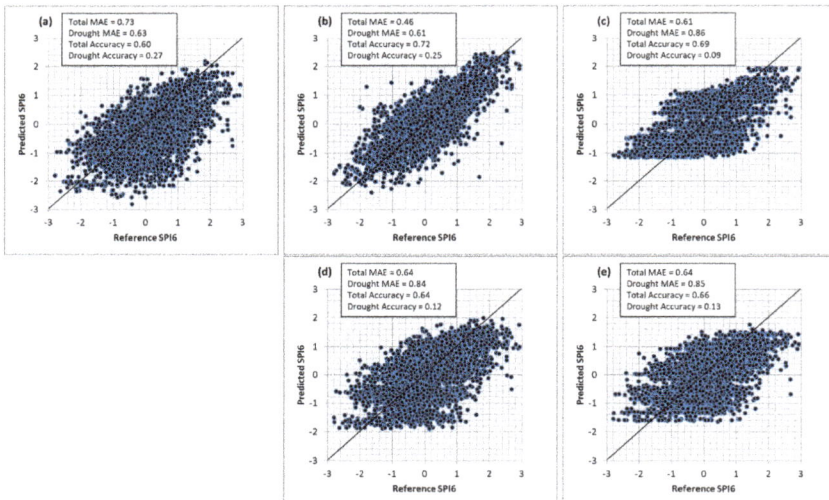

Figure 7. Scatter plots of reference SPI6 vs. 1-month lead SPI6 predictions for testing based on (**a**) FCST_ONLY; (**b**) ERT_WRF; (**c**) Adaboost_WRF; (**d**) ERT_INSITU; and (**e**) Adaboost_INSITU. Reference SPI6 are based on 20% of the WRF model outputs.

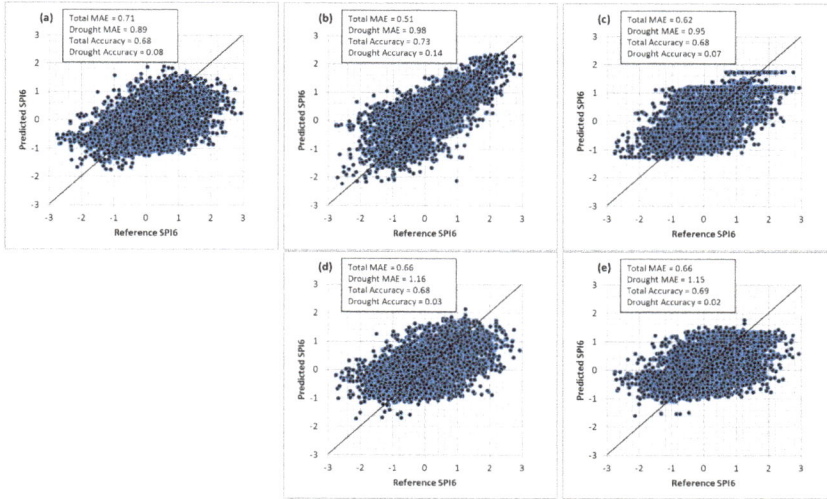

Figure 8. Scatter plots of reference SPI6 vs. 3-month lead SPI6 predictions for testing based on (**a**) FCST_ONLY; (**b**) ERT_WRF; (**c**) Adaboost_WRF; (**d**) ERT_INSITU; and (**e**) Adaboost_INSITU. Reference SPI6 are based on 20% of the WRF model outputs.

5.3. Spatial Distribution Maps of SPI6 Predictions

Spatially distributed maps of 1- to 6-month lead SPI6 predictions based on FCST_ONLY and ERT_WRF were created. Some examples are shown in Figure 9; in order to provide the WRF-based SPI6 map used for training machine learning models as well as in-situ SPI6 map with available data from all four weather stations, 21 months with all data available were identified. Although no extreme drought events were observed in the 21 months, Nadi (91680) station experienced severe droughts in March, June, July, and October 2010.

The WRF-based SPI6 (Figure 9a,b), 1-month lead SPI6 predictions based on FCST_ONLY (Figure 9c,d), and 1-month lead SPI6 predictions based on ERT_WRF (Figure 9e,f) for March and June of 2010 are shown. Four weather stations are also shown with SPI6 based on observation data for March and June, 2010 (Figure 9). Only Nadi station was in severe drought in March and June of 2010 (SPI6 = −1.51 and −1.94, respectively). In March 2010, Udu Point and Suva stations were in moderate drought (SPI6 = −1.37 and −1.05, respectively) while Nabouwalu station was in near normal condition (SPI6 = −0.67). In June, Udu Point station was in moderate drought (SPI6 = −1.48) while Nabouwalu and Suva stations were in near normal condition (SPI6 = −0.45 and −0.52, respectively).

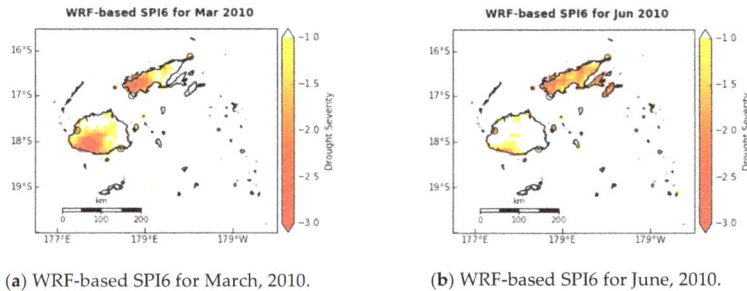

(**a**) WRF-based SPI6 for March, 2010.　　　　(**b**) WRF-based SPI6 for June, 2010.

Figure 9. *Cont.*

155

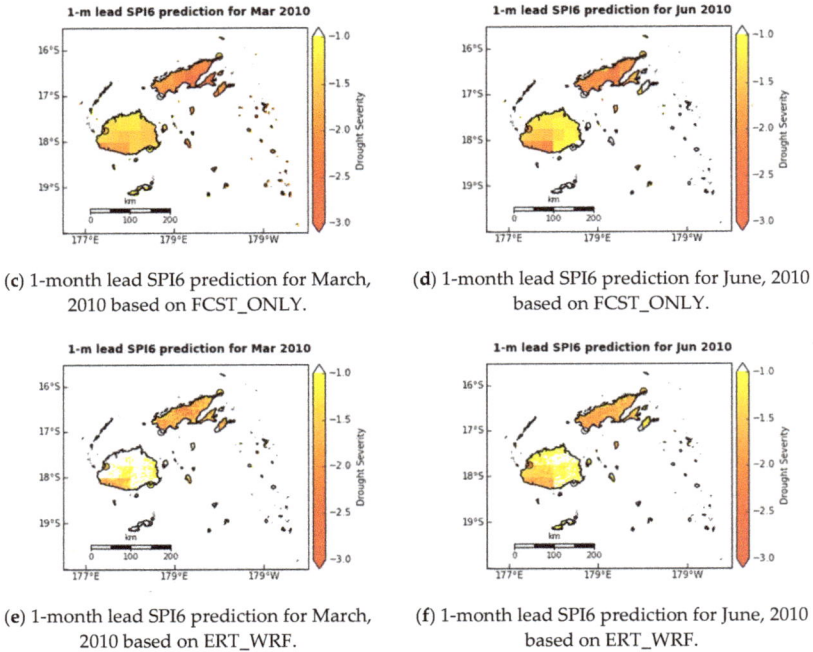

(c) 1-month lead SPI6 prediction for March, 2010 based on FCST_ONLY.

(d) 1-month lead SPI6 prediction for June, 2010 based on FCST_ONLY.

(e) 1-month lead SPI6 prediction for March, 2010 based on ERT_WRF.

(f) 1-month lead SPI6 prediction for June, 2010 based on ERT_WRF.

Figure 9. Spatial distribution maps of 1-month lead SPI6 predictions for March 2010 and June 2010 and WRF-based SPI6.

5.4. Relative Importance of Input Variables to Machine Learning Models

Python modules for machine learning models provide information on the relative importance of input variables. The importance of the most important variable is set to 100% and relative importance scores of other input variables are determined. In all cases, the most important variable was SPI6_RS in this study, and only the scores of other input variables are shown in Figure 10.

When in-situ precipitation data were used for reference data, the relative importance of all other input variables was quite low; the score of the second important variable MEI only ranges between 4% and 8% for ERT_INSITU (Figure 10c). For Adaboost_INSITU, the scores of input variables vary with lead time, but all were below 20% (Figure 10d). The importance of temporal (MONTH) and topographical (ELEV) information as well as large-scale climate indices (SPCZ, MEI) were more obvious when the WRF model outputs were used for reference data (Figure 10a,b). For ERT_WRF, the scores of MONTH, MEI, and SPCZ were higher than other input variables, mostly over 20% (Figure 10a). The scores of those three variables as well as ELEV were higher for Adaboost_WRF; the score for MONTH even reached about 55% (Figure 10b).

Differences in the relative importance of the input variables between the sources of reference data indicate that temporal characteristics of drought occurrences and the effect of ENSO, SPCZ strength, as well as topography of the region could not be adequately applied to the models when in-situ data were used for reference data, because in-situ data from only few stations are available. The use of the WRF model output precipitation data, on the other hand, enabled the use of diverse information from those variables.

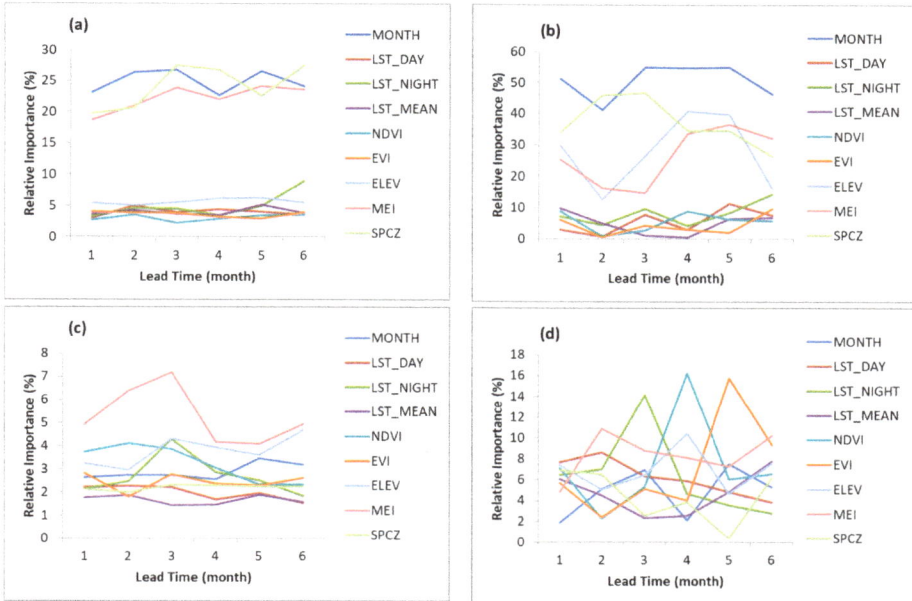

Figure 10. Relative importance scores of input variables to machine learning models for (**a**) ERT_WRF; (**b**) Adaboost_WRF; (**c**) ERT_INSITU; and (**d**) Adaboost_INSITU.

6. Conclusions

We developed hybrid drought prediction models using APCC MME seasonal climate forecasts and machine learning models and examined their performance for the case study area of Fiji. The purpose of the models is to provide spatially distributed detailed drought prediction data of SPI6 for the area. The APCC MME provides up to 6-month lead precipitation forecast data. Remote sensing data were used to bias-correct the forecast data as well as to train machine learning models; machine learning models of ERT and Adaboost were used to provide spatially distributed drought information for ungauged areas. In order to overcome the limitation of sparse monitoring network, dynamic downscaling of historical climate with the WRF model was used to produce reference data.

When compared to the performance of the hybrid models trained based on different reference data, the models trained using the WRF model outputs performed better than the models trained using in-situ data: ERT_WRF outperformed ERT_INSITU in all cases, and Adaboost_WRF outperformed Adaboost_INSITU except for Drought MAE and Drought Accuracy of 1-month lead predictions, Total MAE and Total Accuracy of 2-month lead predictions, and Total Accuracy of 3-month lead predictions. The superiority of the models trained based on the WRF model outputs indicates that the spatial extent of the training data is important because in-situ data are from only four weather stations. The added value caused by the topography is clear, especially in the convergence/divergence field over the islands; this crucially impacted inland and coastal precipitation and caused greater detail in precipitation to be found in the WRF model outputs [24].

The use of the ERT_WRF model produced better results compared to Adaboost_WRF in terms of Total MAE, Total Accuracy, and Drought Accuracy for all lead times, as well as in terms of Drought MAE of 1-month lead predictions. For other lead times, no statistical difference between ERT_WRF and Adaboost_WRF were found (2- to 4-month lead predictions) or ERT_WRF showed larger error than Adaboost_WRF (5- to 6-month lead predictions) in terms of Drought MAE. It shows that the choice of the machine learning model matters; the use of simulated input data with added noise to attain the same numbers of samples between drought categories may have improved the performance of ERT and surpassed the advantage of Adaboost, supporting weak learners.

Compared to FCST_ONLY, ERT_WRF performed better in terms of Total MAE and Total Accuracy for all lead times as well as in terms of Drought Accuracy for 2- to 6-month lead predictions. Although there was no statistically significant difference for 1-month and 3-month lead predictions in terms of Drought MAE and the error of ERT_WRF was larger for 2-month and 4- to 6-month lead predictions, Drought Accuracy of ERT_WRF for 2- to 6-month lead predictions was higher than FCST_ONLY. The hybrid model, especially ERT_WRF, showed good performance compared to simply bias corrected forecasts.

Hybrid models with better performance than simply bias corrected forecasts in most cases for areas with sparse monitoring networks were successfully developed. It should be noted that the performance of the compared methods may be evaluated differently according to the purpose of the study with the appropriate choice of a performance measure. In future studies, the use of more diverse input variables related to drought for machine learning models need to be investigated. Only SPI based on precipitation data was examined in this study; drought prediction based on drought indices considering the effect of evapotranspiration, such as the Standardized Precipitation-Evapotranspiration Index [49] or the Standardized Evapotranspiration Deficit Index [50], may also help to reduce vulnerability to droughts.

Author Contributions: J.R. designed this study, led data analysis and manuscript writing, and served as the corresponding author. H.Y. contributed to the selection of variables, data production, discussion of results, and manuscript writing.

Funding: This research received no external funding.

Acknowledgments: This research was supported by the APEC Climate Center. Authors are thankful to the Fiji Meteorological Service for the provision of data, information, and constructive comments.

Conflicts of Interest: The authors declare no conflict of interest.

References

1. Intergovernmental Panel on Climate Change. *Climate Change 2007: The Physical Science Basis: Working Group I Contribution to the Fourth Assessment Report of the Intergovernmental Panel on Climate Change*; Solomon, S., Qin, D., Manning, M., Chen, Z., Marquis, M., Averyt, K.B., Tignor, M., Miller, H.L., Eds.; Cambridge University Press: Cambridge, UK; New York, NY, USA, 2007; 996p.
2. Hay, J.E.; Mimura, N.; Campbell, J.; Fifita, S.; Koshy, K.; McLean, R.F.; Nakalevu, T.; Nunn, P.; de Wet, N. *Climate Variability and Change and Sea-Level Rise in the Pacific Islands Region: A Resource Book for Policy and Decision Makers, Educators and Other Stakeholders*; South Pacific Regional Environment Programme: Apia, Samoa, 2003; 108p.
3. Trenberth, K.E.; Hoar, T.J. The 1990–1995 El Niño-Southern Oscillation event: Longest on record. *Geophys. Res. Lett.* **1996**, *23*, 57–60. [CrossRef]
4. Mantua, N.J.; Hare, S.R.; Zhang, Y.; Wallace, J.M.; Francis, R.C. A Pacific interdecadal climate oscillation with impacts on salmon production. *Bull. Am. Meteorol. Soc.* **1997**, *78*, 1069–1079. [CrossRef]
5. Mantua, N.J.; Hare, S.R. The Pacific decadal oscillation. *J. Oceanogr.* **2002**, *58*, 35–44. [CrossRef]
6. Bergeron, T. Richtlinien einer dynamischen Klimatologie. *Meteorol. Z.* **1930**, *47*, 246–262.
7. Trenberth, K.E. Spatial and temporal variations of the Southern Oscillation. *Q. J. R. Meteorol. Soc.* **1976**, *102*, 639–653. [CrossRef]
8. Folland, C.K.; Renwick, J.A.; Salinger, M.J.; Mullan, A.B. Relative influences of the interdecadal Pacific oscillation and ENSO on the South Pacific convergence zone. *Geophys. Res. Lett.* **2002**, *29*, 1643. [CrossRef]
9. Hu, Z.-Z.; Huang, B. Interferential impact of ENSO and PDO on dry and wet conditions in the US Great Plains. *J. Clim.* **2009**, *22*, 6047–6065. [CrossRef]
10. Nicholls, N.; Wong, K.K. Dependence of rainfall variability on mean rainfall, latitude, and the Southern Oscillation. *J. Clim.* **1990**, *3*, 163–170. [CrossRef]

11. Deo, R.C. On meteorological droughts in tropical Pacific Islands: Time-series analysis of observed rainfall using Fiji as a case study. *Meteorol. Appl.* **2011**, *18*, 171–180. [CrossRef]

12. Mataki, M.; Koshy, K.C.; Lal, M. Baseline climatology of Viti Levu (Fiji) and current climatic trends. *Pac. Sci.* **2006**, *60*, 49–68. [CrossRef]

13. Kumar, R.; Stephens, M.; Weir, T. Rainfall trends in Fiji. *Int. J. Climatol.* **2014**, *34*, 1501–1510. [CrossRef]

14. Kumar, R.; Stephens, M.; Weir, T. Temperature trends in Fiji: A clear signal of climate change. *S. Pac. J. Nat. Appl. Sci.* **2013**, *31*, 27–38.

15. Risbey, J.S.; Lamb, P.J.; Miller, R.L.; Morgan, M.C.; Roe, G.H. Exploring the structure of regional climate scenarios by combining synoptic and dynamic guidance and GCM output. *J. Clim.* **2002**, *15*, 1036–1050. [CrossRef]

16. Feresi, J.; Kenny, G.J.; de Wet, N.; Limalevu, L.; Bhusan, J.; Ratukalou, I. *Climate Change Vulnerability and Adaptation Assessment for Fiji*; Technical Report; The International Global Change Institute, University of Waikato: Hamilton, New Zealand, 2000; 135p.

17. Agrawala, S.; Ota, T.; Risbey, J.; Hagenstad, M.; Smith, J.; van Aalst, M.; Koshy, K.; Prasad, B. *Development and Climate Change in Fiji: Focus on Coastal Mangroves*; Organisation for Economic Co-operation and Development: Paris, France, 2003; 56p.

18. Intergovernmental Panel on Climate Change. *Climate Change 2013: The Physical Science Basis: Working Group I Contribution to the Fifth Assessment Report of the Intergovernmental Panel on Climate Change*; Stocker, T.F., Qin, D., Plattner, G.-K., Tignor, M., Allen, S.K., Boschung, J., Nauels, A., Xia, Y., Bex, V., Midgley, P.M., Eds.; Cambridge University Press: Cambridge, UK; New York, NY, USA, 2013; 1535p.

19. Terry, J.P.; Raj, R. The 1997–98 El Niño and drought in the Fiji Islands. In *Hydrology and Water Management in the Humid Tropics, Proceedings of the Second International Colloquium, Panama, Republic of Panama, 22–26 March 1999*; UNESCO: Paris, France, 2002; pp. 80–93.

20. ENSO Update. Available online: http://www.met.gov.fj/ENSO_Update.pdf (accessed on 3 November 2017).

21. Yin, H.; Donat, M.G.; Alexander, L.V.; Sun, Y. Multi-dataset comparison of gridded observed temperature and precipitation extremes over China. *Int. J. Climatol.* **2015**, *35*, 2809–2827. [CrossRef]

22. Derrick, R.A. *The Fiji Islands: A Geographical Handbook*; Fiji Government Printing Department: Suva, Fiji, 1957; 68p.

23. Terry, J.P.; Raj, R. Hydrological drought in western Fiji and the contribution of tropical cyclones. In *Climate and Environmental Change in the Pacific*; Terry, J.P., Ed.; School of Social and Economic Development, University of the South Pacific: Suva, Fiji, 1998; pp. 73–85.

24. Rhee, J.; Yang, H. *Development of a Drought Forecast Model for Fiji Based on High-Resolution Dynamic Downscaling of Climate Data and Machine Learning of Long-Range Climate Forecast and Remote Sensing Data*; APEC Climate Center Research Report 2017-04; APEC Climate Center: Busan, South Korea, 2018; 63p.

25. Guttman, N.B. Accepting the standardized precipitation index: A calculation algorithm. *J. Am. Water Resour. Assoc.* **1999**, *35*, 311–322. [CrossRef]

26. McKee, T.B.; Doesken, N.J.; Kleist, J. The relationship of drought frequency and duration to time scales. In Proceedings of the 8th Conference on Applied Climatology, Anaheim, CA, USA, 17–22 January 1993; American Meteorological Society: Boston, MA, USA, 1993; pp. 179–183.

27. Fiji Meteorological Service. Personal communication, 2017.

28. APEC Climate Data Service System (ADSS). Available online: http://adss.apcc21.org (accessed on 7 July 2017).

29. Ashouri, H.; Hsu, K.-L.; Sorooshian, S.; Braithwaite, D.K.; Knapp, K.R.; Cecil, L.D.; Nelson, B.R.; Prat, O.P. PERSIANN-CDR: Daily precipitation climate data record from multisatellite observations for hydrological and climate studies. *Bull. Am. Meteorol. Soc.* **2015**, *96*, 69–83. [CrossRef]

30. Tucker, C.J. Red and photographic infrared linear combinations for monitoring vegetation. *Remote. Sens. Environ.* **1979**, *8*, 127–150. [CrossRef]

31. Liu, H.Q.; Huete, A. A feedback based modification of the NDVI to minimize canopy background and atmospheric noise. *IEEE Trans. Geosci. Remote* **1995**, *33*, 457–465.

32. Borlace, S.; Santoso, A.; Cai, W.; Collins, M. Extreme swings of the South Pacific Convergence Zone and the different types of El Niño events. *Geophys. Res. Lett.* **2014**, *41*, 4695–4703. [CrossRef]

33. Cai, W.; Lengaigne, M.; Borlace, S.; Collins, M.; Cowan, T.; McPhaden, M.J.; Timmermann, A.; Power, S.; Brown, J.; Menkes, C.; et al. More extreme swings of the South Pacific convergence zone due to greenhouse warming. *Nature* **2012**, *488*, 365–369. [CrossRef] [PubMed]

34. Kidwell, A.; Lee, T.; Jo, Y.-H.; Yan, X.-H. Characterization of the variability of the South Pacific convergence zone using satellite and reanalysis wind products. *J. Clim.* **2016**, *29*, 1717–1732. [CrossRef]
35. Vincent, E.M.; Lengaigne, M.; Menkes, C.E.; Jourdain, N.C.; Marchesiello, P.; Madec, G. Interannual variability of the South Pacific Convergence Zone and implications for tropical cyclone genesis. *Clim. Dyn.* **2011**, *36*, 1881–1896. [CrossRef]
36. Mishra, A.K.; Singh, V.P. Drought modeling–A review. *J. Hydrol.* **2011**, *403*, 157–175. [CrossRef]
37. Leilah, A.A.; Al-Khateeb, S.A. Statistical analysis of wheat yield under drought conditions. *J. Arid Environ.* **2005**, *61*, 483–496. [CrossRef]
38. Steinemann, A.C. Using climate forecasts for drought management. *J. Appl. Meteorol. Climatol.* **2006**, *45*, 1353–1361. [CrossRef]
39. Morid, S.; Smakhtin, V.; Bagherzadeh, K. Drought forecasting using artificial neural networks and time series of drought indices. *Int. J. Climatol.* **2007**, *27*, 2103–2111. [CrossRef]
40. Han, P.; Wang, P.X.; Zhang, S.Y.; Zhu, D.H. Drought forecasting based on the remote sensing data using ARIMA models. *Math. Comput. Model.* **2010**, *51*, 1398–1403. [CrossRef]
41. Ribeiro, A.; Pires, C.A. Seasonal drought predictability in Portugal using statistical-dynamical techniques. *Phys. Chem. Earth* **2016**, *94*, 155–166. [CrossRef]
42. Tadesse, T.; Brown, J.F.; Hayes, M.J. A new approach for predicting drought-related vegetation stress: Integrating satellite, climate, and biophysical data over the US central plains. *ISPRS J. Photogramm.* **2005**, *59*, 244–253. [CrossRef]
43. Rhee, J.; Im, J. Meteorological drought forecasting for ungauged areas based on machine learning: Using long-range climate forecast and remote sensing data. *Agric. For. Meteorol.* **2017**, *237*, 105–122. [CrossRef]
44. Jensen, J.R.; Lulla, K. *Introductory Digital Image Processing: A Remote Sensing Perspective*; Taylor & Francis: Milton Park, UK, 1987; 65p.
45. Quan, X.-W.; Hoerling, M.P.; Lyon, B.; Kumar, A.; Bell, M.A.; Tippett, M.K.; Wang, H. Prospects for dynamical prediction of meteorological drought. *J. Appl. Meteorol. Clim.* **2012**, *51*, 1238–1252. [CrossRef]
46. Geurts, P.; Ernst, D.; Wehenkel, L. Extremely randomized trees. *Mach. Learn.* **2006**, *63*, 3–42. [CrossRef]
47. Freund, Y.; Schapire, R.E. A decision-theoretic generalization of on-line learning and an application to boosting. *J. Comput. Syst. Sci.* **1997**, *55*, 119–139. [CrossRef]
48. Rhee, J.; Im, J.; Carbone, G.J. Monitoring agricultural drought for arid and humid regions using multi-sensor remote sensing data. *Remote Sens. Environ.* **2010**, *114*, 2875–2887. [CrossRef]
49. Vicente-Serrano, S.M.; Beguería, S.; López-Moreno, J.I. A Multiscalar drought index sensitive to global warming: The Standardized Precipitation Evapotranspiration Index. *J. Clim.* **2010**, *23*, 1696–1718. [CrossRef]
50. Kim, D.; Rhee, J. A drought index based on actual evapotranspiration from the Bouchet hypothesis. *Geophys. Res. Lett.* **2016**, *43*, 10277–10285. [CrossRef]

water

MDPI

Article

Wavelet-ANN versus ANN-Based Model for Hydrometeorological Drought Forecasting

Md Munir H. Khan [1,2], Nur Shazwani Muhammad [1,]* and Ahmed El-Shafie [3]

[1] Smart and Sustainable Township Research Centre, Faculty of Engineering and Built Environment, Universiti Kebangsaan Malaysia (UKM), Bangi 43600, Selangor Darul Ehsan, Malaysia; shihab.bd@gmail.com

[2] Faculty of Engineering & Quantity Surveying, INTI International University (INTI-IU), Persiaran Perdana BBN, Putra Nilai, Nilai 71800, Negeri Sembilan, Malaysia

[3] Department of Civil Engineering, Faculty of Engineering, University of Malaya (UM), Kuala Lumpur 50603, Malaysia; elshafie@um.edu.my

* Correspondence: shazwani.muhammad@ukm.edu.my; Tel.: +60-3-8921-6226

Received: 4 June 2018; Accepted: 10 July 2018; Published: 27 July 2018

Abstract: Malaysia is one of the countries that has been experiencing droughts caused by a warming climate. This study considered the Standard Index of Annual Precipitation (SIAP) and Standardized Water Storage Index (SWSI) to represent meteorological and hydrological drought, respectively. The study area is the Langat River Basin, located in the central part of peninsular Malaysia. The analysis was done using rainfall and water level data over 30 years, from 1986 to 2016. Both of the indices were calculated in monthly scale, and two neural network-based models and two wavelet-based artificial neural network (W-ANN) models were developed for monthly droughts. The performance of the SIAP and SWSI models, in terms of the correlation coefficient (R), was 0.899 and 0.968, respectively. The application of a wavelet for preprocessing the raw data in the developed W-ANN models achieved higher correlation coefficients for most of the scenarios. This proves that the created model can predict meteorological and hydrological droughts very close to the observed values. Overall, this study helps us to understand the history of drought conditions over the past 30 years in the Langat River Basin. It further helps us to forecast drought and to assist in water resource management.

Keywords: drought analysis; ANN model; drought indices; meteorological drought; SIAP; SWSI; hydrological drought; discrete wavelet

---

1. Introduction

Drought gradually happens with a lack of rainfall for a long period of time (i.e., months or years). This natural disaster is considered to be the most complex and least understood by many scientists. The impact of drought varies with respect to the affected areas. The damage may include impacts on the social and agriculture sectors, and the economy [1]. In 2007, it was reported that, because of the tremendously hot temperature, heat waves, and heavy rainfalls, extreme events would accumulate and become more frequent [2]. Although Malaysia experiences a tropical climate and receives more than 2000 mm of total rainfall annually, over the recent years, the country has experienced several drought episodes. For example, the state of Melaka faced a serious water shortage when water levels in the dams fell under critical levels in 1991, and the Durian Tunggal dam, which serves as a major water supply dam, ran dry [3]. In 1998, an El Nino-related drought severely hit the states of Selangor, Kedah, and Penang, which caused severe social and environmental impacts across the country [3]. This drought caused water rationing and hardship for 1.8 million residents of Kuala Lumpur and other towns in Klang Valley. The Langat River Basin also experienced a rise in temperature nearly 5° higher than usual on many days in March and April 2016 [4]. A research study applied the standardized precipitation index (SPI) to evaluate dry conditions using the data from 10 gauging stations throughout peninsular Malaysia, and found that extreme dry conditions are becoming more frequent than extreme wet conditions [5]. Thus, emphasis should be placed on measures to reduce the impact of dry conditions, although the authorities usually put more focus on reducing extreme wet conditions (i.e., floods).

Drought is generally analyzed by means of drought indices, which are effectually a function of precipitation and other hydrometeorological variables [6]. Different drought indices have been discovered and are used in different nations [6]. Hydrologists have defined four major categories of drought, namely, meteorological drought, agricultural drought, hydrological drought, and socioeconomic drought [1]. Drought monitoring by indices in specific areas must be based on the availability of hydrometeorological data and the capability of the index to dependably detect spatial and temporal differences through a drought event. Nevertheless, no single indicator or index alone can precisely describe the onset and severity of the event. Numerous climate and water supply indices are used to describe the severity of any drought event. Although none of the major indices is inherently superior to the rest in all circumstances, some indices are better suited for certain uses than others [7]. In this study, the first objective was to assess the drought using two drought indices (DIs), the Standard Index of Annual Precipitation (SIAP) and the Standardized Water Storage Index (SWSI), to represent meteorological and hydrological droughts, respectively. The SIAP and SWSI were chosen for their simplicity, and they do not require parameter estimation. Gourabi [8] used SIAP and the dependable rainfall index (DRI) for the recognition of drought years in several areas in Iran, and to analyze the effects on rice yield and water surface. Sing et al. [9] used SIAP and a few other indices to assess the drought spells in the Almora district of Uttarakhand, India. On the other hand, to calculate SWSI, the Standardized Drought Assessment Toolbox (SDAT), developed by Farahmand and AghaKouchak in 2015 [10], is used. The SDAT methodology standardizes the marginal probability of drought-related variables (e.g., precipitation, soil moisture, and relative humidity) using the empirical distribution function of the data. This approach does not require an assumption of the representativeness of a parametric distribution function to describe drought-related variables. Additionally, the nonparametric framework does not require a parameter estimation and goodness-of-fit evaluation, which makes the SDAT framework computationally much more efficient. Wang et al. [11] used four drought indices, including SWSI, in order to assess the intensity and timing of drought events in the upper and middle Yangtze River Basin in China. In the second objective of this study, artificial neural network (ANN)–based models coupled with a wavelet were developed and their performance evaluation was carried out for both SIAP and SWSI models.

Many researchers have developed and applied various models to predict hydrological events, which could be divided into two major types, conceptual models (CM) and data-based models (DDM) [12]. The conceptual models usually incorporate simplified schemes of physical laws and are generally nonlinear, time-invariant, and deterministic, with parameters that are representative of watershed characteristics. However, when they are calibrated to a given set of hydrological signals (time series), there is no guarantee that the conceptual models can predict accurately when they are used to extrapolate beyond the range of calibration or verification experience [13,14]. It was also a bit difficult to understand the nature of these kind of models, so, in order to use such kind of models it was very important that, in order to get better results, one should have all of the knowledge about the models and its parameters [15]. However, DDM, which are basically numerical and based on biological neuron systems, recently known as an artificial brain or intelligence, have received more attention in water related applications because of their ease, fast progress time, and less data necessity. The ANN- or data-driven models have become increasingly popular in hydrologic forecasting because they are effective at dealing with the nonlinear characteristics of hydrological data [16]. Among the various machine learning methods, artificial neural networks (ANNs), which include back-propagation neural network (BPNN), radial basis function (RBF) neural network, generalized regression neural network (GRNN), Elman neural network, and multilayer feed-forward (MLFF) network, are among the most popular techniques for hydrological time series forecasting [17]. Although data driven models have attained high levels in the hydrological field, there is still space present to improve the forecasting methods [18]. Hydrological processes are non-linear and arbitrary. By simply applying such models on an original time series, the facts of alteration are overlooked, so that prediction correctness is reduced [19].

In the last decade, wavelet transform has become a widely applied technique for analyzing variations, periodicities, and trends in time series [20,21]. Wavelet transform, which can produce a good local representation of the signal, in both the time and frequency domains, provides considerable information on the structure of the physical process to be modelled. Discrete wavelet transformation provides a decomposition of original time series. Subseries decomposed by discrete wavelet transform, from original time series, provide detailed information about the data structure and its periodicity [22]. The attributes of each subseries are different. The wavelet components of the original time series improve on a forecasting model by giving useful information on various resolution levels [23]; however, not much research has applied a wavelet for drought forecasting. A major limitation of artificial neural networks (ANNs) is their inability to deal with nonstationary data. To overcome this limitation, researchers have increasingly begun to use a wavelet analysis to preprocess the inputs of the hydrologic data. Shabri [24] proposed a hybrid wavelet–least square support vector machine (WLSSVM) model that combines the wavelet method and the LSSVM model for monthly stream flow forecasting. Belayneh and Adamowski [25] studied drought forecasting using machine learning techniques and found that coupled wavelet neural network models were the most accurate for forecasting three month SPI (SPI 3) and six month SPI (SPI 6) values over lead times of one and three months in the Awash River Basin in Ethiopia. Therefore, in this study, coupling wavelets with ANN was expected to provide significant improvements in the model performance.

2. Materials and Methods

2.1. Standard Index of Annual Precipitation (SIAP)

The SIAP is known for transferring the raw data of precipitation to relative amounts, so that the deviation of rainfall from mean can be divided to standard deviation. Khalili [26] developed the SIAP and applied it to the study the processes of drought and wet conditions in Iran [27]. The values of the SIAP can be computed by Equation (1), provided by Khalili [6,26], as follows:

$$SIAP = \frac{P_i - \overline{P}}{PSD} \tag{1}$$

where SIAP is the drought index, P_i is the annual precipitation, P is the mean of precipitation in the period, and PSD is the standard deviation of the period. SIAP classifies drought intensity into five major categories, namely, extremely wet, wet, normal, drought, and extreme drought. Details on the SIAP classifications are given in Table 1 [9,27].

In this study, SIAP is applied for short-term/monthly drought analysis. The pattern of the raw rainfall data shows a normal distribution, which supports the concept behind using SIAP on a short-term/monthly scale. Hence, Equation (1) is rewritten as follows:

$$SIAP\,(M) = \frac{P_i - \overline{P}}{SD}$$

where SIAP (M) is the drought index on a monthly time scale, P_i is the monthly rainfall in the ith month (i = 1, 2, 3, 4, ... 360), P is the mean of the monthly rainfall data for the whole period of study, and SD is the standard deviation of the monthly rainfall for the duration of the study.

Table 1. Classification of Standard Index of Annual Precipitation (SIAP) values.

Classes of Drought Intensity	SIAP Values
Extremely wet	0.84 or more
Wet	0.52 to 0.84
Normal	−0.52 to 0.52
Drought	−0.52 to −0.84
Extreme drought	−0.84 or less

2.2. Standardized Water Storage Index (SWSI)

The SWSI is used to assess the deficit in the terrestrial water reserves. The SWSI calculation is based on Equation (2). It is calculated by the SDAT toolbox in MATLAB, which was developed by Farahmand and AghaKouchak [10].

$$SWSI_{i,j} = \frac{S_{i,j} - S_{j,mean}}{S_{j,sd}} \qquad (2)$$

where $S_{i,j}$ is the seasonal water level for year i and month j, $S_{j,mean}$ is the mean water level of the corresponding month for the duration of the study, and $S_{j,sd}$ is the standard deviation. The details of SWSI classification are given in Table 2.

Table 2. Classification of the Standardized Water Storage Index (SWSI).

SWSI Values	Classification
2.0 or more	Extremely wet
1.5 to 1.99	Very wet
1.0 to 1.49	Moderately wet
−0.99 to 0.99	Near normal
−1.49 to −1.00	Moderate drought
−1.99 to −1.5	Severe drought
−2 or less	Extreme drought

2.3. Development of Forecasting Model Using ANN

An artificial neural network can be defined as a set of simple processing units working as a parallel distributed processor [28]. These units, which are called neurons, are responsible for storing experimental knowledge for later disposal. The ANNs mimic the biological nervous system, similar to the brain; they learn through examples and have acquired knowledge stored in the connection weights between neurons [29]. The data are introduced in the input layer and the network progressively processes the data through the subsequent layers, producing a result in the output layer. The input neurons are linked to those in the intermediate layer through w_{ji} weights, and the neurons in the intermediate layer are linked to those in the output layer through w_{ki} weights. The symbols i, j, and k represent the ith, jth, and kth neuron in input, hidden, and output layers, respectively. The network maps out the relation between the input data and the output variables based on the nonlinear activation functions. The purpose of training a network is to minimize the error between outputs of the network and the target values. The training algorithm reduces the error by adjusting the weights and biases of the network. In training, the input values are multiplied by the respective connection weights and then the biases are added. The same process is repeated for the output layer, where the output of a hidden layer is used as an input the output layer. The combination of net weighted input and biases net_j to the jth neuron of the hidden layer can be expressed as [30] follows :

$$net_j = \sum_{i=1}^{l} (w_{ji}x_i + b_j) \qquad (3)$$

where x_i is the input value to the ith neuron of the input layer, while w_{ji} is the weight of the jth neuron of the hidden layer connected to the ith neuron of the input layer, and b_j is the bias of the jth hidden neuron. The net value, net_j, is passed through a transfer or activation function in the hidden layer to produce an output from the hidden neuron. The output from the hidden layer can be expressed as [30] follows:

$$y_j = f(net_j) = f_h\left(\sum_{i=1}^{p}(w_{ji}x_i + b_j)\right) \tag{4}$$

where y_j is the output from the jth hidden neuron. The output from the hidden layer, y_j, is used as an input to the output layer, and the same process as in hidden layer is repeated in the output neurons in order to produce an output from the output layer. The net weighted input to the output neuron can be represented by [30] the following:

$$net_k = \sum_{j=1}^{q} f_o(y_j)w_{ki} + b_k \tag{5}$$

Similar to above, the output from the kth neuron in the output layer is given by [30] the following:

$$y_k = f(net_k) = f_o\left(\sum_{j=1}^{q} w_{ki}f_h\left(\sum_{i=1}^{p}(w_{ji}x_i + b_j)\right) + b_k\right) \tag{6}$$

The ANN weights are made and modified iteratively through a procedure called calibration. The ANN models used in this study have a feed-forward multilayer perceptron (MLP) architecture that was trained with the Levenberg–Marquardt (LM) back-propagation algorithm. MLPs have often been used in hydrologic forecasting because of their simplicity. MLPs consist of an input layer, one or more hidden layers, and an output layer. The Levenberg–Marquardt (LM) algorithm is used for training because it is considered one of the fastest methods for training ANNs. The major drawback of feed-forward network models, as used in this study, is their inability to mimic the temporal pattern trend during the model training stage. Therefore, this type of model may not be capable of providing a reliable and accurate forecasting solution [31]. The efficiency of the models may be assessed using several statistical parameters, which describe the adhering degree among the data that are observed and predicted by the model [32,33]. A neuron computes and gives feedback based on the weighted sum of all of its inputs, according to an activation function based on its output [34]. The activation function selected here is the sigmoidal activation function. Standard neural network training procedures adjust the weights and biases in the network to minimize a measure of 'error' in the training cases, which is most commonly the sum of the squared differences between the network outputs and the targets. Finding the weights and biases that minimize the chosen error function is commonly done by using some gradient-based optimization method, with derivatives of the error, with respect to the weights and biases calculated by back-propagation. A detailed theory of the back-propagation algorithm is beyond the scope of this research and can be found in Haykin [28]. In this study, the Neural Networks Toolbox of MATLAB® is used. Figure 1 shows a simple neural network structure.

The performance of the presented models is evaluated based on their correlation coefficient (R) and root mean-square error (RMSE). The estimation of R is done using Equation (7), as follows:

$$R = \frac{\frac{1}{n}\sum_{t=1}^{n}(y_t^o - \overline{y}_t^o)(y_t^f - \overline{y}_t^f)}{\sqrt{\frac{1}{n}\sum_{t=1}^{n}(y_t^o - \overline{y}_t^o)^2}\sqrt{\frac{1}{n}\sum_{t=1}^{n}(y_t^f - \overline{y}_t^f)^2}} \tag{7}$$

where y_t^o and y_t^f are the observed and forecasted values at time t, respectively, and n is the number of data points.

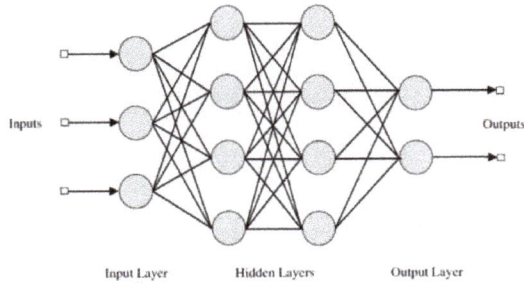

Figure 1. Neural network structure.

The correlation coefficient (R) measures how well the predicted values correlate with the observed values and shows the degree to which the two variables are linearly related. An R value close to unity indicates a satisfactory result, while a low value or one that is close to zero implies an inadequate result.

The RMSE provides information about the predictive capabilities of the model. The RMSE evaluates how close the predictions match the observations, shown in Equation (8), as follows:

$$\text{RMSE} = \sqrt{\frac{1}{n}\sum_{t=1}^{n}\ (y_t^o - y_t^f)^2} \tag{8}$$

The criteria for deciding the best models are based on how small the RMSEs found in training, testing, and validation of the data are.

2.4. Discrete Wavelet

Wavelet analysis is a multi-decomposition analysis that provides information on both the time and frequency domains of the signal, and is the important derivative of the Fourier transform. The wavelet will become an important tool in time series forecasting. The basic objective of wavelet transformation is analyzing the time series data, in both the time and frequency domains, by decomposing the original time series in different frequency bands using wavelet functions. Compared to the Fourier transform, time series are analyzed using sine and cosine functions. Wavelet transformations provide useful decompositions of the original time series by capturing useful information on various decomposition levels.

Nowadays, wavelet analysis is one of the most powerful tools in the study of time series. Wavelet transform can be divided into two categories, continuous wavelet transform (CWT) and discrete wavelet transform (DWT). CWT is not often used for forecasting because of its computational complexity and time requirements [35]. Among the reviewed papers by Nourani et al., [36], only about 20% of the studies used the CWT for decomposing the hydrological time series, and the majority of studies utilized the DWT. This is because real world observed hydrologic time series are measured and gathered in discrete form, rather that in a continuous format [36]. DWT is often used in forecasting applications to simplify numeric solutions. DWT requires less computation time and is simpler to apply. DWT is given by the following:

$$\psi_{m,n}(t) = \frac{1}{\sqrt{s_0^m}}\ \psi\left\{\frac{t - n\tau_0 s_0^m}{s_0^m}\right\} \tag{9}$$

where $\psi\ (t)$ is the mother wavelet, and m and n are integers that control the scale and time, respectively. The most common choices for the parameters are $S_o = 2$ and $\tau_o = 1$. According to Mallat's theory, the original discrete time series can be decomposed into a series of linearity-independent approximation and detail signals, by using the inverse DWT. The inverse DWT is given by Mallat [37], as follows:

$$x(t) = T + \sum_{m=1}^{M} \sum_{t=0}^{2^{M-m}-1} W_{m,n} \, 2^{-\frac{m}{2}} \psi\left(2^{-m} t - n\right). \tag{10}$$

where $W_{m,n} = 2^{-\frac{m}{2}} \sum_{t=0}^{N-1} . \psi(2^{-m} t - n) x(t)$ is the wavelet coefficient for the discrete wavelet at scale $s = 2^m$ and $\tau = 2^m n$.

2.5. W-ANN Model

The W-ANN model is obtained by combining the DWT and ANN models. The W-ANN model uses the subseries obtained from using DWT on original data. The W-ANN model structure developed in this study can be described with the following steps:

1. Decompose the original time series for each input into subseries components (details and approximations) by DWT.
2. Select the most important and effective of each subseries component for each input by the correlation coefficient.
3. Construct a W-ANN model using the new summed series obtained by adding the significant components of details sub-time series and approximations sub-time series for each input as the new input to the ANN, and the original output time series as the output of the ANN. Figure 2 shows a schematic representation of the model.

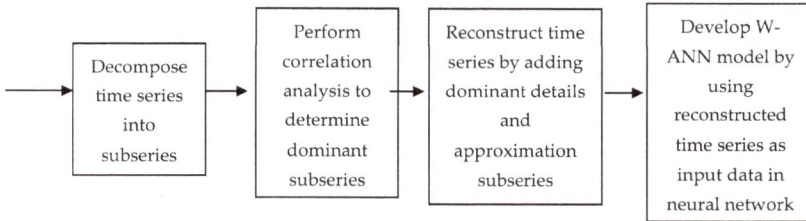

Figure 2. Schematic diagram of wavelet-based artificial neural network (W-ANN) model development.

2.6. Study Area and Data Collection

2.6.1. Langat River Basin

The Langat River is situated in the state of Selangor, Malaysia. This river basin is located near Kuala Lumpur, the capital city of Malaysia. Therefore, the study area has been rapidly developed, which makes it dependent on the Langat River for water supply [38]. The Langat River has an estimated total catchment area of 1817 km^2 and is located at latitude 2°40′152″ N to 3°16′15″ N and longitude 101°19′20″ E to 102°1′10″ E [30], and the main river is 141 km in length. The Beranang River, Semenyih River, and Lui River are the main tributaries of the Langat River, as shown in Figure 3. There are two reservoirs in the Langat River Basin, Hulu Langat and Semenyih. The northeastern part of the river basin has a reduced level (RL) of 960 m above the mean sea level and is mountainous. The temperature of the area varies from 23.5 °C to 33.5 °C all year, and the comparative humidity ranges from 63% to 95%, with an average of 81%. Heavier rainfall happens in the month of November, with a monthly average rainfall of 270 mm; the average annual rainfall of the study area is about 2400 mm [38]. The area also sometimes experiences rainstorms, and these usually occur in the early evening through the year, and are usually of a short duration with a high intensity.

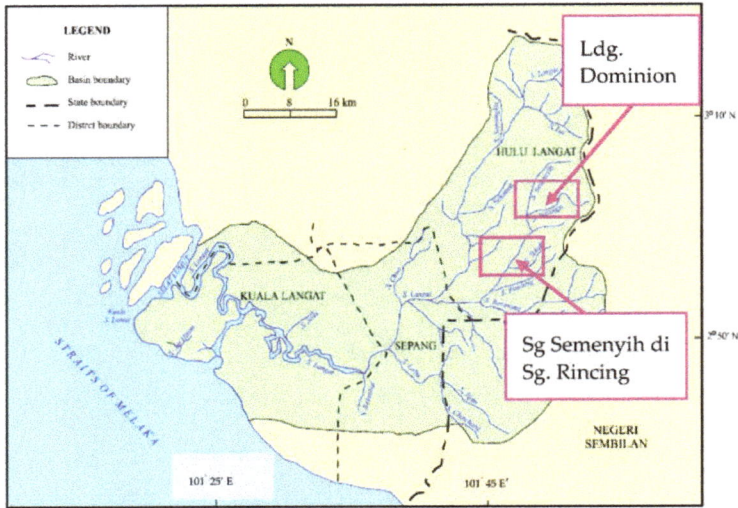

Figure 3. Langat River Basin.

2.6.2. Data Collection

Thirty years (1986–2016) of rainfall and water level data of stations were collected from the Department of Irrigation and Drainage (DID), Malaysia. Table 3 gives details of the gauging stations, including the station name, station number, coordinates (latitude and longitude), data availability, and percentage of missing data.

Table 3. Details on rainfall and water level gauging stations.

Station	Station Name	Station No.	Coordinates		Data Availability (Years)	Missing Data (%)
			Latitude (N)	Longitude (E)		
1	Sg. Semenyih di Sg. Rincing	WL 2918401	02°54′55″	101°49′25″	1986–2016	5.4%
2	Ldg. Dominion	RF 3018107	03°00′13″	101°52′55″		6.5%

For the simplicity of naming the stations, water level (WL) station Sg. Semenyih di Sg. Rincing (WL 2918401) will be referred as station 1, and rainfall station (RF) Ldg. Dominion (RF 3018107) as station 2. The missing rainfall data of station 2 is estimated using the normal ratio method from the observations of rainfall at some of the other stations, as close to and as evenly spaced around the station with the missing record as possible [39].

2.6.3. Distribution of Rainfall and Water Level

The mean and median values were estimated for 30 years of raw data, from October 1986 to September 2016, and are presented in Figure 4. The most likely time for drought to happen is when the rainfall is low. It can be seen that, for the distribution of the rainfall data for station 2, the highest mean and median were in November, at 369.4 mm and 317.0 mm, respectively. The lowest rainfall was in January, with a mean of 138.4 mm and median of 126.3 mm, followed by June, July, and February. There are basically three different seasons in the Langat River Basin of Malaysia. The wet period of the year is from October through to the beginning of January, and the dry months are generally observed from January to March, and June to September. October and November are the wettest months, with an average rainfall of 321.5 mm and 369.4 mm, respectively.

Figure 5 presents the water level data of station 1, and it can be seen that the water level steadily decreased for the second half of the duration of this study.

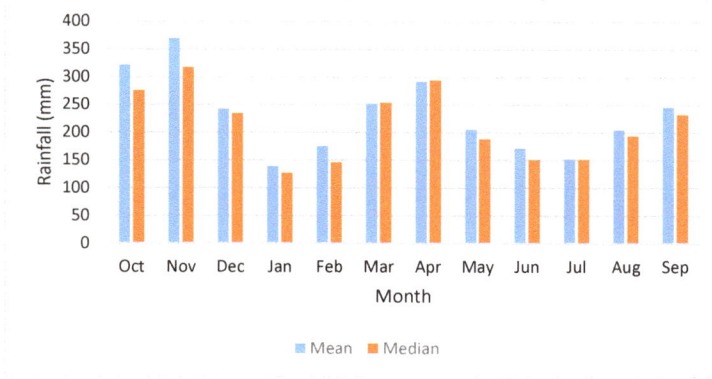

Figure 4. Monthly rainfall distribution at station 2, estimated using data from 1986 to 2016.

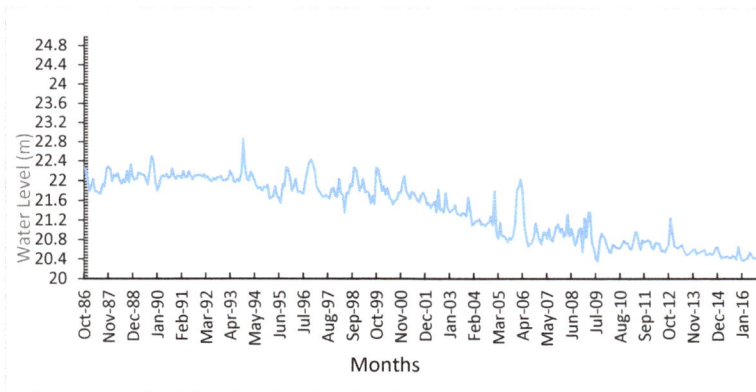

Figure 5. Water level data for 30 years at station 1 (1986–2016).

3. Results and Discussion

3.1. Assessment Using Standard Index of Annual Precipitation (SIAP)

As illustrated in Figure 6, the highest SIAP value was 4.921 (October 1994) and the lowest value was −1.591 (August 1990). In 1988, the drought period was 11 months; followed by 1990, with a drought period of 10 months; and then 2015, with a nine month dry period. Figure 6 also shows that in 1988, there was a 10 month dry period from March to December.

Figure 7 shows a categorization of the results for the five different classes of drought. It shows that 17% of the months were extremely wet, 7% were wet, 39% were normal, 17% had drought, and 20% had very severe drought. Overall, drought happened during 37% of the total months, and wet periods occurred during 24% of the total months. Table 4 shows a summary of the drought classifications.

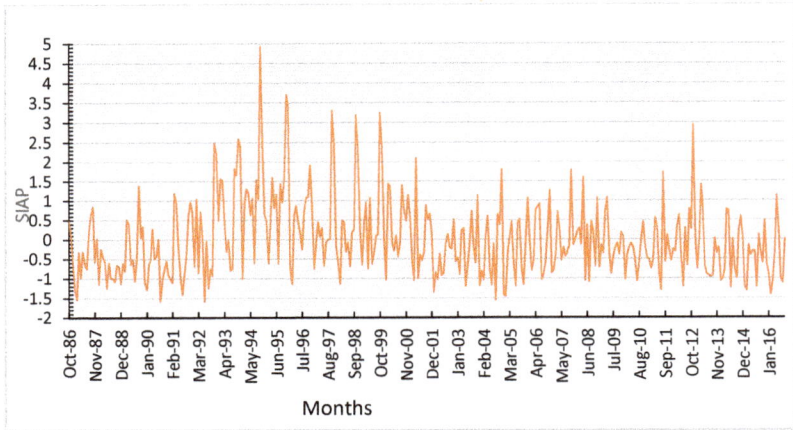

Figure 6. Standard Index of Annual Precipitation (SIAP) values for 30 years at station 2.

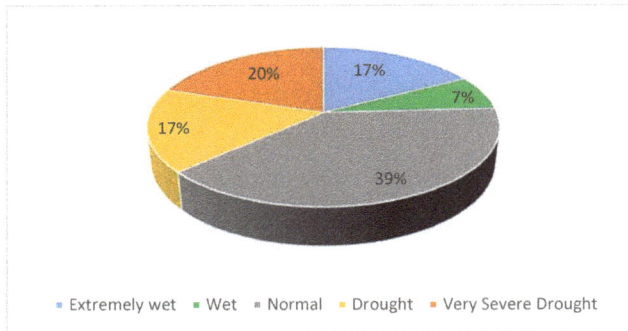

Figure 7. Distribution of SIAP values into classes (station 2).

Table 4. Summary of drought classifications for station 2.

Category	Number of Months	Percentage (%)
Extremely wet	62	17
Wet	25	7
Normal	140	39
Drought	61	17
Very severe drought	72	20
Total	360	100

Artificial Neural Network (ANN) Model

The ANN architecture does not have a systematic way to establish suitable architecture. Networks that are too small and simple can lead to underfitting, while networks that are too complex tend to overfit the training pattern [40]. Usually, nonlinear sigmoidal activation functions are used, as reported in the literature, which were also adopted in this study. The inputs to the ANN model were normalized and kept within the range of 0.1 to 0.9. Normalization or scaling is not really a functional requirement for the NNs to learn, but it significantly helps as it transposes the input variables into the data range that the sigmoid activation functions lie in (i.e., 0.1). The learning rate and momentum coefficient are influential parameters that control the convergence rate, but optimize them for the best output. Here, the two parameters were kept constant at 0.4 and 0.6, respectively, throughout the network structure for various numbers of hidden neurons. The network input models that were tested for the forecasting were based on the SIAP at station 2, shown by Equations (11) and (12). The input combinations consisted of lagged data of the rainfall and drought index, and the output was kept as a single drought index variable.

$$SIAP\ (t) = f\ (R_{t-1}, R_{t-2}, R_{t-3}) \quad \text{Input model number 1} \tag{11}$$

$$SIAP\ (t) = f\ (SI_{t-1}, SI_{t-2}, SI_{t-3}) \quad \text{Input model number 2} \tag{12}$$

where SIAP or SI is the drought index; R is the precipitation; n is the time lag, which is effectively the lead time of the forecast; and t is time in months. The input models based on the main parameter, rainfall, in calculating the index or a drought index itself as input, performed better in the forecasting using ANN [41]. The same study also illustrated a lack of impact of the secondary parameters on the performance of the networks. In the ANN model stated above, there are three classifications of samples, training, which was kept at 70% (252 samples); validation, 15% (54 samples); and testing, 15% (54 samples). In the majority of the cases, data division is carried out on an arbitrary basis. However, the way the data divided can have a significant effect on the model performance. Shahin et al. [42] investigated the issue of data division and its impact on the ANN model performance for a case study of predicting the settlement of shallow foundations on granular soils. The results indicated that the statistical properties of the data in the training, testing, and validation sets need to be taken into account in order to ensure that the optimal model performance is achieved [42]. During training, it adjusts the network according to its final measured error. The validation process was used at the end of training as an extra check on the performance of the model. If the performance of the network was found to be consistently good on both the test and the validation samples, then it was reasonable to assume that the network would generalize well on unseen data. For testing, this does not affect the training part, but it provides an independent measure of the network performance during and after training. Each MLP was trained with 5 to 15 hidden neurons in a single hidden layer, as shown in Table 5, to select the most effective model by analyzing the performance. The three best-performing combinations are shown for each input model.

Table 5. Correlation coefficient (R) of artificial neural network (ANN) network structure (SIAP).

Input Model Number	Number of Neurons	R			
		Training	Validation	Testing	Overall
1	10	0.907	0.865	0.908	0.899
1	15	0.803	0.845	0.758	0.800
1	12	0.796	0.765	0.801	0.783
2	8	0.712	0.813	0.705	0.741
2	9	0.737	0.799	0.782	0.770
2	10	0.882	0.875	0.851	0.868

For the comparison between the output and target, it was found that for input model number 1, for training, validation, testing, and overall, the R values were 0.907, 0.865, 0.909, and 0.899, respectively. An R value of 1 means a close relationship, 0 means no relationship. So, this means that the relationship between the two (output and target) are close and related, which is shown in Figure 8. The errors in the training, validation, and testing stages are illustrated in Figure 9.

Figure 10 displays a section of time series from January 1987 to December 1989 of the SIAP observed values against the forecasted ones, by using input model number 1. The results effectively exemplify the high accuracy of the short-range forecasts of the droughts at station 2. Such studies may be a way to identify the operational accuracy of forecasts, and have been used by others for similar purposes [43].

The forecasted and actual index values were similar, so the model can be said to be reliable. Therefore, this ANN model can be used to predict short- to medium-term drought occurrences in Malaysia. In addition, SIAP is an effective index for the assessment of drought monitoring and the characteristics of drought conditions in the Langat River Basin. Authorities can render early warnings for the timely implementation of preparedness based on predictions.

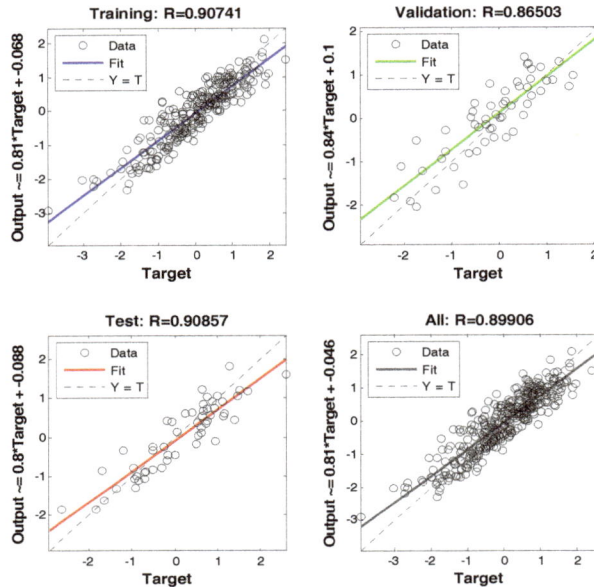

Figure 8. Neural network training regression for input model 1.

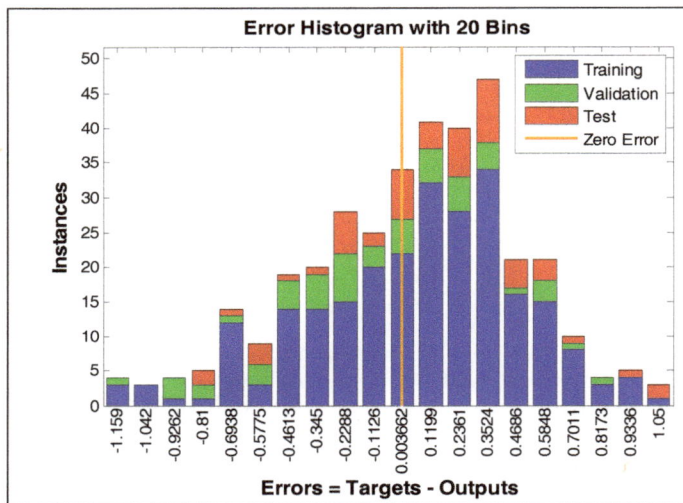

Figure 9. Error histogram of input model number 1.

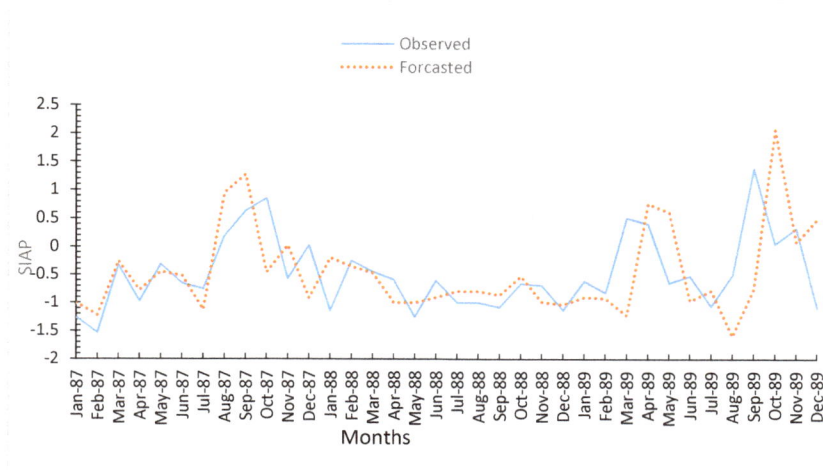

Figure 10. Comparison of observed and forecasted SIAP at station 2 of input model number 1.

3.2. Assessment Using SWSI for Hydrological Drought

Figure 11 shows the time series created using SWSI for the 30 years of data at station 1. The data used for the analysis was the water level of the river. Initially, the values are seen to be classified as very wet or above, then, they slowly change to near normal. The trend of SWSI in Figure 11 is similar to the raw input water level data, seen in Figure 5, which shows that almost the first half of the whole period was very wet or above normal, and the second half was below normal or had droughts. Climate change, rapid urbanization, environmental degradation, and industrial development may have resulted in water and related resources within the basin becoming increasingly stressed. A study conducted in Malaysia highlighted that extreme dry conditions are becoming more frequent than extreme wet conditions [5]. With reference to Figure 5, the time series starts with one month of moderately wet conditions, which follows 12 months of near-normal conditions. From November 1987 to May 1994 (months 14 to 92), the conditions are classified as very wet, extremely wet, or moderately wet. After this wet period, the values are observed to decrease gradually, from very wet conditions to near normal conditions. From June 1994 to November 2008 (months 93 to 266), the conditions were near normal. However, there were few months that were moderately wet, very wet, or extremely wet. The first drought occurred in December 2008 (month 267), with an index value of −1.39. Drought started to occur more frequently from this point onward. Near-normal conditions are observed from January 2009 (month 268) to February 2013 (month 317). However, within this period, the months with drought increased. From March 2013 (month 318) to September 2016 (month 360), all of the months experienced drought. The most frequent type of drought was moderate drought, followed by extreme drought. The number of occurrences of severe drought is less than that of moderate and extreme drought.

Table 6 shows the number of months that each drought occurred, with percentages varying from 2.50% to 66.67%. The most observed condition within the period of study was near normal. Near normal conditions occurred for 240 months, about 66.67%. Except for the near normal, all of the other conditions were below 11%. Moderate drought occurred for 37 months (10.28%). Moderately wet conditions occurred for a similar number of months (35 months; 9.72%). Very wet conditions were observed in 16 months (4.44%), followed by extremely wet conditions in 12 months (3.33%). Extreme drought occurred in 11 months (3.06%). The least frequent condition was severe drought, which occurred in 9 months (2.5%).

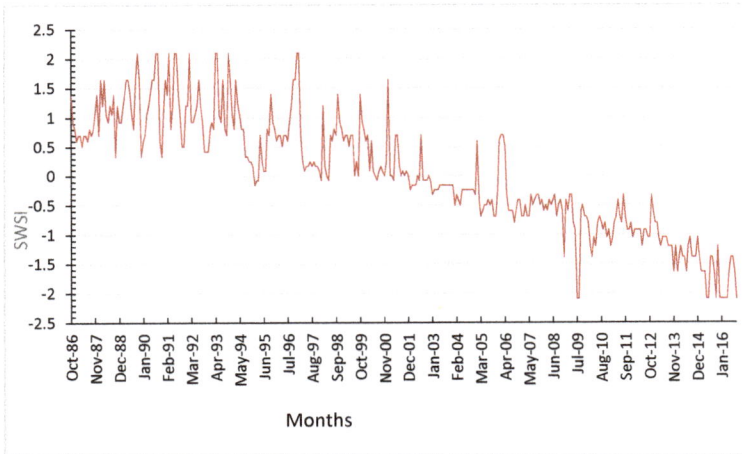

Figure 11. Standardized Water Storage Index (SWSI) values for station 1 for 30 years (360 months).

Table 6. SWSI drought percentage from station 1.

Drought Classification	Condition	Number of Months	Percentage (%)
Extremely wet	>2	12	3.33
Very wet	1.5 to −2	16	4.44
Moderately wet	1.0 to 1.5	35	9.72
Near normal	−1.0 to 1.0	240	66.67
Moderate drought	−1.5 to −1.0	37	10.28
Severe drought	−2.0 to −1.5	9	2.50
Extreme drought	<−2	11	3.06

Artificial Neural Network Model for Hydrological Drought

The network input models that were tested are based on SWSI at station 1, shown by Equations (13) and (14), as follows:

$$\text{SWSI (t)} = f\,(W_{t-1}, W_{t-2}, W_{t-3}) \qquad \text{Input model number 3} \qquad (13)$$

$$\text{SWSI (t)} = f\,(SW_{t-1}, SW_{t-2}, SW_{t-3}) \quad \text{Input model number 4} \qquad (14)$$

where SWSI or SW is the drought index, W is the water level, and *n* is the time lag, which is effectively the lead time of the forecasted SWSI model developed for station 1. Similar to the SIAP ANN model, in this case as well, each MLP was trained with 5 to 15 hidden neurons in a single hidden layer, as shown in Table 7, in order to select the most effective model by analyzing performance. The three best-performing combinations are shown for each input model.

Table 7. Correlation coefficient (R) of ANN network structure (SWSI).

Input Model Number	Number of Neurons	R			
		Training	Validation	Testing	Overall
3	10	0.968	0.967	0.969	0.968
3	11	0.898	0.908	0.951	0.918
3	7	0.767	0.801	0.822	0.796
4	15	0.901	0.855	0.835	0.865
4	10	0.911	0.899	0.853	0.888
4	6	0.751	0.772	0.811	0.779

The output, which is the forecasted results, is plotted together with the observed results in Figure 12, using the SWSI input model (number 3). The dotted line shows the forecasted values and the solid line shows the observed results, which were calculated by SWSI. In general, the two plots are not very different. The forecasted values have only minor differences.

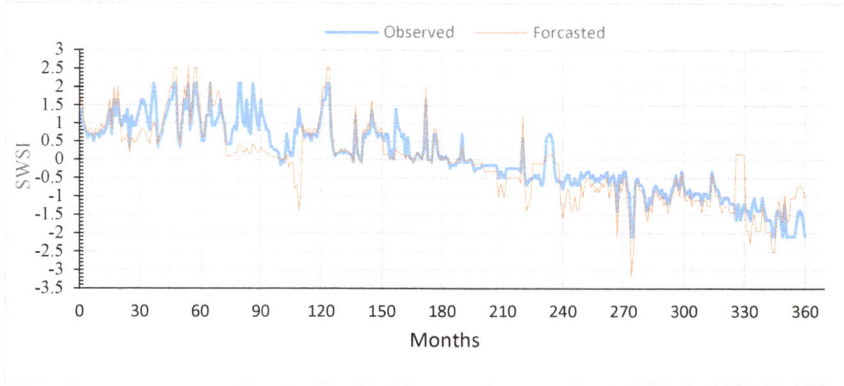

Figure 12. SWSI observed and forecasted values (360 months) of station 1 for input model number 3.

An error histogram for input model number 3 of SWSI was also plotted, and is shown in Figure 13. The error histogram assists in authenticating the performance of the network. The blue part represents the training data and the green part represents the validation data. The biggest portion of data is surrounding the zero line. The zero line offers a way to confirm the outliers to determine if the data contains errors. It can also confirm that those data features are not like the leftovers of the dataset [44].

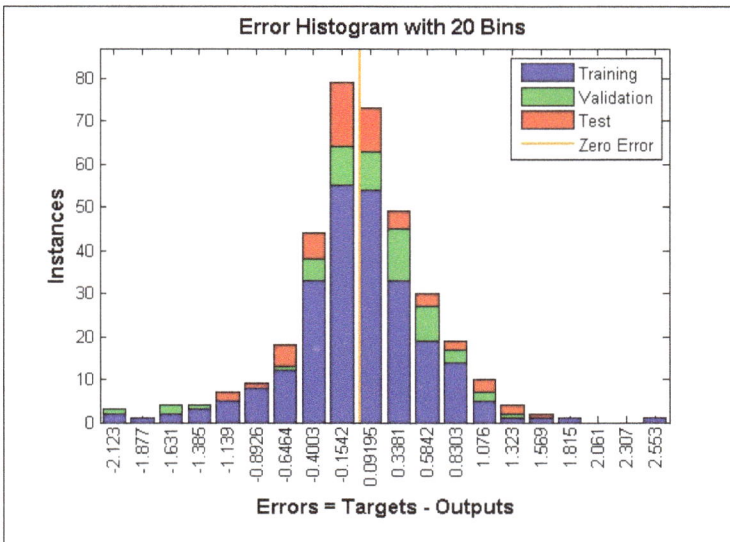

Figure 13. Error histogram for SWSI input model number 3.

The R value shown in Figure 14 concludes the connection between the output or target values of the artificial neural network models. The R value is also known as the correlation coefficient. Strong and random connections were identified when the R value was 1 and 0, respectively [45]. The line must be at a 45° angle toward 1 to be a perfect fit. The 45° line means that the output value is equal to the input target values. For SWSI, the R values for training, validation, testing, and overall, are the same at 0.96. This indicates a strong correlation in the prediction of drought, based on the observed values and developed model [44].

A time series of the calculated indices plotted shows that drought is not increasing gradually, but occurs irregularly. The water level decreased and drought increased gradually every year. SWSI considers values between +1 and −1 as near normal, whereas other hydrological drought indices (e.g., SDI) consider all of the values below 0 as drought.

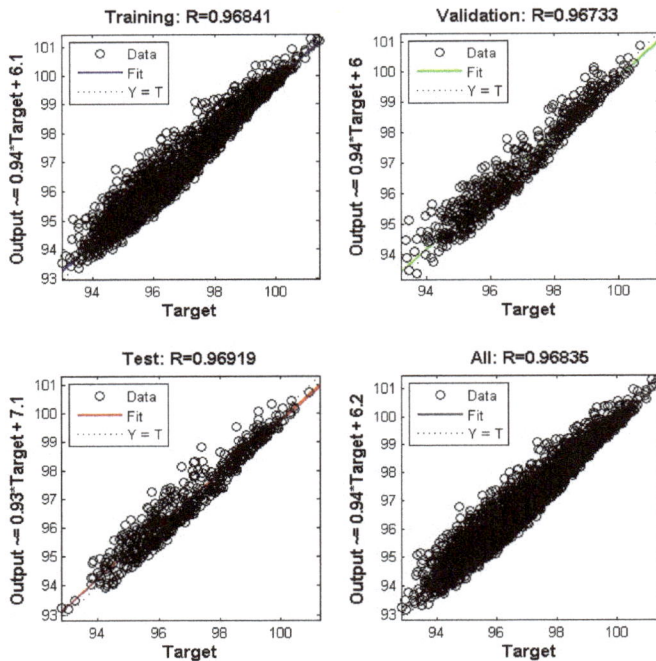

Figure 14. Correlation coefficient for SWSI at station 1 for input model number 3.

3.3. W-ANN Model

As seen in Table 8, there is a correlation between the DWT wavelet components D1, D2, D3, D4, D5, D6, D7, and D8 of the SIAP, SWSI, rainfall, and water level series, with the original series. It can be observed, in the case of SIAP, SWSI, and rainfall, that D1, D2, D3, and D8 show significantly higher correlations than the average of correlations among them compared to the D4, D5, and D6 components. However, for the water level, only D7 and D8 subseries show higher than average correlations. According to the correlation analysis, the effective components were selected as the dominant wavelet components, as stated above. Afterwards, the significant wavelet components and the approximation (A8) component were added to constitute the new series.

Table 8. The correlation coefficient between each sub-time series and original drought indices/raw input series.

Discrete Wavelet Components (db3)	Correlation between Detailed Sub-Time Series and Observed Drought Index/Rainfall/Water Level Data				
	SIAP	SWSI	Rainfall	Water Level *	Dominant
D1	0.5291	0.1996	0.5291	0.1335	√
D2	0.5732	0.2221	0.5732	0.1522	√
D3	0.3645	0.1900	0.3645	0.1409	√
D4	0.2130	0.1322	0.2130	0.1120	×
D5	0.1616	0.1519	0.1616	0.1333	×
D6	0.2576	0.1885	0.2576	0.0694	×
D7	0.2015	0.0820	0.2015	0.3151 *	
D8	0.3201	0.3923	0.3201	0.4303 *	√
Average	0.3274	0.1948	0.3274	0.1858	

* Only D7 and D8 subseries show higher than average correlations. √ indicates that the subseries is dominant; × indicates that the subseries is not dominant.

Secondly, the W-ANN models were developed for monthly drought prediction, using wavelet subseries. The most important part of this wavelet-based model is the selection of inputs for its formation. The summed wavelet components (the new series) instead of the original data were employed as inputs of the W-ANN model for drought prediction. Four different models based on combinations of different input data (SIAP, SWSI, rainfall, and water level) were evaluated. The forecasting performance of the wavelet–neural network models are presented in Table 9, in terms of RMSE and R. The table shows that the W-ANN model has a significant positive effect on the monthly drought forecast. As seen from the table, model number 4, with three months of previous SWSI data, has the lowest RMSE and the highest correlation coefficients among all of the wavelet–neural network models. For meteorological drought prediction, while the highest correlation coefficient (R) obtained by the ANN model is 0.899, with the wavelet-ANN model, this value increased to 0.940. Similarly, for the case of hydrological drought, while the R obtained by the ANN model is 0.968, with the wavelet-ANN model, this value increased to 0.973. The application of wavelet in the ANN model achieved higher correlation coefficients for all of the models, except for input model number 3. In both types of drought forecasting, it was found that the models based on preceding drought index values as inputs performed better than the models developed with raw data, such as rainfall or water level as inputs. This proves that the created models can improve hydrologic and meteorological drought prediction close to the observed values.

Table 9. Root mean-square error (RMSE) and R statistics of different W-ANN models.

Input Model (After Wavelet Decomposition)	RMSE (Validation)	R (Overall)	Hidden Neurons
1 (R_{t-1}, R_{t-2}, R_{t-3})	0.38	0.932	8
1 (R_{t-1}, R_{t-2}, R_{t-3})	0.41	0.931	10
1 (R_{t-1}, R_{t-2}, R_{t-3})	0.38	0.901	15
2 (SI_{t-1}, SI_{t-2}, SI_{t-3})	0.40	0.922	8
2 (SI_{t-1}, SI_{t-2}, SI_{t-3})	0.42	0.931	10
2 (SI_{t-1}, SI_{t-2}, SI_{t-3})	0.39	0.940	15
3 (W_{t-1}, W_{t-2}, W_{t-3})	0.40	0.902	8
3 (W_{t-1}, W_{t-2}, W_{t-3})	0.43	0.901	10
3 (W_{t-1}, W_{t-2}, W_{t-3})	0.38	0.910	15
4 (SW_{t-1}, SW_{t-2}, SW_{t-3})	0.19	0.971	10
4 (SW_{t-1}, SW_{t-2}, SW_{t-3})	0.17	0.972	13
4 (SW_{t-1}, SW_{t-2}, SW_{t-3})	0.21	0.973	15

Table 10 shows the performance improvement of the W-ANN models, and it can be seen that the models for meteorological drought forecasting improved by 3.67% and 8.29%; however, for the hydrological drought forecasting models, there was a decrease of R value by 5.99% for input model number 3. Input model number 4 performed better than the other models that were considered in this study, with a performance improvement of 9.57%.

Table 10. Performance improvement of R statistics of different W-ANN models.

Input Model	R (With Wavelet Decomposition)	R (Without Wavelet Decomposition)	Performance Improvement (%)
1 (R_{t-1}, R_{t-2}, R_{t-3})	0.932	0.899	+3.67
2 (SI_{t-1}, SI_{t-2}, SI_{t-3})	0.940	0.868	+8.29
3 (W_{t-1}, W_{t-2}, W_{t-3})	0.910	0.968	−5.99
4 (SW_{t-1}, SW_{t-2}, SW_{t-3})	0.973	0.888	+9.57

Figure 15 shows a scatter plot using model number 4 (SWSI), and it shows that the W-ANN forecasts approximate the general behavior of the observed data more satisfactorily for the drought months.

Figure 16 shows a scatter plot using model number 3 (SIAP), and it shows that the W-ANN forecasts do not linearly approximate the general behavior of the observed data, but the correlation coefficient is 0.940.

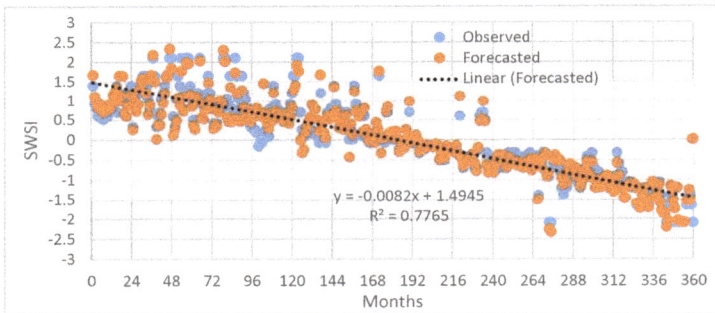

Figure 15. Scatter plot comparing observed and forecasted hydrological drought using W-ANN models.

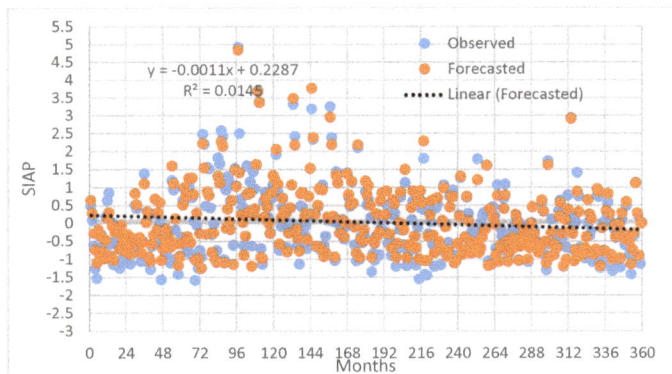

Figure 16. Scatter plot comparing observed and forecasted meteorological drought using W-ANN models.

4. Conclusions

Drought occurrences in the Langat River catchment of peninsular Malaysia were characterized using meteorological and hydrological drought indices, SIAP and SWSI, respectively. Overall, SWSI and SIAP were found to be effective indices for the assessment of drought. The occurrence of hydrological and meteorological droughts was found to be around 16% and 37% by SWSI and SIAP, respectively. Two neural network-based models and two wavelet-based ANN models were developed using the values of SIAP and SWSI. For SWSI and SIAP, correlation coefficients of 0.96 and 0.90, respectively, were calculated. Therefore, it is concluded that both of the models are found to be reliable. However, with the W-ANN model, these values increased to 0.940 and 0.973 for meteorological and hydrological drought forecasting, respectively. This proves that the proposed models are able to predict hydrologic and meteorological drought very close to the observed values. This study can help in the drought assessment and the prediction of drought occurrence in the study area. Authorities can issue an early warning for the timely implementation of preparedness, based on predictions.

Author Contributions: M.M.H.K. designed the research, carried out the analysis, and wrote the article. N.S.M. and A.E.-S. helped to conceive the research, review the overall results, and initiate the article. All of the authors contributed substantially to the work reported.

Funding: Fundamental Research Grant Scheme (Grant No. FRGS/2/2014/TK02/UKM/03/2), provided by the Ministry of Education Malaysia

Acknowledgments: The authors would like to acknowledge support from the Ministry of Education (MOE) and the Universiti Kebangsaan Malaysia (UKM) for research grant FRGS/2/2014/TK02/UKM/03/2. They also acknowledge contributions from Adil Rassam Timimi and Ahmed Ali Jabir at the University of Nottingham, Malaysia, for help with the analysis. Appreciation is due to the Department of Irrigation and Drainage (DID) for the data supplied to conduct this research study.

Conflicts of Interest: The authors declare no conflict of interest.

References

1. Basics, D.; Drought, T. Types of Drought. Available online: http://drought.unl.edu/DroughtBasics/TypesofDrought.aspx (accessed on 14 October 2017).
2. Deni, S.; Jemain, A.; Ibrahim, K. The spatial distribution of wet and dry spells over Peninsular Malaysia. *Theor. Appl. Climatol.* **2008**, *94*, 163–173. [CrossRef]
3. Jamalluddin, S.A.; Low, K.S. Droughts in Malaysia: A Look at Its Characteristics, Impacts, Related Policies and Management Strategies. In Proceedings of the Water and Drainage 2003 Conference, Kuala Lumpur, Malaysia, 28–29 April 2003.
4. Staff, S.; Staff, S. Drought beyond India: Malaysia Faces a Massive Water Crisis as South East Asia Swelters. Available online: http://scroll.in/article/806812/drought-beyond-india-malaysia-faces-a-massive-water-crisis-as-south-east-asia-swelters (accessed on 25 November 2017).
5. Wan, W.Z.; Ibrahim, K.; Jemain, A.A. Evaluating dry conditions in Peninsular Malaysia using bivariate copula. *Anziam J.* **2010**, *51*, 555–569.
6. Morid, S.; Smakhtin, V.; Moghaddasi, M. Comparison of seven meteorological indices for drought monitoring in Iran. *Int. J. Climatol.* **2006**, *26*, 971–985. [CrossRef]
7. Karavitis, C.A.; Alexandris, S.; Tsesmelis, D.E.; Athanasopoulos, G. Application of the Standardized Precipitation Index (SPI) in Greece. *Water* **2011**, *3*, 787–805. [CrossRef]
8. Gourabi, B.R. The Recognition of Drought with Dri and Siap Method and its Effects on Rice Yield and Water Surface in Shaft, Gilan, south Western of Caspian Sea. *Aust. J. Basic Appl. Sci.* **2010**, *4*, 4374–4378.
9. Arvind Singh, T.; Bhawana, N.; Lokendra, S. Drought spells identification with indices for Almora district of Uttarakhand, India. *Am. Int. J. Res. Sci. Technol. Eng. Math.* **2015**, *12*, 1–5.
10. Farahmand, A.; AghaKouchak, A. A generalized framework for deriving nonparametric standardized drought indicators. *Adv. Water Resour.* **2015**, *76*, 140–145. [CrossRef]
11. Wang, W.; Wang, P.; Cui, W. A comparison of terrestrial water storage data and multiple hydrological data in the Yangtze River basin. *Adv. Water Sci.* **2015**, *26*, 759–768.

12. Nourani, V.; Komasi, M.; Mano, A. A Multivariate ANN-Wavelet Approach for Rainfall–Runoff Modeling. *Water Resour. Manag.* **2009**, *23*, 2877. [CrossRef]

13. Panagoulia, D. Hydrological modeling of a medium-sized mountainous catchment from incomplete meteoro-logical data. *J. Hydrol.* **1992**, *137*, 279–310. [CrossRef]

14. Panagoulia, D. Artificial neural networks and high and low flows in various climate regimes. *Hydrol. Sci. J.* **2006**, *51*, 563–587. [CrossRef]

15. Govindaraju, R. Artificial neural networks in hydrology. II: hydrologic applications. *J. Hydrol. Eng.* **2000**, *5*, 124–137.

16. Peng, T.; Zhou, J.; Zhang, C.; Fu, W. Streamflow Forecasting Using Empirical Wavelet Transform and Artificial Neural Networks. *Water* **2017**, *9*, 406. [CrossRef]

17. Zhou, J.; Peng, T.; Zhang, C.; Sun, N. Data Pre-Analysis and Ensemble of Various Artificial Neural Networks for Monthly Streamflow Forecasting. *Water* **2018**, *10*, 628. [CrossRef]

18. Remesan, R.; Shamim, M.A.; Han, D.; Mathew, J. Runoff prediction using an integrated hybrid modelling scheme. *J. Hydrol.* **2009**, *372*, 48–60. [CrossRef]

19. Tayyab, M.; Zhoua, J.; Adnana, R.; Zenga, X.; Zenga, X. Application of Artificial Intelligence Method Coupled with Discrete Wavelet Transform Method. *Procedia. Comput. Sci.* **2017**, *107*, 212–217. [CrossRef]

20. Zhou, T.; Wang, F.; Yang, Z. Comparative Analysis of ANN and SVM Models Combined with Wavelet Preprocess for Groundwater Depth Prediction. *Water* **2017**, *9*, 781. [CrossRef]

21. Seo, Y.; Choi, Y.; Choi, J. River Stage Modeling by Combining Maximal Overlap Discrete Wavelet Transform, Support Vector Machines and Genetic Algorithm. *Water* **2017**, *9*, 525.

22. Wang, D.; Ding, J. Wavelet network model and its application to the prediction of hydrology. *Nat. Sci.* **2003**, *1*, 67–71.

23. Kim, T.W.; Valdes, J.B. Nonlinear model for drought forecasting based on a conjunction of wavelet transforms and neural networks. *J. Hydrol. Eng.* **2003**, *6*, 319–328. [CrossRef]

24. Shabri, A. A hybrid model for stream flow forecasting using wavelet and least Squares support vector machines. *Jurnal Teknologi* **2015**, *73*, 89–96. [CrossRef]

25. Belayneh, A.; Adamowski, J. Drought forecasting using new machine learning methods. *J. Water Land Dev.* **2013**, *18*, 3–12. [CrossRef]

26. Khalili, A.; Bazrafshan, J. Assessing the efficiency of several drought indices in different climatic regions of Iran. *Nivar J.* **2003**, *48*, 79–93.

27. Najjar, S.; Rouhollah, R.Y. Studying & Comparing the Efficiency of 7 Meteorological Drought Indices in Droughts Risk Management (Case Study: North West Regions). *Appl. Math. Eng. Manag. Technol.* **2015**, *3*, 131–142.

28. Haykin, S. *Neural Networks: A Comprehensive Foundation*, 2nd ed.; Prentice Hall: Upper Saddle River, NJ, USA, 1999.

29. Demuth, H.; Beale, M. *Neural Network Toolbox: For Use with Matlab*; The MathWorks, Inc.: Natick, MA, USA, 2005.

30. Bishop, C.M. *Neural Networks for Pattern Recognition*; Oxford University Press: Oxford, UK, 1995; p. 482.

31. El-Shafie, A.; Noureldin, A.; Taha, M.; Hussain, A. Dynamic versus static neural network model for rainfall forecasting at Klang River Basin, Malaysia. *Hydrol. Earth Syst. Sci. Discuss.* **2011**, *8*, 6489–6532. [CrossRef]

32. Dawson, C.W.; Abrahart, R.J.; See, L.M. HydroTest: A web-based toolbox of statistical measures for the standardised assessment of hydrological forecasts. *Environ. Modell. Softw.* **2007**, *27*, 1034–1052. [CrossRef]

33. Napolitano, G.; Serinaldi, F.; See, L. Impact of EMD decomposition and random initialisation of weights in ANN hindcasting of daily stream flow series: An empirical examination. *J. Hydrol.* **2011**, *406*, 199–214. [CrossRef]

34. Barua, S.; Ng, A.; Perera, B. Artificial Neural Network—Based Drought Forecasting Using a Nonlinear Aggregated Drought Index. *J. Hydrol. Eng.* **2012**, *17*, 1408–1413. [CrossRef]

35. Kisi, O. Wavelet regression model as an altrnative to neural networks for river stage forecasting. *Water Resour. Manag.* **2011**, *25*, 579–600. [CrossRef]

36. Nourani, V.; Baghanam, A.H.; Adamowski, J.; Kisi, O. Applications of hybrid wavelet-Artificial Intelligence models in hydrology: A review. *J. Hydrol.* **2014**, *514*, 358–377. [CrossRef]

37. Mallat, S.G. A theory for multi decomposition signal decomposition: The wavelet representation. *IEEE Trans. Pattern Anal. Mach. Intell.* **1989**, *11*, 674–693. [CrossRef]

38. Hasan, M.; Begum, M.; Al Mamun, A.; Haque Khan, Z. Selection of extreme drought event for Langat Basin and its consequence on salinity intrusion through Langat River System. *IOSR J. Mech. Civ. Eng.* **2014**, *11*, 62–69. [CrossRef]

39. Singh, V.P. *Elementary Hydrology*; Prentice Hall of India: New Delhi, India, 1994.

40. Dawson, C.W.; Abrahart, R.J.; Shamseldin, A.Y.; Wilby, R.C. Flood estimation at ungauged sites using artificial neural networks. *J. Hydrol.* **2006**, *319*, 391–409. [CrossRef]

41. Morid, S.; Smakhtin, V.; Bagherzadeh, K. Drought forecasting using artificial neural networks and time series of drought indices. *Int. J. Climatol.* **2007**, *27*, 2103–2111. [CrossRef]

42. Shahin, M.A.; Maier, H.R.; Jaksa, M.B. Data Division for Developing Neural Networks Applied to Geotechnical Engineering. *J. Comput. Civ. Eng.* **2004**, *18*, 105–114. [CrossRef]

43. Khadr, M. Forecasting of meteorological drought using Hidden Markov Model (case study: The upper Blue Nile river basin, Ethiopia). *Ain Shams Eng. J.* **2016**, *7*, 47–56. [CrossRef]

44. Claudio, G.; Joseph, Q.; Carmine, T.; Svetoslav, I.; Silviya, P. Public Transportation Energy Consumption Prediction by means of Neural Network and Time Series Analysis Approaches. In Proceedings of the 6th International Conference on Automotive and Transportation Systems, Slerno, Italy, 27–29 June 2015; pp. 64–70.

45. Amit, K.Y.; Hasmat, M.; Chandel, S.S. Selection of most relevant input parameters using WEKA for artificial Neural network based solar radiation prediction models. *Renew. Sustain. Energy Rev.* **2014**, *31*, 509–519.

water

MDPI

Article

Improved Forecasting of Extreme Monthly Reservoir Inflow Using an Analogue-Based Forecasting Method: A Case Study of the Sirikit Dam in Thailand

Somchit Amnatsan [1,2,*], Sayaka Yoshikawa [2] and Shinjiro Kanae [2]

[1] Water Management and Maintenance Division, Regional Irrigation Office 2, Lampang 52100, Thailand
[2] Department of Civil and Environmental Engineering, School of Environment and Society,
   Tokyo Institute of Technology, 2-12-1-M1-6 Ookayama, Meguro-ku, Tokyo 152-8552, Japan;
   yoshikawa.s.ad@m.titech.ac.jp (S.Y.); kanae@cv.titech.ac.jp (S.K.)
* Correspondence: somnatsan@yahoo.com; Tel.: +66-86-859-6488

Received: 24 September 2018; Accepted: 5 November 2018; Published: 9 November 2018

Abstract: Reservoir inflow forecasting is crucial for appropriate reservoir management, especially in the flood season. Forecasting for this season must be sufficiently accurate and timely to allow dam managers to release water gradually for flood control in downstream areas. Recently, several models and methodologies have been developed and applied for inflow forecasting, with good results. Nevertheless, most were reported to have weaknesses in capturing the peak flow, especially rare extreme flows. In this study, an analogue-based forecasting method, designated the variation analogue method (VAM), was developed to overcome this weakness. This method, the wavelet artificial neural network (WANN) model, and the weighted mean analogue method (WMAM) were used to forecast the monthly reservoir inflow of the Sirikit Dam, located in the Nan River Basin, one of the eight sub-basins of the Chao Phraya River Basin in Thailand. It is one of four major dams in the Chao Phraya Basin, with a maximum storage of 10.64 km^3, which supplies water to 22 provinces in this basin, covering an irrigation area of 1,513,465 hectares. Due to the huge extreme monthly inflow in August, with inflow of more than 3 km^3 in 1985 and 2011, monthly or longer lead time inflow forecasting is needed for proper water and flood control management of this dam. The results of forecasting indicate that the WANN model provided good forecasting for whole-year forecasting including both low-flow and high-flow patterns, while the WMAM model provided only satisfactory results. The VAM showed the best forecasting performance and captured the extreme inflow of the Sirikit Dam well. For the high-flow period (July–September), the WANN model provided only satisfactory results, while those of the WMAM were markedly poorer than for the whole year. The VAM showed the best capture of flow in this period, especially for extreme flow conditions that the WANN and WMAM models could not capture.

Keywords: reservoir inflow forecasting; artificial neural network; wavelet artificial neural network; weighted mean analogue; variation analogue

1. Introduction

Reservoirs are manmade structures that are widely used in water resource management, and are recognized as some of the most efficient infrastructure components in integrated water resource management and development [1]. Reservoirs are among the major solutions to water demand and water-related problems, including irrigation, hydropower, urban and industrial water supply, conservation of ecology, and flood control. Nevertheless, there are several factors that affect the performance of the reservoir system, for example, the reservoir sedimentation [2] and the reservoir operation. In reservoir operation, care is required, especially for multipurpose reservoirs where there may be a number of potentially conflicting objectives. For water supply, operations should keep reservoirs as full as possible, whereas flood control requires reservoirs to be kept as empty as possible to allow the capture of flood water [3]. Reservoirs should be neither partially empty at the end of the rainy season nor full at the time of a series of peak floods that lead to heavy releases, causing floods in downstream areas [4]. Due to its complexity, reservoir operation is a challenging problem for water resource planners and managers. To optimize operating rules, many optimization and simulation models have been developed and applied over the past several decades [5–9]. However, these operating rules are not easy to implement, as appropriate reservoir operations depend on the accuracy of inflow forecasting and the operating time horizon [10]. Accurate inflow prediction is not only an important non-engineering measure to ensure flood-control safety and increase water resource use efficiency, but also can provide guidance for reservoir planning and management, because streamflow is the major input into reservoirs [11,12].

Due to its importance, several models and methodologies for reservoir inflow forecasting have been developed and applied in real-world situations [13]. One method that is widely used to forecast reservoir inflow is the artificial neural network (ANN) model. Although this is a black-box model in which the internal structure of the process involved cannot be understood, it has many advantages from the viewpoint of practical application. First, it is able to recognize the relation between the input and output variables without explicit physical consideration [14]. Second, it is very convenient to review the model when the data of interest are suspected as having changed. It can be recalculated as soon as new data are available with low cost and time requirements. Third, once the model is developed, it can be adapted very flexibly to other areas or for other purposes. In addition to these advantages, ANN models have been shown to be applicable to hydrology, including reservoir inflow prediction [14,15]. There have been several reports of the application of the ANN model for predicting short-term reservoir inflow at hourly and daily time scales [16,17]. Most studies have concluded that the ANN model provides satisfactory forecasting results. The ANN model can also be applied to forecast long-term and seasonal reservoir inflow as reported in several studies [18–20]. Some studies attempted to improve forecasting results by incorporating sea-surface temperature (SST) and climatic indices as inputs of the ANN model [21]. Most studies reported the good prediction results and the incorporation of SST provide improved predictions relative to the same model using only reservoir inflows.

Although ANNs have been used successfully in various fields, the precision of the results has still required improvement in many cases. Several hybrid ANN models have been proposed to fulfill this requirement. Kim and Valdés [22] developed a model for drought forecasting in the Conchos River Basin in Mexico, making use of the ability of neural networks to model and forecast nonlinear and non-stationary time series and the ability of wavelet transforms to provide useful decompositions of an original time series. The results indicated that the conjunction model significantly improved the ability of neural networks to forecast the index regional drought. A similar study which indicated the successful integration of the ANN and wavelet analysis to predict water levels in the Nan River, Thailand, can be found in the work of Amnatsan et al. [23].

Another technique that has been widely used in forecasting is the analogue method (AM), which was first introduced by Lorenz in 1969 to predict the evolution of the states of a dynamic system [24]. This is the simplest statistical technique that can establish nonlinear relationships between variables in a straightforward manner [25]. The analogue forecasting approach is based on the hypothesis that two relatively similar synoptic situations may produce similar local effects [26]. This approach has two main advantages and has been commonly used in weather prediction. First, the use of observed weather patterns helps to maintain the local-scale weather in the simulated field. Second, it is easy to construct scenarios for non-normally distributed variables, such as daily precipitation, because the AM does not assume the form of probability distribution of downscaled variables [27]. There have been many reports of successful implementation of the AM in weather prediction [25,26,28,29]. However, there have been few reports regarding its application to streamflow forecasting. Bellier et al. [30] evaluated probabilistic flood forecasting on the Rhone River using ensemble- and analogue-based precipitation forecasts. They reported that forecasting performance of the two methods for the peak amplitude and peak timing of floods was very similar. Svensson [31] performed flow forecasting based on flow persistence and historical flow analogues. The river flows at one and three months in the future at 93 individual river flow stations across the United Kingdom were forecast using two historical AMs, i.e., the weighted mean method and the shifted weighted mean method. The results indicated that forecasts based on persistence of the previous month's flow generally outperformed the analogue approach, particularly for slowly responding catchments with large underground water storage in aquifers. For the weighted-mean AM, the forecasting performance was increased with the length of historical flow records. The considerable success used of the weighted-mean in an interlayer forward validated scheme was reported in Panagoulia [32].

In this study, the wavelet artificial neural network (WANN) and the weighted mean AM (WMAM) were used to forecast the monthly reservoir inflows of the Sirikit Dam in Thailand. Monthly and seasonal inflow forecasting are very important for proper management of this multipurpose dam, which has a large catchment area of 13,130 km^2 and maximum storage of 10.64 km^3. This is one of four major dams that supply water to 22 provinces in the Chao Phraya Basin, covering an irrigation area of 1,513,465 hectares. Difficulty in the operation of this dam occurs mainly in the monsoon season, especially in July to September, the months which account for about 50% of the annual inflow. During this period, the dam managers have to decide whether to keep or release water. They have to retain sufficient water to supply demand in the next dry season, but for downstream flood control they must not keep too much water. As the capacity of the downstream river is limited, large amounts of water cannot be released in too short a time. An incorrect decision due to lack of an accurate and timely inflow forecast will lead to excessive release in a short time, resulting in flooding in downstream areas. On the other hand, a long forecast lead time will allow dam managers to release water gradually. Therefore, monthly or seasonal weather and reservoir inflow forecasting are crucial for proper management of this dam [33].

In addition to the WANN and WMAM methods, a forecasting method designated as the variation analogue method (VAM) was developed and employed to forecast the reservoir inflow of this dam. This study was performed to evaluate the performance of different forecasting methods in predicting the reservoir inflow, especially with regard to predicting extreme flow. Many researchers have reported that ANN-based models cannot predict extreme values in river flow [34,35]. The WMAM, which was found to show good predictive performance for a low-response watershed [31], may not be able to forecast the peak flow for the high-response catchment of the Sirikit Dam.

Several previous studies have indicated that SSTs and ocean indices are associated with the seasonal and interannual climate of Thailand [36–39], and therefore the variability of rainfall and reservoir inflows may be associated with SST anomalies. Manusthiparom [40] reported that adding SSTs as ANN inputs significantly improved the results of monthly rainfall and runoff forecasting for the Chao Phraya River Basin. In this study, we incorporated SSTs and ocean indices into the WANN and the VAM to improve the performance of inflow forecasting. Their forecasting performance was compared using four indicators: the root mean square error (RMSE), the correlation (R), the Nash–Sutcliffe efficiency index (EI), and the coefficient of determination (CD).

2. Study Area and Methods

2.1. Study Area and Data

The Sirikit Dam is located in the Nan River Basin, one of the eight sub-basins of the Chao Phraya River Basin in Thailand, as shown in Figure 1. It is the largest earth-filled dam in Thailand, with a catchment area of 13,130 km^2 and a maximum storage of 10.64 km^3. The main functions of this dam are flood prevention, water supply for domestic use, ecological conservation, agriculture, industry, fishing, and as an important tourist attraction.

The sources of data used in this study are listed in Table 1. The monthly reservoir inflow data of Sirikit Dam used in this study were obtained from the Electricity Generating Authority of Thailand (EGAT). The data were for the period from January 1974 to December 2014.

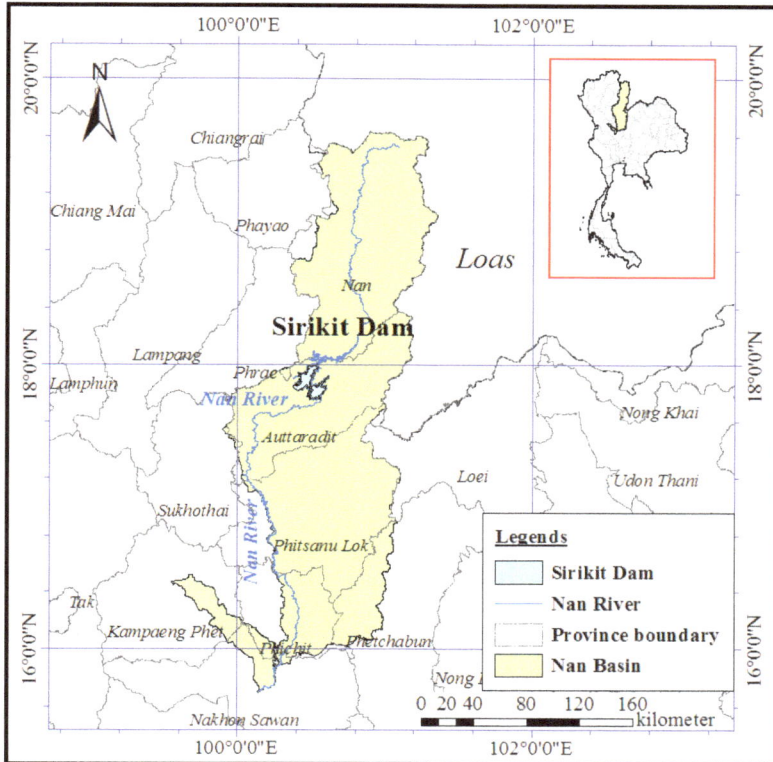

Figure 1. Location of the Sirikit Dam.

Table 1. Sources of data used in this study.

Data Used	SST Regions/Ocean Index Name	Source
Sea-surface temperature (SST)	Niño 1 + 2 Niño 3 Niño 3.4 Niño 4 Pacific Ocean South China Sea Andaman Sea	The U.S. National Oceanic and Atmospheric Administration (NOAA)
Ocean index	Southern Oscillation Index (SOI)	
	Dipole Mode Index (DMI)	Japan Agency for Marine-Earth Science and Technology (JAMSTEC)
Monthly reservoir inflow		Electricity Generating Authority of Thailand (EGAT)

The SSTs and the El Niño/La Niña Southern Oscillation (ENSO) indices in Niño 3, Niño 4, Niño 1 + 2, and Niño 3.4 regions, including the Southern Oscillation Index (SOI), were used in this study. The data were taken from the U.S. National Oceanic and Atmospheric Administration (NOAA), available at http://www.cpc.ncep.noaa.gov/. These are monthly data from January 1950 to December 2014.

In addition, the Dipole Mode Index (DMI) was also used as another ocean index. The DMI, which represents the intensity of the Indian Ocean Dipole (IOD), shows the anomalous SSTs between the western Indian Ocean and the southeastern Indian Ocean. The DMI data were obtained from the Japan Agency for Marine-Earth Science and Technology (JAMSTEC) website (http://www.jamstec.go. jp).

Additional SST data for the Pacific Ocean, the South China Sea, and the Andaman Sea were also used in this study to improve the accuracy of reservoir-inflow forecasting. The Extended Reconstructed Sea Surface Temperature (ERSST) version 3b dataset, a global monthly SST analysis derived from the International Comprehensive Ocean-Atmosphere Dataset with missing data filled in by statistical methods, was taken from the U.S. National Oceanic and Atmospheric Administration (NOAA), available at http://www.ncdc.noaa.gov.

For the WANN forecasts, data from January 1974 to December 2004 were used for training, from January 2005 to December 2010 for validation, and from January 2011 to December 2014 for testing of the models. For the WMAM and VAM forecasts, the inflow data from January 1974 to December 2004 were used as historical analogues for forecasting inflow from January 2005 to December 2014. This forecasting period corresponded to the validation and testing periods in the WANN models.

2.2. Wavelet Artificial Neural Network

The WANN is a hybrid version of the ANN model in which wavelet analysis is used as a data pre-processing technique to improve accuracy. According to the investigation of the American Society of Civil Engineers (ASCE) Task Committee on Application of Artificial Neural Networks in Hydrology [14] that a feed-forward network with a single hidden with an arbitrary number of sigmoidal hidden nodes can approximate any continuous function, a multilayer perceptron (MLP) feedforward network with one hidden layer was adopted in this study. The network was trained in a supervised manner with an error back-propagation algorithm. As suggested in Panagoulia et al. [41], input variables should be "first stage" selected, depending on their robustness, from an inclusive set which influences the physical model underlying the ANN structure with the constraint of minimizing redundancy and noise. In the second stage of selection, an association via statistics must be established to determine those first stage input variables that are maximally and distinctly connected to the major internal model variables. In this study, the reservoir inflow was selected and the autocorrelation analysis between different lag versions of those inflows was performed in an initial experiment. After that, other inputs were selected and cross-correlation analyses between the different lag versions of those inputs and the inflows were performed. Trials with changes in activation function, learning rate, number of hidden neurons, and momentum of the ANN network were also performed to obtain the best forecasting results. After obtaining the best forecasting results for each input dataset, the original input data were then decomposed into their detailed (high frequency) and approximated (low frequency) components by a discrete wavelet transform. Based on the study of Wang et al. [42], using different mother wavelets in the wavelet neural network affected the accuracy of prediction results. In this study, the Haar wavelet, the simplest and oldest of all wavelet functions [43], was used. This wavelet function provided a good prediction result in the study of Wang et al. [42]. The simplicity of this wavelet function facilitated the decomposition process and consequently supported practical implementation. Only one level of decomposition was used in this study. After decomposition, the decomposed data were used as the input for the ANN model. The architecture of the WANN model used in this study is shown in Figure 2.

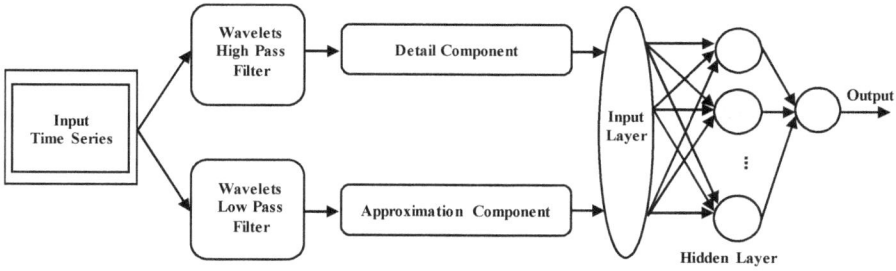

Figure 2. Architecture of the wavelet artificial neural network (WANN) model.

2.3. Weighted Mean Analogue Method

The full details of the WMAM were presented by Svensson [31]. In the present study, the forecast was obtained by first calculating the reservoir inflow anomalies. In the calculation, monthly reservoir inflows were transformed to the log form to ensure that the distribution was similar to a normal distribution, and when assessing the similarity of the analogues to the recent past, the highest inflows became less extreme. After log transformation, standardized reservoir inflow anomalies were calculated for use in the analysis as follows. For each of the 12 calendar months (mon), the mean reservoir inflow (m_{mon}), and standard deviation (s_{mon}) were calculated from the log-transformed monthly reservoir inflow (q_t). A series of standardized monthly anomalies (a_t) was then calculated as:

$$a_t = \frac{q_t - m_{mon}}{s_{mon}} \tag{1}$$

where t denotes the serial number of the month, starting from January 1974, and mon refers to the calendar month corresponding to t. Forecasts could be made once the observed data for the latest month were received by comparing the monthly anomalies of the most recent past months with all possible historical sequences of anomalies covering the same months of the year. From this annual series of potential analogues based on the RMSE, the N_{ana} historical analogues most similar to those of the recent past were selected. Then, we used the inverse of these RMSEs to weight the inflow anomalies in the months following the analogues to obtain the WMAM forecast. The RMSE was calculated for each potential analogue in the observed record as follows:

$$\text{RMSE} = \sqrt{\frac{1}{D_{ana}} \sum_{k=1}^{D_{ana}} \left(a_p(k) - a_r(k)\right)^2} \tag{2}$$

where $a_p(k)$ is the inflow anomaly for each month k in the potential analogue of duration D_{ana}, and $a_r(k)$ is the corresponding inflow anomaly in the recent past. The RMSEs for the selected N_{ana} analogues were used to calculate the weight, w, for each analogue as follows:

$$w(n) = \frac{1}{\text{RMSE}(n)} \Big/ \sum_{n=1}^{N_{ana}} \frac{1}{\text{RMSE}(n)} \tag{3}$$

where $n = 1, \dots, N_{ana}$ is the rank of the ordered RMSEs (the potential analogue $a_p(n)$ with the smallest RMSE had rank $n = 1$). The weighted mean forecast anomalies, $a_f(m)$, for each month $m = 1, \dots, D_f$ in the forecast duration, D_f, formed the last part of the constructed analogue, a_c, and were calculated as:

$$a_f(m) = a_c(D_{ana} + m) = \sum_{b-1}^{N_{ana}} w(b) a_{p,b}(D_{ana} + m), \tag{4}$$

where $a_{p,n}$ is the vector of inflow anomalies for the potential analogue with rank n. The D_{ana}, N_{ana}, and D_f were set to 5, 5, and 1, respectively.

2.4. Variation Analogue Method

The VAM was developed and used to forecast the reservoir inflow in this study. The idea behind this method emerged from the concept of a force system. Consider a force system in which two objects are located at different locations and subjected to different forces. If we observe both objects through a small window and notice that they move to the same location at the same time, we cannot assume that the next location of these objects after some time interval will be the same. This is because they are subjected to different forces and start moving from different initial locations. This is similar to most forecasting methods in hydrology and meteorology that try to compare historical amounts of rainfall, discharge, runoff, or inflow to forecast future values of the data of interest. If similarities in the values of the data of interest are the result of different forcing factors and different initial conditions, the forecasting result may be worse than expected. Compare this to another force system in which two objects are located at different locations but are subjected to the same force. In this system, at the same time interval, the objects will again move to different locations. However, although the new locations of the two objects are different, their displacement will be equal. If we compare the displacements of the objects instead of their locations, we can predict that in the next time interval of interest the two objects will have the same displacement. Consequently, the locations of the two objects can be calculated from their predicted displacement. Based on this concept, the forecasting method known as VAM was developed in this study. This method compares the variation (displacement) in standardized inflows instead of comparing standardized inflows as in the WMAM. It replaces data points by their successive differences so that the model target is shifted towards prediction based on differences rather than absolute positioning. By this method, a measure of chronological stability around a suitably chosen statistical quantity is established based on long-term data calculation. Considering the standardized monthly anomalies (a_t) as calculated in Equation (1), the variation (v_t) in a_t can be calculated as:

$$v_t = a_t - a_{t-1} \tag{5}$$

Once the observed data for the latest month have been received and the variation (v_t) has been calculated, this variation is compared to all possible historical sequences of variations covering the same months of the year. From this annual series of potential variation analogues (v_{ana}), the N_{ana} historical variation analogues most similar to the recent variation are selected. The variation for the next month can then be forecast as:

$$v_{t+1} = \frac{1}{N_{ana}} / \sum_{n=1}^{N_{ana}} \left[v_t + v_{ana(t+1)} - v_{ana(t)} \right]. \tag{6}$$

Then, the forecast standardized monthly anomaly for the next month can be calculated as:

$$a_{t+1} = v_{t+1} + a_t \tag{7}$$

Comparison of the variation in standardized inflows is similar to comparison of the displacement of objects subjected to a force—if the objects have the same properties and are subjected to the same force, their displacement will be the same, regardless of their initial locations. Building on this concept, if the displacement of one of these objects is known, it is possible to predict the displacement of the other objects. Applying this to inflow forecasting, if the variation in inflow in the current month of the current year is similar to the variation in inflow in the same month of a historical year, the inflows of the two years are inferred to occur due to similar forcing factors. If it is assumed that these forcing factors persist, the variation in inflow in the next month of the current year can be forecast from the variation in inflow of the historical year. Using this method, reservoir inflows are standardized as in the WMAM forecasts. Then, the variation in standardized inflows between successive months is calculated and used in forecasting as described above.

3. Results

The forecasting using the WANN model in this study was begun by finding the input parameters of the ANN model that produced the best forecast. After several trials, the best forecasting results were obtained from a model with 22 input parameters, as shown in Table 2. The activation function of this model in both the hidden and output layers was a hyperbolic function. The number of hidden neurons, learning rate, and momentum that provided the best results were 10, 0.0001, and −0.5, respectively. After obtaining the best forecast from the ANN model, all input parameters were decomposed into their detailed (high frequency) and approximated (low frequency) components. Then, all decomposed components were fed into the neural network model. The performance indicators of the WANN model in each model period are shown in Table 3.

Table 2. Input parameters of the artificial neural network (ANN) model that produced the best forecast. SOI: Southern Oscillation Index; DMI: Dipole Mode Index.

Input Parameter	SST Region/Ocean Index Name	Lag Used (Month)
	Niño 1+2	5, 17, 18
	Niño 3	4, 16, 17
Sea-surface temperature	Niño 3.4	5, 15, 16
	Pacific Ocean	6, 7, 18
	South China Sea	6, 18, 19
	Andaman Sea	7, 18, 19
Ocean index	SOI	5
	DMI	16
Reservoir inflow	-	1, 12

Table 3. Performance indicators of the wavelet artificial neural network (WANN) model. CD: coefficient of determination; EI: efficiency index; RMSE: root mean square error; R: correlation.

Model Period	Model Performance Indicators			
	RMSE	R	EI	CD
Training	179.85	0.95	0.90	0.89
Validation	248.68	0.90	0.81	0.85
Testing	210.80	0.95	0.89	0.77

For the forecasting using the WMAM and VAM methods, reservoir inflow data from 1974 to 2004 were used to forecast the inflow of the years 2005 to 2014. Therefore, there were at least 31 years of monthly records for use as historical analogues. For the WMAM method, the selection of potential historical analogues was based on calculation of the RMSE as described in the Methodology section. Figure 3 shows an example of reservoir inflow forecasting for March 2013. Five historical analogues gave the minimum root mean square values selected for the forecast. After selection, the weights for each analogue were calculated, and these weights were then used to calculate the forecast standardized value and converted to obtain the forecast inflow for March 2013. The yellow broken line and the yellow solid line are the forecast and observed standardized inflows in March 2013, respectively. The forecast standardized inflows were converted to inflows in a normal form and used to calculate the performance indicators.

Figure 4 shows an example of the variation values plotted against standardized inflow values from February to January of the following year. Assuming that the most current month is December 2005, we can forecast the inflow in January 2006. The variation from March to January of previous years is plotted alongside the variation from March to December of 2005. Then, the most similar variation analogue is selected by comparing the variation vectors from November to December. In this example, the most similar analogue is the plot for 1993–1994, as shown in Figure 4. Thus, the variation in January 2006 is calculated from the variation in December 1993 and January 1994. Then, the forecast standardized value for January 2006 is calculated from this forecast variation value, as shown in Figure 4. The plot of standardized inflows from October 2005 to January 2006 is very similar to the plot from October 1993 to January 1994. It is evident from the plot that VAM forecasting has the advantage that it allows determination of similar patterns among inflow events even if they occurred in different zones. This is different from the WMAM, in which selection of potential historical dialogues depends on the RMSE between inflows of previous years and the current year. The selection will include all nearby inflow patterns even if they are not similar to the inflow pattern of the current year, while similar patterns in different zones, as in this example, will not be selected.

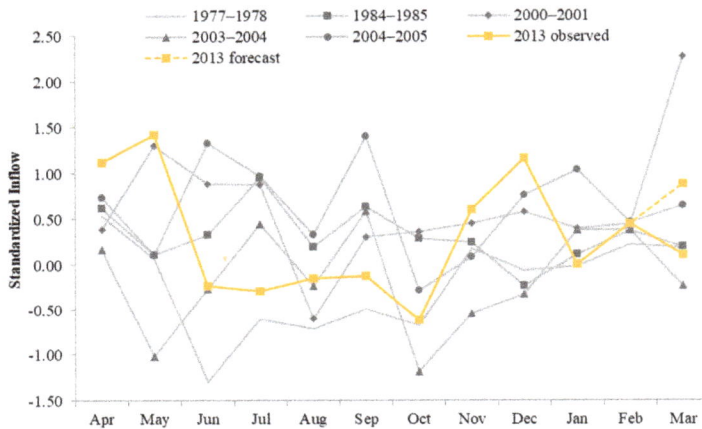

Figure 3. Forecasting of the reservoir inflow in March 2013 using the weighted mean analogue method.

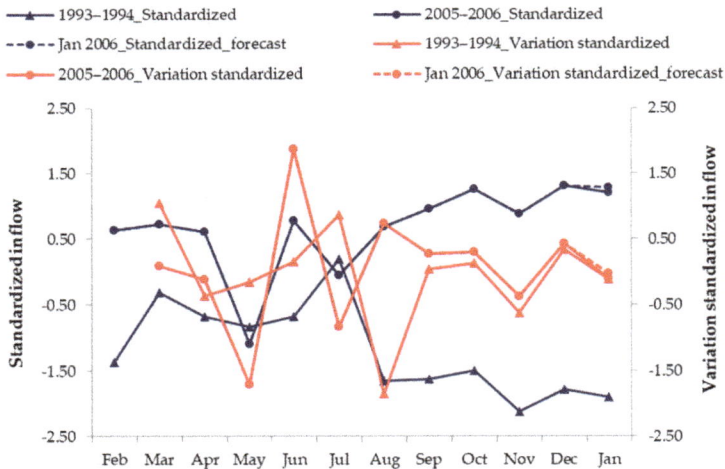

Figure 4. Variation and standardized values of inflows used for forecasting inflow in January 2006.

To improve forecasting results of the VAM, additional processes to assist selection of the most similar analogue were investigated. Consider the predicted inflow of the Sirikit Dam in August 1995, which was the most extreme Sirikit Dam inflow on record, shown in Figure 5. The plot of the variation in standardized inflow for June–July 1995 is very similar to the plot for the same period in 1992. The forecast variation in standardized inflow in August 1995 should follow the red dotted line. Nevertheless, the actual variation followed the dark red line, which is very far from the forecast result. These observations indicate that the VAM still has some weaknesses and requires further improvement.

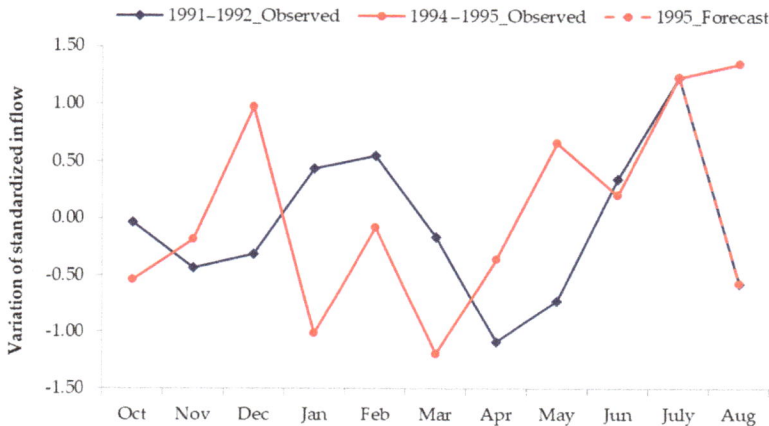

Figure 5. Plot of variation in standardized inflow forecasts for August 1995.

Several previous studies have indicated that SSTs and climatic indices are associated with climate and rainfall in Thailand. Moreover, several previous studies have reported that incorporating SSTs and climatic indices into river-flow forecasts can improve results. The incorporation of SSTs and climatic indices as inputs to the WANN in this study confirmed that these data can improve reservoir inflow forecasts, implying that SST and climatic indices are also associated with reservoir inflow. Based on this assumption, years with similar patterns of standardized inflow variation should have similar SST and climatic-index patterns. These similar SST and climatic-index patterns can then act as guidelines in the selection of the most potentially useful historical analogues. Therefore, cross-correlation analysis between SSTs and the climatic indices used as the inputs of the WANN model and standardized inflow was performed. The correlation values between SSTs, climatic indices, and standardized inflow were calculated for each month. The SSTs and climatic indices with correlation values exceeding the threshold for significance (0.304 for 41-year inflow data in this study) [44] were considered significant SSTs and indices in the selection of historical analogues of the corresponding month. As examples, Appendix A lists the significant SSTs and climatic indices for the Sirikit Dam inflows in January and August; the number −1 behind a month indicates the month in the previous year compared to the year of inflow. For example, Niño 3 (Jan-1) refers to the Niño 3 index in January of the previous year compared to the year of the inflow to be forecast. These significant SSTs and climatic indices will be used to decide the most useful potential historical analogue for forecasting the inflow of the current year.

An example of forecasting the inflow of the Sirikit Dam in January 2008 is presented. In this case, the most current month is December 2007, and the inflow to forecast is that of January 2008. The steps of forecasting the inflow are as follows.

1. The variations in standardized inflows for March to December 2007 are plotted along with the variations in standardized inflows for March to January of the available analogues. Potential

analogues with variation patterns similar to that of December 2007 are selected. In this case, there are three historical analogue candidates: December 1976, December 1988, and December 1989 (Figure 6).

2. To forecast the inflow for January 2008, the significant SSTs and climatic indices for inflows in January 2008, January 1977, January 1989, and January 1990 are plotted to assist in selection of the best potential analogue (see Appendix B). In this case, most of the significant climatic indices and SSTs for January 2008 are very similar to those for January 1989, and therefore January 1989 is selected as the best potential analogue.

3. The forecast variation for January 2008 is calculated from the variation in January 1989 using Equation (6) and plotted as the red-dotted line in Figure 6.

4. After obtaining the variation for January 2008, the standardized inflow is calculated using Equation (7) and converted to the normal form of inflow. The forecast inflow values calculated from this method and the observed inflow in January 2008 are 138.29 and 136.84 million cubic meters, respectively.

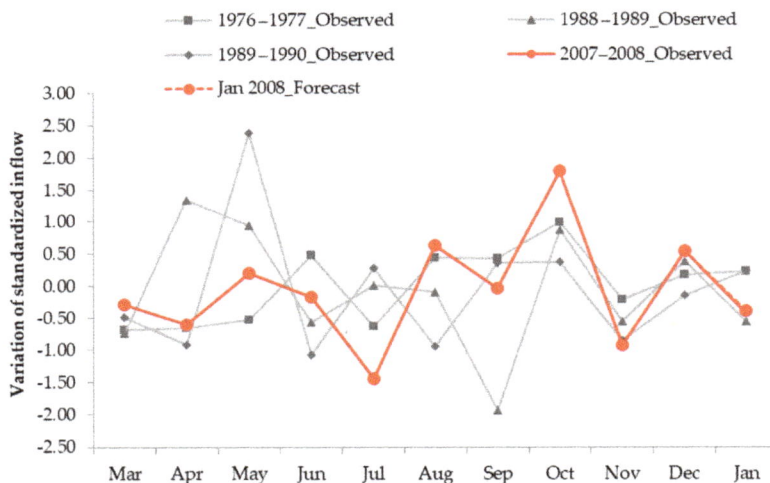

Figure 6. Plot of variation in forecasts of the Sirikit Dam inflow in January 2008.

For a greater understanding of forecasting using the VAM with the consideration of SSTs and climatic indices (the VAM-improved), the readers can read the examples of forecasting for the inflow in August 1995 and August 2011, which were the most extreme inflows on record (Appendices C and D, respectively).

Based on the results described above, the improved VAM that considers climatic indices was used to forecast July, August, and September, which are high-flow periods in Sirikit Dam inflow. The forecasting performance of all methods for the whole-year and high-flow periods was evaluated and compared. To compare the performance of the WANN with other methods, the forecasting results of the WANN in the validation and testing periods were combined and the performance indicators were recalculated to match the forecasting period of the WANN with that of the WMAM and VAM. The performance indicators of all methods in predicting the reservoir inflow of the Sirikit Dam from January 2005 to December 2014 are shown in Table 4. Plots of forecast and observed inflows for the whole-year and high-flow periods are shown for comparison in Figures 7 and 8, respectively.

Table 4. Performance indicators of all forecasting methods for forecasting the Sirikit Dam inflow in 2005–2014. VAM: variation analogue method; WANN: wavelet artificial neural network; WMAM: weighted mean analogue method.

Forecasting Period	Method	Model Performance Indicator			
		RMSE	R	EI	CD
Whole year (January–December)	WANN	234.20	0.92	0.85	0.81
	WMAM	335.45	0.84	0.69	0.62
	VAM	186.33	0.95	0.90	1.01
	VAM-improved	115.55	0.98	0.96	0.92
High flow (July–September)	WANN	366.78	0.83	0.67	0.60
	WMAM	627.42	0.37	0.04	0.37
	VAM	305.59	0.84	0.84	0.87
	VAM-improved	215.55	0.95	0.92	0.86

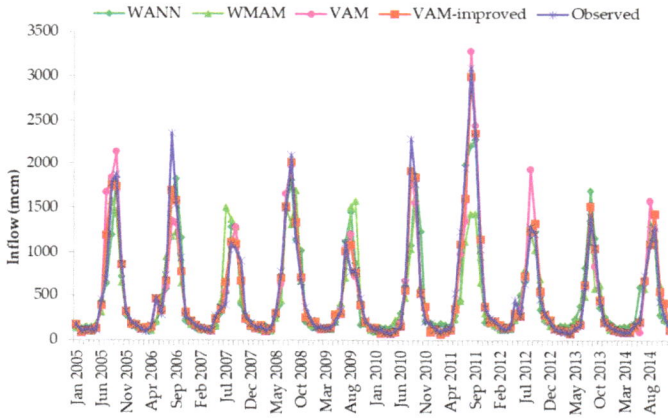

Figure 7. Comparison of plots of forecast and observed inflows for whole-year periods from 2005 to 2014.

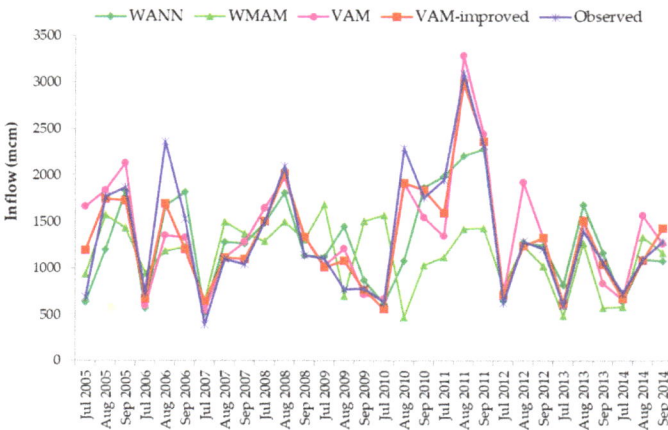

Figure 8. Comparison of plots of forecast and observed inflows in high-flow periods from 2005 to 2014.

For the whole-year forecasting period, which included both low-flow and high-flow patterns, the WANN model provided good forecasting results as all performance indicators were above 0.80. The WMAM provided only satisfactory forecasting results, as the EI and CD values were <0.70 and the RMSE was higher than for the other methods. The forecasting performance of the VAM was superior to that of other methods as all performance indicators were >0.9 and the RMSE had the lowest value. It can also be seen from the comparison plot in Figure 8 that the VAM forecast captured the extreme inflow of the Sirikit Dam reservoir.

For the high-flow period, the forecasting performance of all methods significantly worsened. The WANN method, which produced good results for the overall period, provided only satisfactory results for this period. This was not unexpected; poor performance in predicting peak flow is a common weakness of ANN methods, as noted in several previous reports (e.g., Sudheer [34]; Yang et al. [35]). The forecasting performance of the WMAM was markedly lower in the high-flow period compared to the whole year, indicating that this method is not suitable for prediction of inflow in a high-response watershed, especially for the high-flow season, as in this case. This weakness of the WMAM was also noted by Svensson [31], who reported a high degree of uncertainty in the historical analogue approach, particularly in catchments with a rapid response. The VAM captured flow best in this period comparing to the WANN and the WMAM, especially peak flow. Taking SSTs and climatic indices into consideration significantly improves forecasting results of the VAM. The VAM-improved performance indicators were the best in both the whole-year and high-flow forecasting periods; this model provided very good performance indicators even in high-flow periods, with all indicators having values above 0.85. The improvement can be seen in Figure 8, where most of the VAM-improved forecast values are closer to the observed values than those of the standard VAM (i.e., the VAM without consideration of SSTs and climatic indices).

Based on the very high reliability and low uncertainty of the improved VAM indicated by the results, this method can be used for management of the Sirikit Dam, especially in high-flow periods. It provides very good forecasting results, with all performance indicators above 0.85, and its uncertainty as defined by RMSE values is less than the reservoir surcharge storage (998,000,000 m^3). Moreover, in testing, it predicted extreme flow such as occurred in August 2011, whereas other methods did not. However, in forecasting based on historical analogues, there may be high-return-period events for which no suitable analogues are available. Therefore, scenarios based on forecasts with various uncertainty values should be modeled. The most suitable scenario for the current month can then be selected for dam operation. For example, if the water level in the dam is very low in August, well below the upper curve of the reservoir operation rule, and low inflow is forecast in September, then a scenario with less inflow than that forecast is selected to retain the water for the coming dry season. On the other hand, if the water level in the dam is very high in August, close to the upper curve of the reservoir operation rule, and high inflow is forecast in September, then a scenario with more inflow than that forecast is selected. The dam operator can then decide to release water gradually to make space for the expected inflow. For the low-flow season, the forecasting results of all of the methods examined in this study were acceptable for dam operation, as their uncertainties were close enough to observed values. Moreover, most of the inflow in the low-flow period is kept for water supply, so uncertainty is less important.

4. Discussion and Conclusions

This study compared three forecasting methods—the WANN method, the WMAM, and the VAM—for use in forecasting the monthly reservoir inflow of the Sirikit Dam in Thailand. The results indicate that for whole-year forecasting, which includes both low- and high-flow seasons, the WANN method provided good results, while those obtained using the WMAM were only satisfactory. The performance of the VAM was superior to that of the other methods and accurately predicted extreme inflow. For the high-flow period (July–September), the VAM predicted flow best, especially in the case of peak flow. However, the performance of all of the methods was significantly lower for the high-flow period. The WANN method, which produced good results for whole-year forecasts, provided only satisfactory results for the high-flow period. The performance of the WMAM method was markedly worse in the high-flow period compared to the whole-year period. Based on the results of this study, the following conclusions were reached with regard to the methodologies and application to Sirikit Dam in Thailand.

1. The WANN model, a hybrid of ANN and wavelet analysis, produced good results in forecasting the monthly reservoir inflow of Sirikit Dam. However, for the high-flow period it provided only satisfactory results. This indicated that the WANN model is weak in forecasting peak flows because such flows are rare compared to low- and moderate-flow events. As ANN-based models rely on learning from past events, the number of peak flow events is insufficient for ANN models to learn and produce good forecasting results. This characteristic of ANN-based models is a common issue that has been reported in the literature (e.g., Sudheer [34]; Yang et al. [35]). Wavelet analysis, a data pre-processing technique, generally improved the forecasting results, but the improvement was not enough to predict peak flow. In conclusion, the WANN method has a poor ability to forecast peak flows.

2. The WMAM provided only reasonably satisfactory predictions for the whole-year period and its performance was markedly worse in the high-flow period. This may have been because the forecasting is dependent on the RMSE between historical and current inflows. The selection of historical analogues based on RMSE may result in the inclusion of all recent inflow patterns, even if they are not similar to the pattern of the current year. This leads to incorrect selection of analogues, especially for a high-response catchment such as the Sirikit Dam. This characteristic of analogue-based methods such as the WMAM was reported previously by Svensson [31], who concluded that the uncertainty of the historical analogue approach can be large, especially in catchments with a fast response.

3. The developed VAM provided excellent predictions of the monthly reservoir inflow of the Sirikit Dam. Its ability to forecast extreme peak flow represented an advantage over the other methods. However, it has the drawback that it relies on past observation data. Therefore, in the absence of a similar historical analogue it may not provide good results. This is especially important in the case of events with return periods that may be longer than the record length, making rare situations that have not been observed in the past very difficult to forecast. The example of this situation is the case of forecasting the inflow in August 1995 described earlier. In addition, changes in land use, urbanization processes, or changes in the morphology of the rivers may affect the discharge arriving to the reservoirs. The study of the effect of those changes should be further conducted to clarify this issue.

4. The incorporation of SSTs and climatic indices in the WANN model and the VAM significantly improved forecasts. In the WANN model, SSTs and climatic indices were used as an input to the model. In the VAM, significant SSTs and climatic indices for the inflow each month were plotted and compared to aid in selecting appropriate historical analogues. The idea behind investigating use of SSTs and climatic indices as guidelines for selection of the most suitable historical analogues was derived from the results of several previous studies that indicated that SSTs and climatic indices were associated with the Thai climate and rainfall (e.g., Singharattna et al. [36]; Bejranonda and Koch [37]; Chansaengkrachang [38]; Bridhikitti [39]; and Manusthiparom [40]). The improvement in forecasts in this study after incorporation of SSTs and climatic indices supports these previous reports. However, future studies should clarify the individual contributions of SSTs and the climate indexes.

5. Although the VAM could provide excellent predictions of the reservoir inflow of the Sirikit Dam, the leading time of prediction in this study is only one month, which may not be enough for the open large reservoir where prediction times of longer than one month are often needed. The further study of longer lead-time prediction using the VAM is hence needed for better reservoir operation.

Author Contributions: Conceptualization, S.A. and S.K.; Methodology, S.A.; Validation, S.A., S.Y. and S.K.; Formal analysis, S.A.; Data curation, S.A.; Writing—original draft preparation, S.A.; Writing—review and editing, S.A., S.Y., and S.K.; Supervision, S.K.; Project administration, S.K.; Funding acquisition, S.K.

Funding: The first author was provided a research grant by the Agricultural Research Development Agency (ARDA) of Thailand. This research was tremendously supported by Advancing Co-design of Integrated Strategies with Adaptation to Climate Change (ADAP-T) Project through the Science and Technology Research Partnership for Sustainable Development (SATREPS) by the Japan Science and Technology Agency and the Japan International Cooperation Agency (JST-JICA). It was also supported by the Japan Society for the Promotion of Science (JSPS) KAKENHI Grant Number JP16H06291.

Conflicts of Interest: The authors declare no conflict of interest.

Appendix A

Table A1. Significant SSTs and climatic indices for Sirikit Dam Inflow in January and August. SCS: South China Sea; AO: Arctic Oscillation.

January		August	
Significant SSTs/Indices	Correlation	Significant SSTs/Indices	Correlation
Andaman (MAR-1)	−0.404	Andaman (AUG)	−0.321
Andaman (APR-1)	−0.366	Andaman (MAY-1)	−0.372
Andaman (MAY-1)	−0.400	Andaman (JUN-1)	−0.336
Andaman (JUN-1)	−0.320	AO (FEB-1)	−0.345
Andaman (JUL-1)	−0.342	AO (DEC-1)	−0.308
Andaman (AUG-1)	−0.556	AO (MAY)	−0.356
Andaman (SEP-1)	−0.499	DMI (APR)	0.312
Andaman (OCT-1)	−0.303	DMI (MAR-1)	0.325
Andaman (NOV-1)	−0.371	DMI (APR-1)	0.421
Andaman (DEC-1)	−0.325	DMI (MAY-1)	0.324
Niño 3 (JAN-1)	−0.375	SOI (MAR)	0.448
Niño 3 (FEB-1)	−0.422	Niño 1 + 2 (MAR)	−0.316
Niño 3 (MAR-1)	−0.391	Niño 1 + 2 (APR)	−0.440
Niño 3 (APR-1)	−0.388	Niño 1 + 2 (MAY)	−0.404
Niño 3.4 (Jan-1)	−0.398	Niño 1 + 2 (JUN)	−0.342
Niño 3.4 (FEB-1)	−0.403	Niño 1 + 2 (JUL)	−0.307
Niño 3.4 (MAR-1))	−0.422	Niño 1 + 2 (AUG)	−0.334
Niño 3.4 (APR-1)	−0.459	Niño 3 (MAR)	−0.330
Niño 3.4 (MAY-1)	−0.327	Niño 3 (APR)	−0.421
Niño 4 (JAN-1)	−0.327	Niño 3 (MAY)	−0.439
Niño 4 (FEB-1)	−0.321	Niño 3 (JUN)	−0.357
Niño 4 (MAR-1)	−0.322	Niño 3 (JUL)	−0.310
Pacific (FEB-1)	−0.339	Niño 3 (AUG)	−0.329
Pacific (MAR-1)	−0.327	Niño 3.4 (APR)	−0.318
Pacific (APR-1)	−0.325	Niño 3.4 (MAY)	−0.353
Pacific (MAY-1)	−0.394	Pacific (JUL)	−0.409
Pacific (JUN-1)	−0.487	Pacific (AUG)	−0.355
Pacific (JUL-1)	−0.591		
Pacific (AUG-1)	−0.544		
Pacific (SEP-1)	−0.354		
SCS (APR-1)	−0.334		
SCS (MAY-1)	−0.320		
SCS (JUN-1)	−0.346		
SCS (JUL-1)	−0.361		
SCS (AUG-1)	−0.340		
SOI (JAN-1)	0.362		
SOI (FEB-1)	0.412		
SOI (MAR-1)	0.531		
SOI (APR-1)	0.398		

Appendix B

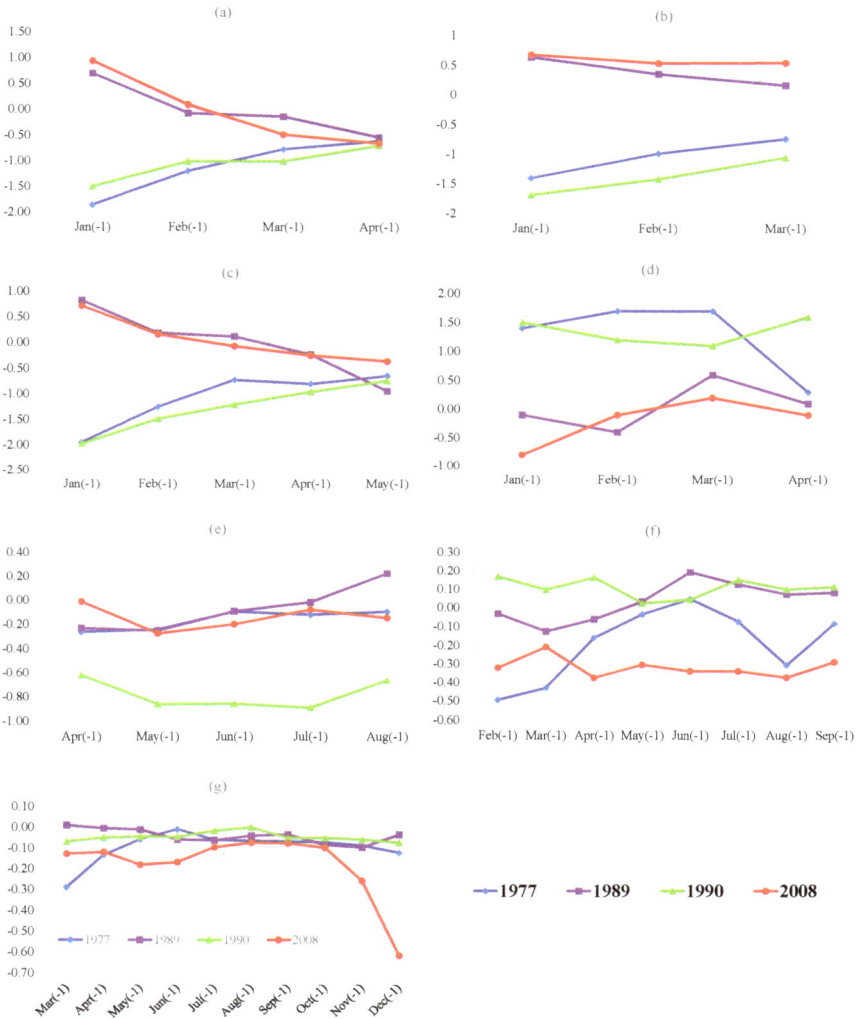

Figure A1. Plots of significant SSTs and climatic indices for forecasting Sirikit Dam inflow in January 2008 (**a**) Niño 3; (**b**) Niño 4; (**c**) Niño 3.4; (**d**) SOI; (**e**) Standardized SST in the South China Sea; (**f**) Standardized SST in the Pacific Ocean (**g**) Standardized SST in the Andaman sea.

Appendix C

The inflow forecasts for August 1995 using the VAM.

In the inflow forecasts for August 1995 described in the Results section, if we assume that all inflow data from 1974 to 2014 were available for prediction of inflow in August 1995 (which is impossible because 1996–2014 was in the future in August 1995), the potential variation analogues would be as plotted as in Figure A2. When considering the plots of significant SSTs and climatic indices for the inflow in August 1995 and all candidate analogues (see Figure A3), it is evident from the plots that most of the significant SSTs and climatic indices for August 1995 are similar to those for August 2001 and August 2013. It is also evident that the most significant SSTs and climatic indices for August 1992 differ from those for August 1995. Therefore, the variation in standardized inflow in August 1995 does not align with the variation in August 1992, even though it is the most visually similar among the available candidates. In this case, the variation in standardized inflow in August 1995 could be calculated from the variation in August 2001 and August 2013 using Equation (6). The forecast variation in August 1995 is shown by the red dotted line in Figure A2, which is closer to the observed value than before. This is the strength of the improved VAM that considers significant SSTs and climatic indices to predict extreme peak flow accurately.

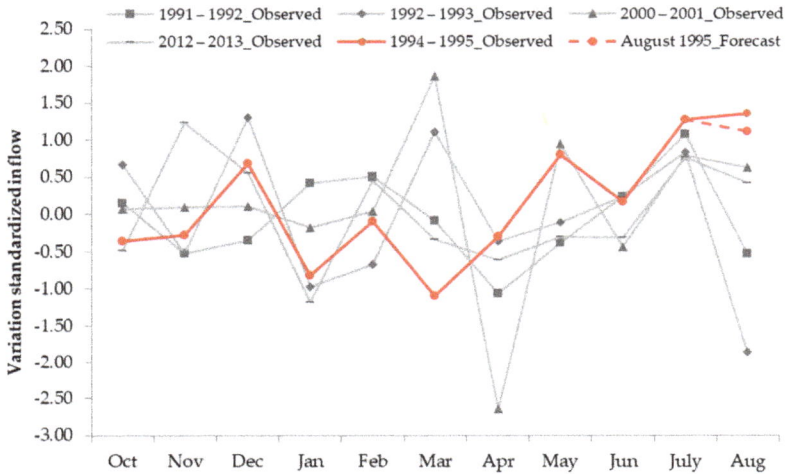

Figure A2. Plot of variation in forecasts of Sirikit Dam inflow in August 1995.

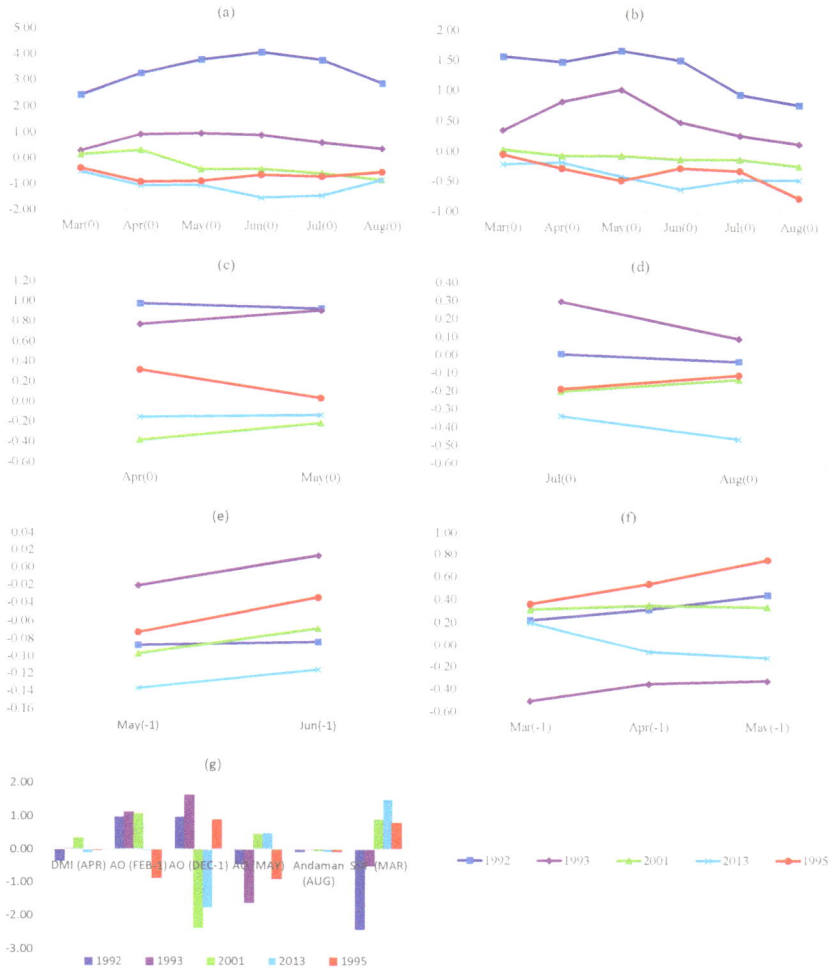

Figure A3. Plots of Significant SSTs and climatic indices for forecasting Sirikit Dam inflow in August 1995 (**a**) Niño 1 + 2; (**b**) Niño 3; (**c**) Niño 3.4; (**d**) Standardized SST in the Pacific Ocean; (**e**) Standardized SST in the Andaman Sea; (**f**) DMI; (**g**) Other SSTs and indices as stated in Appendix A.

Appendix D

The inflow forecasts for August 2011 using the VAM.

The inflow in August 2011 is another extreme event in the record. To confirm the ability of the VAM to predict extreme inflow, the forecast procedure for this extreme event is discussed. The prediction follows the steps described for the January 2008 inflow. Here, the most current month is July 2011, so we require a similar pattern of variation in July in the optimal analogue. Visual examination of the variation plots produces six potential analogues with patterns similar to that of July 2011 (see Figure A5). The significant SSTs and climatic indices for inflow in August 2011 and all candidate analogues are plotted to determine which analogue is the best potential candidate for forecasting (see Figure A6). The August 2011 patterns of SSTs and significant indices are not very similar to any one or two historical analogues, unlike the previous cases in the Results section. Therefore, the climate-forcing factors affecting the rainfall and inflow of the Sirikit Dam Basin in August 2011 are a mix of the forcing factors in the candidate analogue years. Based on this assumption, the variation in standardized inflow in August 2011 is the average of the variation in August of all candidate analogues and can be calculated using Equation (6). The results are shown by the red dashed line, which is very close to the observed value depicted by the dark red line (see Figure A5). This result confirms the strength of the improved VAM in forecasting extreme peak flows. In other methods that attempt to compare or search for historical inflow information, finding candidate analogues is very difficult, especially in the case of extreme events such as in August 2011, because the events occur in a very high-flow zone and similar events are very rare (see Figure A4).

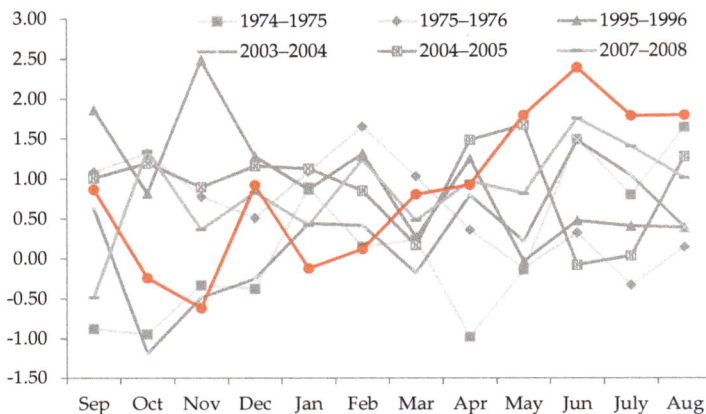

Figure A4. Plots of standardized inflow for forecasting Sirikit Dam inflow in August 2011.

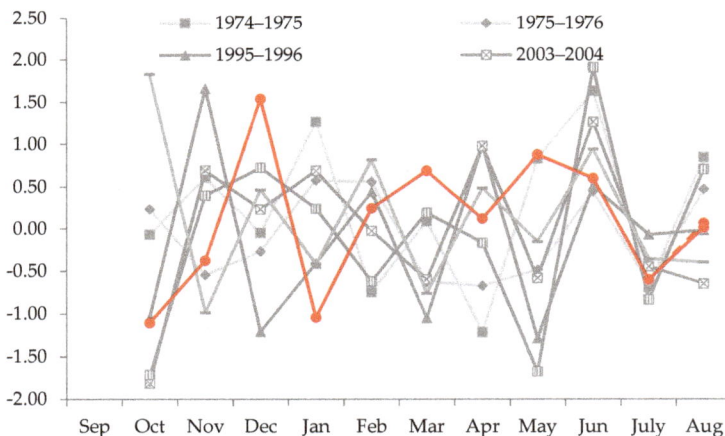

Figure A5. Plots of variation-standardized inflow for forecasting Sirikit Dam inflow in August 2011.

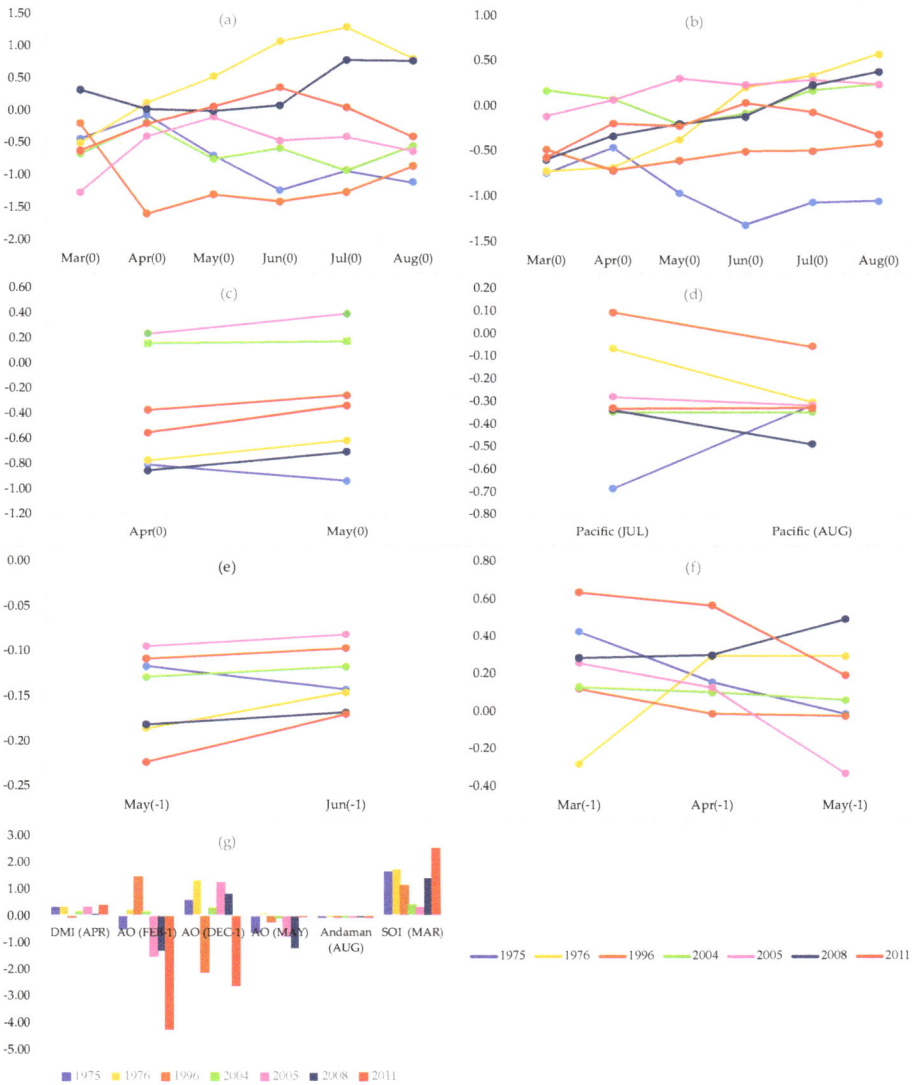

Figure A6. Plots of significant SSTs and climatic indices for forecasting Sirikit Dam inflow in August 2011 (**a**) Niño 1 + 2; (**b**) Niño 3; (**c**) Niño 3.4; (**d**) Standardized SST in the Pacific Ocean; (**e**) Standardized SST in the Andaman Sea; (**f**) DMI; (**g**) Other SSTs and indices as stated in Appendix A.

References

1. Anand, J.; Gosain, A.K.; Khosa, R. Optimisation of multipurpose reservoir operation by coupling SWAT and genetic algorithm for optimal operating policy (case study: Ganga River basin). *Preprints* **2018**. [CrossRef]
2. Schleiss, A.J.; Franca, M.J.; Juez, C.; De Cesare, G. Reservoir sedimentation vision paper. *J. Hydraul. Res.* **2016**. [CrossRef]
3. US Army Corps of Engineers. *Forecast-Based Advance Release at Folsom Dam: Effectiveness and Risk-Phase 1*; PR-48; Hydrologic Engineering Center: Davis, CA, USA, 2002; pp. 1–111.

4. Kumar, T. *Operations Research in Water Quality Management*; Harvard Water Resources Group: Cambridge, UK, 2013.
5. Lin, S.-C.; Wu, R.-S.; Chen, S.-W. Study on optimal operating rule curves including hydropower purpose in parallel multireservoir systems. *WIT Trans. Ecol. Environ.* **2005**, *83*, 151–160.
6. Mateo, C.M.R.; Hanasaki, N.; Komori, D.; Yoshimura, K.; Kiguchi, M.; Champathong, A.; Yamazaki, D.; Sukhapunnaphan, T.; Oki, T. A simulation study on modifying reservoir operation rules: Tradeoffs between flood mitigation and water supply. *IAHS Publ.* **2013**, *362*, 33–40.
7. Mythili, B.; Devi, U.G.; Raviteja, A.; Kumar, P.S. Study of optimizing techniques of reservoir operation. *Int. J. Eng. Res. Gen. Sci.* **2013**, *1*, 2091–2730.
8. Heydari, M.; Othman, F.; Taghieh, M. Optimization of multipurpose reservoir system operations by using matrix structure (case study: Karun and Dez Reservoir Dams). *PLoS ONE* **2016**, *10*, e0156276. [CrossRef] [PubMed]
9. Mohammed, J.M. Recalibration of Kainji Reservoir operating rules for optimal operation. *Ann. Rev. Res.* **2018**, *1*, 1–7.
10. Sivapragasam, C.; Vasudevan, G.; Vincent, P. Effect of inflow forecast accuracy and operating time horizon in optimizing irrigation releases. *Water Resour. Manag.* **2007**, *21*, 933–945. [CrossRef]
11. Dong, X.; Dohmen-Janssen, C.M.; Booij, M.; Hulscher, S. Effect of flow forecasting quality on benefits of reservoir operation. A case study for the Geheyan reservoir (China). *Hydrol. Earth Syst. Sci. Discuss.* **2006**, *3*, 3771–3814. [CrossRef]
12. Wang, Y.; Guo, S.; Chen, H.; Zhou, Y. Comparative study of monthly inflow prediction methods for the Three Gorges Reservoir. *Stoch. Environ. Res. Risk Assess.* **2013**, *28*, 555–570. [CrossRef]
13. Mohammadi, K.; Eslami, H.R.; Dardashti, S.D. Comparison of regression, ARIMA and ANN models for reservoir inflow forecasting using snowmelt equivalent (a case study of Karaj). *J. Agric. Sci. Technol.* **2005**, *7*, 17–30.
14. ASCE Task Committee on Application of Artificial Neural Networks in Hydrology. Artificial neural networks in hydrology. I: Preliminary concepts. *J. Hydrol. Eng.* **2000**, *5*, 115–123. [CrossRef]
15. ASCE Task Committee on Application of Artificial Neural Networks in Hydrology. Artificial neural networks in hydrology. II: Hydrologic applications. *J. Hydrol. Eng.* **2000**, *5*, 124–137. [CrossRef]
16. Kim, T.; Choi, G.; Heo, J.-H. Inflow Forecasting for real-time reservoir operation using artificial neural network. In Proceedings of the World Environmental and Water Resources Congress, Kansas City, MO, USA, 17–21 May 2009; pp. 4947–4955.
17. Zhang, J.; Cheng, C.-T.; Liao, S.-L.; Wu, X.-Y.; Shen, J.-J. Daily reservoir inflow forecasting combining QPF into ANNs model. *Hydrol. Earth Syst. Sci. Discuss.* **2009**, *6*, 121–150. [CrossRef]
18. Othman, F.; Naseri, M. Reservoir inflow forecasting using artificial neural network. *Int. J. Phys. Sci.* **2011**, *6*, 434–440. [CrossRef]
19. Valipour, M.; Banihabib, M.E.; Behbahani, S.M.R. Monthly inflow forecasting using autoregressive artificial neural network. *J. Appl. Sci.* **2012**, *12*, 2139–2147. [CrossRef]
20. Vijayakumar, N.; Vennila, S. Reservoir inflow forecasting at Krishnagiri Reservoir Project using artificial neural network. In Proceedings of the National Conference on Contemporary Advancements in Civil Engineering Practices—CACEP, P.S.G College of Technology, Coimbatore, India, 23 May 2016.
21. Muluye, G.Y.; Coulibaly, P. Seasonal reservoir inflow forecasting with low-frequency climatic indices: A comparison of data-driven methods. *Hydrol. Sci. J.* **2007**, *52*, 508–522. [CrossRef]
22. Kim, T.-W.; Valdés, J.B. A nonlinear model for drought forecasting based on conjunction of wavelet transforms and neural networks. *J. Hydrol. Eng.* **2003**, *8*, 319–328. [CrossRef]
23. Amnatsan, S.; Kuribayashi, D.; Jayawadena, A.W. Application of artificial neural networks and wavelet analysis in prediction of water level in Nan River of Thailand. In Proceedings of the Annual Conference 23rd Annual Conference, Hosei University, Tokyo, Japan, 7 September 2010.
24. Zhao, Z.; Giannakis, D. Analog forecasting with Dynamics-Adapted Kernels. *Nonlinearity* **2016**, *29*, 2888–2939. [CrossRef]
25. Zhao, B.; Zhai, P. A new forecast model based on the analog method for persistent extreme precipitation. *Weather Forecast.* **2016**, *31*, 1325–1341. [CrossRef]
26. Horton, P.; Obled, C.; Jaboyedoff, M. The analogue method for precipitation prediction: Finding better analogue situations at a sub-daily time step. *Hydrol. Earth Syst. Sci.* **2017**, *21*, 3307–3323. [CrossRef]

27. Matulla, C.; Zhang, X.; Wang, X.L.; Wang, J.; Zorita, E.; Wagner, S.; Storch, H.V. Influence of similarity measures on the performance of the analog method for downscaling daily precipitation. *Clim. Dyn.* **2008**, *30*, 133–144. [CrossRef]
28. Obled, C.; Bontron, G.; Garçon, R. Quantitative precipitation forecasts: A statistical adaptation of model outputs through an analogue sorting approach. *Atmos. Res.* **2002**, *63*, 303–324. [CrossRef]
29. Diomede, T.; Nerozzi, F.; Paccagnella, T.; Todini, E. The use of meteorological analogues to account for LAM QPF uncertainty. *Hydrol. Earth Syst. Sci.* **2008**, *12*, 141–157. [CrossRef]
30. Bellier, J.; Zin, I.; Siblot, S.; Bontron, G. Probabilistic flood forecasting on the Rhone River: Evaluation with ensemble- and analogue-based precipitation forecasts. *E3S Web Conf.* **2016**, *7*, 1–11. [CrossRef]
31. Svensson, C. Seasonal river flow forecasts for the United Kingdom using persistence and historical analogues. *Hydrol. Sci. J.* **2016**, *61*, 19–35. [CrossRef]
32. Panagoulia, D. Artificial neural networks and high and low flows in various climate regimes. *Hydrol. Sci. J.* **2006**, *51*, 563–587. [CrossRef]
33. Komori, D.; Nakamura, S.; Kiguchi, M.; Nishijima, A.; Yamazaki, D.; Suzuki, S.; Kawasaki, A.; Oki, K.; Oki, T. Characteristics of the 2011 Chao Phraya River flood in Central Thailand. *HRL* **2012**, *6*, 41–46. [CrossRef]
34. Sudheer, K.P.; Nayak, P.C.; Ramasastri, K.S. Improving peak flow estimates in artificial neural network river flow models. *Hydrol. Process.* **2003**, *17*, 677–686. [CrossRef]
35. Yang, T.; Asanjan, A.A.; Welles, E.; Gao, X.; Sorooshian, S.; Liu, X. Developing reservoir monthly inflow forecasts using artificial intelligence and climate phenomenon information. *Water Resour. Res.* **2017**, *53*, 2786–2812. [CrossRef]
36. Singhrattna, N.; Babel, M.S.; Perret, S.R. Hydroclimate variability and its statistical links to the large-scale climate indices for the Upper Chao Phraya River Basin, Thailand. *Hydrol. Earth Syst. Sci. Disscuss.* **2009**, *6*, 6659–6690. [CrossRef]
37. Bejranonda, W.; Koch, M. The role of ocean state indices in seasonal and inter-annual climate variability of Thailand. In Proceedings of the 1st Symposium on Sustainable Water Management and Climate Change, Nakhon Pathom, Thailand, 16–17 June 2010.
38. Chansaengkrachang, K. Empirical orthogonal function analysis of rainfall over Thailand and its relationship with the Indian Ocean Dipole. *Chiangmai Univ. Int. Conf.* **2011**, *1*, 47–54.
39. Bridhikitti, A. Connections of ENSO/IOD and aerosols with Thai rainfall anomalies and associated implications for local rainfall forecasts. *Int. J. Climatol.* **2013**, *33*, 2836–2845. [CrossRef]
40. Manusthiparom, C. Hydroclimatic Prediction for Integrated Water Resources Management in the Chao Phraya River Basin in Thailand. Ph.D. Thesis, University of Tokyo, Tokyo, Japan, September 2003.
41. Panagoulia, D.; Tsekouras, G.J.; Kousiouris, G. A multi-stage methodology for selecting input variables in ANN forecasting of river flows. *Glob. Nest J.* **2017**, *19*, 49–57.
42. Wang, G.; Guo, L.; Duan, H. Wavelet Neural Network Using Multiple Wavelet Functions in Target Threat Assessment. *Sci. World J.* **2013**, *2013*. [CrossRef] [PubMed]
43. Vidakovic, B.; Mueller, P. *Wavelets for Kids: A Tutorial Introduction*; Duke University: Durham, NC, USA, 1991; pp. 1–28.
44. Boslaugh, S. *Statistics in a Nutshell*, 2nd ed.; O'Reilly Media, Inc.: Sebastopol, CA, USA, 2013; pp. 173–192. ISBN 978-1-449-31682-2.

water

MDPI

Article

Applicability of ε-Support Vector Machine and Artificial Neural Network for Flood Forecasting in Humid, Semi-Humid and Semi-Arid Basins in China

Thabo Michael Bafitlhile * and Zhijia Li

College of Hydrology and Water Resources, Hohai University, Nanjing 210098, China; zjli@hhu.edu.cn
* Correspondence: thabo.bafie@hhu.edu.cn or thabo_bafitlhile@yahoo.com; Tel.: +86-25-83787955

Received: 23 November 2018; Accepted: 1 January 2019; Published: 6 January 2019

Abstract: The aim of this study was to develop hydrological models that can represent different geo-climatic system, namely: humid, semi-humid and semi-arid systems, in China. Humid and semi-humid areas suffer from frequent flood events, whereas semi-arid areas suffer from flash floods because of urbanization and climate change, which contribute to an increase in runoff. This study applied ε-Support Vector Machine (ε-SVM) and artificial neural network (ANN) for the simulation and forecasting streamflow of three different catchments. The Evolutionary Strategy (ES) optimization method was used to optimize the ANN and SVM sensitive parameters. The relative performance of the two models was compared, and the results indicate that both models performed well for humid and semi-humid systems, and SVM generally perform better than ANN in the streamflow simulation of all catchments.

Keywords: streamflow; artificial neural network; simulation; forecasting; support vector machine; evolutionary strategy

---

1. Introduction

Timely flood forecasting with high accuracy and excellent reliability is very critical, because human societies are facing a precarious situation of recurring natural disasters such as floods due to the increase in community economy, which brings about an increase in urbanization. Hydrological models have contributed significantly to modern flood forecasting because of their ability to simulate the natural hydrological processes based on physical and empirical laws. Hydrological models are classified into two groups: conceptually or physically based models, and data-driven models (DDMs). Recently, DDMs have gained increasing attention from hydrologists as a complementary technology for modeling complex physical hydrologic processes.

Hydrological modeling can be a complicated process because of the many underlying factors that are involved in the generation of runoff and river flow. Moreover, complications arise because of nonlinearity, and the high degree of spatial and temporal variability resulting from various factors, such as catchment, storm, geomorphologic and climate characteristics. The impediments and complexities encountered when using hydrological models require several processes to be involved in the generation of runoff or streamflow, including evapotranspiration, infiltration rate, antecedent soil moisture content, land use, and land cover. Therefore, it is challenging to use models that demand more input variables like physical models due to limited data, or even in any area, environment or situation where availability of data can be challenging, such as in semi-arid and arid zones. Therefore, DDMs attract attention from hydrologists because of their proficiency in establishing the relationship between rainfall and runoff without any underlying physical processes.

The viability of DDM depends on the disposal of recorded environmental observational data that can help in predictive analytics. Therefore, use of DDMs in hydrological forecasts has become prevalent because of its ability to find a relationship between rainfall and runoff without any other underlying processes, such as evapotranspiration, drainage, and so forth, and also due to the increasing availability of data. In hydrology, DDMs are commonly used for flood forecasting, rainfall-runoff simulation, and water quality prediction. The most used DDMs for prediction and classification are the Support Vector Machine (SVM), Artificial Neural Network (ANN), Fussy rule-based system, and Model Trees (MT) [1].

DDMs are based on computer intelligence (CI) algorithms typically associated with learning from data [2]. They induce causal relationships or patterns between sets of input and output time series data in the form of a mathematical device, which is generally not related to the physics of real-world simulations [3]. They can be used for mathematical prediction problems, reconstructing highly nonlinear functions, performing classification, grouping data, and building rule-based systems [4]. In the hydrological cycle, since DDMs operate with only a limited number of assumptions about the physical behavior of the system, they require pairs of input-output training data to capture the nonlinearity relationships of a rainfall-runoff process.

The following areas have contributed to the development of DDM: artificial intelligence (AI), data mining (DM), knowledge discovery in databases (KDD), CI, machine learning (ML), intelligent data analysis (IDA), soft computing (SC), and pattern recognition. All these areas overlap, often with similar focuses and application areas. The most popular DDMs used in hydrological systems include statistical-like methods, e.g., autoregressive moving average (ARMA), multiple linear regression (MLR), and autoregressive integrated moving average (ARIMA) are popular flood frequency analysis (FFA) methods for modeling flood prediction [5]. Also, ML methods like ANN, SVM and Neuro-fuzzy (NF) have been proven to be useful for both long- and short-term flood forecast. Among popular CI methods are also genetic algorithms (GA); they are not, however, modeling paradigms or function approximation methods, but constitute an optimization approach used in model calibration or model structure optimization [4].

China has invested much time studying rainfall-runoff since the early 1960s [6]. Many years ago, hydrologists focused on developing flood forecasting models for humid areas in the southern part of the Yangtze River, China, because of frequent severe flood events [7]. Further developments due to climate change and an increase in the economy contributed to an increase in runoff. There are increasingly urgent demands for flood forecasting in semi-arid and arid areas, and these have become a severe issue in water science, since flood forecasting is entirely different from that of humid areas [7]. Modeling hydrologic processes of semi-arid and arid basins is challenging due to the specific characteristics of these basins [8]. There is a variability of runoff that sub-basins bring about both in space and time, resulting in a highly complicated rainfall-runoff relationship, and there are also lapses in storage excess runoff generation mechanisms [7,8]. Furthermore, in arid and semi-arid areas, few models are considered adequate due to the difficulty in effectively modeling infiltration-excess runoff processes as the dominant generation mechanism [9,10].

Streamflow is ephemeral under these conditions because of there being only few runoff events each year, and hence generally no hydrologic response at the outlet of the basin. Hydrological research is inadequate in semi-arid and arid zones because of insufficient hydrological and meteorological data [11]. Compared to humid regions, channel flow is perennial, and information on the internal state of the basin is obtained from streamflow records, with most models performing well, because the dominant runoff generation mechanism is saturation excess runoff [8,12]. Semi-arid and arid areas experience flash floods where rainfall intensity is usually very high, and rainfall duration is low [13,14]; there are high flood peaks and rapid flows, and substantial loss of life and property [11,14].

In semi-humid areas, saturation excess and infiltration excess runoff coexist. Consequently, the hydrological prediction is more challenging than for humid regions. Numerous types of research have been carried out to improve the hydrological model for semi-humid and semi-arid regions using conceptual models, physically based models and data-driven models. Seven hydrological models were used to simulate flood events in 3 semi-humid catchments: Xinanjiang (XAJ), Top model, SAC-SMA,

Green-Ampt, Xinanjiang-Green-Ampt, Hebei, and Xinanjiang-Haihe. The averaging method improved the Bayesian model for flood prediction, and the automatic optimization method combined with the manual optimization method calibrated hydrological models. Infiltration excess flow was combined with the surface runoff calculated using Green-Ampt (G-BMA). The results showed that models with saturation-excess mechanisms perform well in semi-humid catchments. It was found that the physically based G-BMA approach outperformed all the other models, including BMA for semi-humid regions, with a high ratio of infiltration-excess surface flow [15]. Ref. [7] also used conceptual models: mix runoff (MIX), Xinanjiang, and Northern Shaanxi were applied to three humid, three semi-humid and three semi-arid watersheds. The results indicate that it is more complicated to model drier regions than wetter watersheds. Simulation results show that all models perform satisfactorily in humid watersheds, and only Northern Shaanxi (NS) is applicable in the arid basin. In semi-humid semi-arid watersheds, XAJ and MIX performed better than NS.

SVM has proven to be robust in hydrological modeling [16]. Ref. [17] adopted the SVM model and the SVM + Ensemble Kalman Filter (SVM + EnKF) model for streamflow forecasting, and the results show that SVM overestimated flood peaks and the SVM + EnKF model provided the best results, indicating that data assimilation (DA) improves the model structure and enhance performance. ASVM estimated model streamflow using rainfall and evaporation as model inputs [18]. The results show that SVMs generalize better by successfully predicting streamflow on test data better than ANN. Ref. [19] developed a simulation framework using SVM coupled with base flow separation to reduce the lag relationship between streamflow and meteorological time series, and it helped to improve the simulation performance.

Ref. [20] employed least square SVM (LSSVM) for daily and monthly streamflow forecasting using temperature, rainfall, and streamflow input data; LSSVM outperformed Fuzzy Genetic Algorithm (FGA) and M5 Model Tree in forecasting daily streamflows. A Gamma Test (GT) derived the best input combination, SVM was employed to predict flood discharge for 2, 5, 10, 25, 50, and 100 year return periods. The SVM model performed better than ANN, adaptive neuro-fuzzy inference system (ANFIS), and nonlinear regression (NLR) [21]. ANN and SVM forecasted streamflow, and SVM successfully forecasted monthly streamflow better than ANN [22]. Ref. [23] applied SVM for real-time radar-derived rainfall forecasting. Ref. [23] used the antecedent grid-based radar-derived rainfall, grid position, and elevation as input variables and radar-derived rainfall as the output variable. The single-mode forecasting model (SMFM) and multiple-mode forecasting models (MMFM) were constructed based on the random forest (RF) and SVM to forecast 1–3-h rainfall for all grids in a catchment and concluded that the performance of SVM-based SMFM exceeds that of RF-based SMFM.

Genetic Algorithm (GA), Grid system and particle swarm optimization (PSO) methods optimized SVM in the prediction of monthly reservoir storage, and GA-based SVM performed better than the SVM optimized with other optimization methods [24]. Ref. [25] also applied GA-SVM for modeling daily reference evaporation in a semi-arid mountain area, and the results show that GA-SVM is superior to the artificial neural network (ANN) in the simulation of evaporation. Ref. [26] compared ANN and linear regression to model the rainfall-runoff relationship, and ANN showed better ability to model streamflow for semi-arid catchment than the linear regression model (LRM). Ref. [27] also used ANN validated by GR2M for simulation of streamflow in an arid region, and ANN performed well in prediction of streamflow compared to GR2M.SVM with other data-driven hydrological models, including ANN and adaptive neuro inference system (ANFIS), were used for hydrological modeling in semi-arid and humid regions, and the results show that there are no substantial variations in the performance of the models, although SVM performed better than the other models [28,29].

Neural fuzzy logic model forecasted downstream water level using upstream hourly telemetrics, and from the results, the efficiencies of the developed model show an acceptable degree of performance according to the tested performance indicators [30]. Ref. [31] compared ANFIS to the ANN model for forecasting monthly river inflow, and the results show that the ANFIS model provided higher inflow forecasting accuracy, especially during extreme flow events, compared with the ANN model.

Also, Ref. [32] compared ANFIS with ANN optimized by GA, and ANFIS still outperformed Genetic Algorithm ANN (GA-ANN). Researchers have proposed both conceptual and DDM hydrological models for different climatic and environmental conditions. However, these models are still not able to represent all the typical geo-climatic characteristics of the vast and diverse territory of China, e.g., Xiananjiang performs better in the humid region and Northern Shaanxi for the semi-arid region in China.

This study aims to gain knowledge of how DDMs, specifically SVM and ANN, perform under different geo-climatic conditions for streamflow simulation and forecasting. Many Evolutionary algorithms (EAs), like genetic algorithms (GAs), evolutionary programming (EP), differential evolution (DE), particle swarm optimization (PSO), have been applied in the field of hydrology for optimization of hydrological models. Evolutionary Strategies (ES), as one of the EAs, has not been utilized in hydrology. Therefore, this study endeavors to explore the ES approach for optimization of SVM and ANN to improve flood prediction in humid, semi-humid and semi-arid areas. This paper applied ε-SVM and ANN for streamflow simulation and forecasting of three different catchments: Changhua, Chenhe, and Zhidan; from humid, semi-humid, semi-arid regions, respectively. This research expected the ES optimization method to fine-tune the sensitive parameter of the ε-SVM and ANN to improve the performance of the models to successfully simulate and forecast streamflow for all catchments, including a semi-arid region which is complicated to model. Measures of performance evaluated and statistically tested the performance of the model, and the results show that the models successfully simulated and forecasted the streamflow of humid and semi-humid areas, and poorly forecasted the streamflow of semi-humid areas; however, SVM performed better than ANN.

2. Back-Propagation Learning Algorithm

The back-propagation algorithm, a mentor learning algorithm using the gradient descent method, is a supervised learning method divided into two phases: propagation and weight update. The two phases repeat until the performance of the network is good enough. Firstly, the inputs and outputs are both provided, the initial estimation of the weight is performed randomly to avoid a zero gradient error if initialized at zero, because it will result in no change in the network. The network then processes the inputs by propagating them forward, through every node except the input nodes, sums the product of the inputs and the weight coming in, and passes the signal through an activation function. The output of every node becomes the input of the nodes in the next layer. The output values of the model are then compared with the desired output to determine the network error [33,34]. The network error gradient is computed and then propagated backward through each weight in the network, causing the system to adjust every weight parameter in the network to reduce the value of the error function by some small amount. The process will go through many iterations, as the weights are continually adjusting, while the network is recurrently learning the target function. The set of data which enables training is 'training data', the data is processed many times as the network tries to find the right model to match the desired output.

$$x_j^l = \theta(S_j^l) = \theta\left(\sum_{i=0}^{d^{(l-1)}} w_{ij}^l \times x_i^{l-1}\right) \tag{1}$$

First, apply x to $x_1^0 \dots x_{d^0}^0$ $x_1^L = h_{(x)}$; an output of each layer until the last layer. $W = \{w_{ij}^l\}$, weights used to determine (h_x) = network output.

Then an error is obtained on a single example (x_n, y_n) is $e\left(h_{(x_n)}, y_n\right) = e_w$.
To implement stochastic gradient descent, we need the gradient of the residual error e_w.

$$\nabla e_{(w)} : \frac{\partial e_w}{\partial w_{ij}^l} \text{ for all } i, j, l$$

$$\text{Computation of } \frac{\partial e_w}{\partial w_{ij}^l} \tag{2}$$

$\frac{\partial e_w}{\partial w_{ij}^l} = \frac{\partial e_w}{\partial S_j^l} \times \frac{\partial S_j^l}{\partial w_{ij}^l}$, we have $\frac{\partial S_j^l}{\partial w_{ij}^l} = x_i^{(i-l)}$; therefore, we have the weight and the value of input in the previous layer.

We only need $\frac{\partial e_w}{\partial S_j^l} = \delta_j^l$, the intermediary quantity signal.

S_j^l is simply the $\sum x_i w_{ij}^l \dots$; δ_j^l is the signal.

For the final layer $\{l = L$ and $j = 1\}$.

$\delta_1^L = \frac{\partial e_w}{\partial S_1^L}$; $e_{(w)} = e(h(x_n), y_n) = e(x_1^L, y_n)$, the error of the neural network in its current state.

Mean Square Error: $e_{(w)} = (x_1^L - y_n)^2$.

$x_1^L = \theta(S_1^l)$, is the output of the network after being passed through an activation function.

Back-propagation of δ.

$$\delta_j^{l-1} = \frac{\partial e_w}{\partial S_i^{(l-1)}} = \sum_{j=1}^{d^l} \frac{\partial e_w}{\partial S_j^l} \times \frac{\partial S_j^l}{\partial x_i^{(l-1)}} \times \frac{\partial x_i^{(l-1)}}{\partial S_i^{(l-1)}} \tag{3}$$

$$\delta_j^{l-1} = \sum_{j=1}^{d^l} \delta_j^l \times w_{ij}^l \times \theta'(S_i^{(l-1)}); \ \delta_i^{(l-1)} = \left(1 - \left(x_i^{(l-1)}\right)^2\right) \sum_{j=1}^{d^l} w_{ij}\delta_j^{(l)}$$

As the error is propagated backward through the network to each node, the connection weights are adjusted correspondingly, based on Equation (3).

$$\Delta w_{ij}(n) = \varepsilon^* \frac{\partial e_w}{\partial w_{ij}} + a^* \Delta w_{ij}(n-1) \tag{4}$$

where $\Delta w_{ij}(n)$ and $\Delta w_{ij}(n-1)$ = weight increment between node i and j during the nth and $(n-1)$th pass, or epoch; ε and α denote learning rate and momentum respectively.

Ref. [35] used a back-propagation neural network (BPNN) for time series forecasting and employed adaptive differential evolution (ADE), differential evolution (DE) and genetic algorithm (GA) for optimization of BPNN; ADE_BPNN outperformed teh other BPNN techniques. Ref. [36] used output weight optimization-hidden weight optimization (OWO-HWO) to optimize the initial weights of the connections, GA was also used for optimizing the network, and GA was found to have tune the parameters of the network better that OWO-HWO. An emotional ANN (EANN) trained by a modified back-propagation algorithm and conventional feed-forward neural network (FFNN) were employed to model the rainfall-runoff process of two watersheds with two distinct conditions. The results showed that EANN outperformed the FFNN model, especially in the estimation of runoff peak values. EANN also performed better than FFNN in multi-step ahead forecasting [37].

ANN techniques, namely, radial basis function (RBF), FFNN and generalized regression neural network (GRNN) forecated streamflow using monthly flow data from two stations. GRNN performed better than FFNN and RBF technique in one-month-ahead streamflow forecasting. Likewise, RBF performed better than FFNN. However, RBF and FFNN simulated streamflow better than GRNN [38]. Both [39,40] confirmed that the back-propagation algorithm improves the performance of the network.

3. Support Vector Machine

SVM was developed in the early 1990s by Vapnik and his collaborators [41,42]. SVM embodies the structural risk minimization (SRM) principle, which minimizes the expected error of a learning model, reduces the problem of overfitting, and enables better generalization [43]. SVM can be applied to regression problems using an alternative loss function to draw the nonlinearity of the observed data x in a high-dimensional feature space, and then to implement a linear regression in the feature space [18,44]. SVM has been productively applied in several hydrologic studies and streamflow forecasting, as well as in groundwater monitoring and runoff prediction problems. SVM operates with the help of kernels. Radial basis function (RBF) has proved to be the best kernel function, and has been further explored in hydrology applications, together with a linear function [18]. The SVM regression function relates the input x to the output $\hat{y}$ as follows:

$$f(x) = w^T \vartheta(x) + b = \hat{y} \qquad (5)$$

where $\vartheta(x)$ is a nonlinear function mapping the input vector to a high-dimensional feature space. w and b are weight vector and bias term, respectively, and can be estimated by minimizing the following structural risk function

$$R = \frac{1}{2}w^T w + C \sum_{i=1}^{N_d} L_\varepsilon(\hat{y}_i) \qquad (6)$$

where N_d is the sample size; C represents the tradeoff between the model complexity and the empirical error; increase in the value of C will increase the relative importance of the empirical risk concerning the regularization term [45]; and L_ε is the Vapnik's ε-insensitive loss function. Both C and ε are user-defined parameters. Vapnik transformed the SVM as an optimization problem

$$\text{Maximize}: \sum_{i=1}^{N_d} y_i(a_i - a_i') - \varepsilon \sum_{i=1}^{N_d}(a_i - a_i') - \frac{1}{2}\sum_{i=1}^{N_d}\sum_{j=1}^{N_d}(a_i - a_i')(a_j - a_j')\vartheta(x_i)^T\vartheta(x_j)^T$$

$$\text{Subject to}: \sum_{i=1}^{N_d}(a_i - a_i') = 0 \quad 0 \le a_i, a_i' \le C, \ i = 1,\ 2,\ 3,\dots N_d \qquad (7)$$

where a_i and a_i' are dual Lagrange multipliers. The solution to Equation (3) is guaranteed to be unique and globally optimal, because the objective function is a convex function. The optimal Lagrange multipliers a_i^* are solved by the standard quadratic programming algorithm. Then the regression function can be rewritten as

$$f(x) = \sum_{i=1}^{N_d} a_i^* \vartheta(x_i)^T \vartheta(x)^T + b = \sum_{i=1}^{N_d} a_i^* K(x_i, x) + b \qquad (8)$$

where $K(x_i, x)$ is the Kernel function. The most used kernel function is the RBF, and this is adopted herein. Some of the solved Lagrange multipliers $a_i - a_i'$ are zero, and should be eliminated from the regression function. The regression function involves the nonzero Lagrange multipliers and the corresponding input vectors of the training data, which are referred to as support vectors (SV). The final regression can be written as:

$$f(x) = \sum_{i=1}^{N_{sv}} a_k K(x_k, x) + b \qquad (9)$$

where x_k denotes the kth support vector and N_{sv} is the number of SV. Herein, the parameter C, which is the tradeoff between the model complexity and the empirical error, is set to 1. This means that the model complexity is as important as the empirical error. In addition, it is acceptable to set the error tolerance ε to 1% for flow forecasting [46].

In general, there are different types of SVM, i.e., linear SVM, LSSVR, v-SVM, and ε-SVR with various kinds of kernel functions, i.e., linear, polynomial and RBF. The most used kernel function is the RBF, and is as follows:

$$K(x, x_i) = e^{\left(\frac{-||x-x_i||^2}{2\sigma^2}\right)} \qquad (10)$$

The SVM model has the following specifications: (1) a global optimal solution is to be found; (2) it avoids overtraining; (3) the solution will be sparse, and only a limited set of training points will contribute to the solution; and (4) nonlinear solutions can be calculated efficiently because of the usage of inner products [46].

4. Evolutionary Strategy

Evolutionary Strategy (ES) is inspired by the natural evolution of species in natural systems. I. Rechenberg pioneered and developed ES in the early 1960s, and published the first paper about ES in 1964; later, H. P. Schwefel also contributed to the improvement of ES [47]. (1+1)-ES is the original ES, because each generation consists of one child, and the best individual is chosen from between the parent and the child to be the individual in the next generation. One ancestor and one descendant per generation, and mutations created by subtracting two numbers drawn from a binomial distribution, comprised the first experiments. Its offspring replaced the ancestor if the latter was not worse than the former [48]. The first generalization of (1+1)-ES is (μ+1)-ES, also called the steady state. In (μ+1)-ES, μ parents are used in each generation, where μ is a user-defined parameter. Each parent also has an associated σ vector that controls the magnitude of mutations. The parents combine to form a single child, and then the child is mutated. The best μ individuals are chosen from among the μ parents and the child, and they become the μ parents of the next generation. Hence, its best individual never gets worse from one generation to the next (elitist), and this could be called extinction of the worst, because of the removal of one individual from the overall population at the end of each generation.

The next ES generalization strategy was (μ+λ)-ES. (μ+λ)-ES starts with a population size of μ, and mutation for each generation generates λ offspring. After the generation of children, we have (μ+λ), and the total population is sorted according to the objective function values—finally, the best μ of the total population are selected as the parents of the next generation [47,49,50]. ES is a commonly used strategy; there are μ parents and λ offspring generated by mutation. Here, none of the μ parents survive to the next generation. Since selection takes place between the λ, the best of the λ members generated become the μ parents of the next generation. The (μ, λ)-ES often works better than the (μ+λ)-ES when the fitness function is noisy or time-varying [47]. In (μ+λ)-ES, a given individual (x, σ)-ES may have a good fitness, but be unlikely to improve due to an inappropriate σ. Therefore, the (x, σ)-ES individual may remain in the population for many generations without improving, which wastes a place in the population. The (x, σ)-ES solves this problem by forcing all individuals out of the population after one generation and allowing only the best children to survive. It helps restrict survival in the next generation to those children with a good σ, which is a σ that results in a mutation vector that allows improvement in x [51]. Combining the two generalization strategies, (μ+λ)-ES and (x, σ)-ES, results in (μ, k, λ, p)-ES [52]. The population of the (μ, k, λ, p)-ES has μ parents, each has a maximum lifetime of k generations, and each generation produces λ children, each of whom has p parents.

The ES algorithms discussed above do not give options for adjusting the standard deviation σ_{kj} of the mutation. Only the adaptive (1+1)-ES algorithm can, by examining all λ of the mutation at each generation and monitoring them in terms of how they contribute to improvements. To find an optimum σ, the elements $\{\sigma_i\}$ of the standard deviation vector have to mutate as follows:

$$\sigma_i' \leftarrow \sigma_i' e^{(\tau' \rho_0 + T \rho_i)} \tag{11}$$

$$x_i' \leftarrow x_i' + \sigma_i' r^i$$

For $i \in [1, n]$, where ρ_0, ρ_i *and* r_i, are scalar random variables taken from N (0, 1), and τ and τ' are tuning parameters. The factor $\tau' \rho_0$ allows for a general change in the mutation rate of x_i', and the factors $T \rho_i$ allow for changes in the mutation rates of specific elements of x_i'. The form of the σ_i' mutation guarantees that σ_i' remains positive. Note that ρ_0 and ρ_i are equally likely to be positive as they are to be negative. This means that the exponential in Equation (11) is equally likely to be greater than one as it is to be less than one. This, in turn, means that σ_i' is just as likely to increase as it is to decrease. Schwetel suggest that this mutation is robust to changes in τ and τ', but he suggests setting them as follows (Equation (8)).

$$\tau = P_1 \left(\sqrt{2\sqrt{n}} \right)^{-1}$$ (12)

$$\tau' = P_2 \left(\sqrt{2n} \right)^{-1}$$

where n is the problem dimension, and P_1 and P_2 are proportional constants that are typically equally to 1.

Firstly, mutate σ', followed by x'. This is because σ' needs to be used to mutate x', so that the fitness of x' indicates, as accurately as possible, the appropriateness of σ'. These ideas lead to the self-adaptive (μ, λ) and $(\mu+\lambda)$ evolutionary strategies.

ES was compared with different methods of GA and penalty function for the optimization of a single-layer sound absorber, in particular with regard to frequency, and using an arbitrary frequency band. The results showed that ES outperformed other optimization methods [53]. Hierarchical ES was proposed for the construction and training of the neural network for fault diagnostics of the rotor bearing system, and the results show that ES is a feasible and effective method for solving classification problems [54].

5. Study Area and Data

In this study, three different catchments in China were selected to evaluate the performance of ε-SVM and ANN, namely, the Changhua, Chenhe and Zhidan catchments, in humid, semi-humid and semi-arid regions, respectively. The total area of the Changhua river basin is 3442 km^2, with a mainstream length of 1624 km, and an overall drop of 965 m. It is a subtropical monsoon climate with abundant rainfall and significant rainfall variation, with an annual rainfall of 1638.2 mm. During the spring season from March to early April, the southeasterly wind prevails upon the ground surface, and the amount of precipitation gradually increases. During the period from May to July, the frontal surface often stagnates or swings over the watershed, resulting in continuous rainfall with high rainfall intensity and long rainy seasons. During the summer months of July and September, the weather is hot, with prevailing southerly thunderstorm and typhoon rainfalls. From October to November, the weather is mainly sunny; from December to February, temperatures are low, with rain and snow weather.

Chenhe basin is located in the northern temperate zone, Shanxi province in China and belongs to the continental monsoon climate. The annual average precipitation is 700–900 mm. The local rainstorm is the primary cause of the flood. The average runoff depth is 100–500 mm, and the runoff coefficient is 0.2–0.5. It is a relatively high runoff yield area, with an erosion modulus of 100–200 t/km^2.

Zhidan hydrologic station is located in Chengguan Town, Zhidan County, Shaanxi province, China. It is in the longitude of 108°46′ E, 36°49′ N. The topographic distribution of the upper reaches is comprised of high mountains, gorges, and barren beaches, with substantial slope changes, sparse vegetation, and severe soil erosion. The station catchment area is 744 km^2, the river length is 81.3 km, and the distance from the estuary is 31 km. The regional climate features a moderate temperate semi-humid semi-arid zone, which is cold and dry in winter, and dry and windy in spring, with droughts and floods in summer, and which is cool and humid in autumn. The average annual temperature, precipitation, sediment transport, and discharge are 7.8 °C, 509.8 mm, 102 million tons, and 2610 m^3/s, respectively. Floods are caused by heavy rains, with rapid fluctuations, sharp peaks, and short duration. The relationship between water level and discharge is generally poor.

This study used seven rainfall stations and one hydrological station for the Changhua catchment (Figure 1a) and eleven flood events between 07/04/1998 and 24/06/2002, nine rainfall stations and one hydrological station for the Changhua catchment (Figure 1b) and eleven flood events between 26/09/2003 and 30/09/2012, and seven rainfall stations and one hydrological station for the Zhidan catchment (Figure 1c) and fifteen flood events for the period between 27/07/2000 and 13/08/2010 for the development of the hydrological models using hourly data.

(a)

(b)

(c)

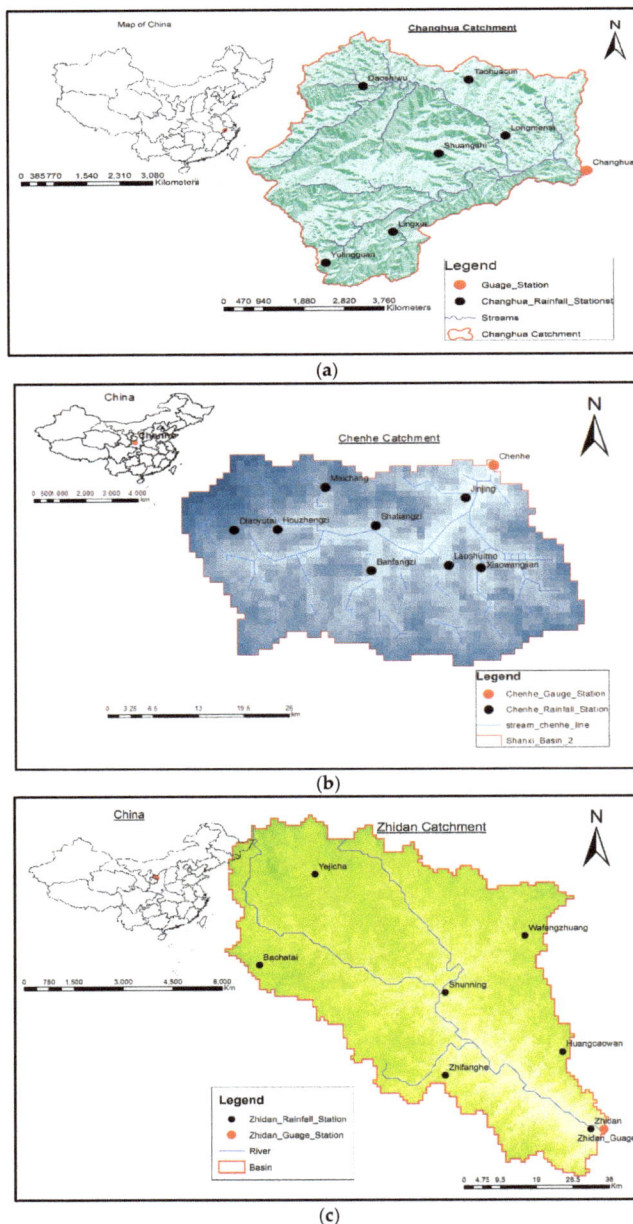

Figure 1. (**a**) Changhua catchment, (**b**) Chenhe catchment, (**c**) Zhidan catchment.

This research applied the vector autoregressive (VAR) method to determine the correlation over time and periodicities in the time series. VAR is one of the most useful, flexible, and easy-to-use models for analyzing the dynamic input of random disturbances on a system of variables [55]; Ref. [56] used VAR for streamflow sequence analysis. Ref [57] analyzed rainfall and groundwater level using VAR, and the results show that there is a significant influence of rainfall on groundwater level. Ref. [58] used VAR for rainfall forecasting; VAR accurately detected the correlation between rainfall and the coordinates of the isohyets; VAR successfully forecasted rainfall, and even outperformed the ARIMA model. Ref. [59] used monthly rainfall and streamflow data to develop streamflow trends using rainfall variability and determined causality between streamflow and rainfall for forecasting. Equation (13) shows a basic VAR model.

$$y_t = Ay_{t-1} + \cdots A_p y_{t-p} + Cx_t + \varepsilon_t \tag{13}$$

where $y_t = (Ay_{t-1} + \cdots A_p y_{t-p})$ is the $K \times 1$ vector of the observable endogenous variables, x_t is a d $\times 1$ vector of the endogenous variables, $A_1 \ldots A_p$ are $K \times K$ matrices of lag coefficients to be estimated, C is a matrix of the exogenous variable coefficient to be estimated, ε_t is white noise. Different criteria are used for optimal lag selection, including the Akaike Information Criterion (AIC), the Schwarz Information Criterion (SC), and the Hannan-Quinn information criterion (HQ). This research adopted the SC criterion for selecting the optimal lag time of each variable, and the auto correlation function is plotted to show the significant lags in the time series of each variable.

Parameter optimization of the model plays a crucial role in the performance of the model. For the ANN model, learning rate, momentum value, and above all the network architecture were optimized using the logistic function and linear function as the activation function and output function, respectively. The optimized parameters for ε-SVM are the cost constant C and error tolerance (ε), and parameter ε controls the width of the e-insensitive loss function. Large ε-values result in a flatter estimated regression function. Parameter σ controls the RBF width, which reflects the distribution range of x-values of training data. Parameters have commonly been determined by a trial and error process, which is inefficient and makes it difficult to achieve a favorable set of parameters that will provide a better-performing model—usually by means a costly grid search, which scales exponentially with the number of parameters used for finding optimal hyperparameters. Nonetheless, for effective optimization of parameters, the model should be nested with an automated, efficient optimization strategy for hyperparameters. Fortunately, the availability of advanced metaheuristic algorithms helps in providing the best solution for the multi-objective optimization problem.

This research adopted the ε-SVM and ANN models. SVM was trained by the RBF kernel function to transform a nonlinear problem into a linear function by mapping the input data into a hypothetical, high-dimensional feature space, while the back-propagation algorithm trained the ANN model. The data was standardized by the two models to remove periodicities present in the time series, and was divided into two datasets—training data set and testing data set—in a ratio of 68% and 32%, respectively. The windowing operator transformed the series data into features that describe the history for the current time point by taking a cross-section of data in time, followed by the application of a sliding window validation operator on the windowed data with a nested model algorithm inside for training and backtesting the hypothesis. When the model was finally developed, the model parameters, including (C, σ, ε for SVM) for SVM, (ε, α, network architecture) for ANN and cross-section, training size, and testing size, were finally optimized, and the model was set for streamflow prediction.

The performance of the models developed in this study was evaluated using seven different statistically different statistical measures of performance:

Root Mean Square Error (RMSE) measures overall performance across the entire range of the dataset. It is sensitive to small differences in the model performance and, being a squared measure, exhibits marked sensitivities to the larger errors that occur at higher magnitudes

$$RMSE = \sqrt{\frac{\sum_{i=1}^{N} \hat{y}_i - y_i}{N}} \tag{14}$$

Coefficient of determination (R^2) describes the proportion of the total statistical variance in the observed dataset that can be explained by the model.

$$R^2 = \left(\frac{\sum_{i=1}^{N}(\hat{y}_i - y_i)(\hat{y}_i - y_i)}{\sqrt{\sum_{i=1}^{N}(y_i - \bar{y}_i)^2}\sqrt{\sum_{i=1}^{N}(\hat{y}_i - \bar{\hat{y}}_i)^2}} \right)^2 \tag{15}$$

Nash Sutcliffe Efficiency (NSE) coefficient is sensitive to extreme values and might yield sub-optimal results when the dataset contains large outliers. Furthermore, it quantitatively describes the accuracy of model outputs other than the discharge.

$$NSE = 1 - \frac{\sum_{i=1}^{N}(y_i - \hat{y}_i)^2}{\sum_{i=1}^{N}(y_i - \bar{y}_i)^2} \tag{16}$$

Mean Square Relative Error (MSRE) provides a relative measure of model performance, the use of squared values makes it far more sensitive to the larger relative errors that will occur at lower magnitudes. It will, in consequence, be less critical of the larger absolute errors that tend to occur at higher magnitudes and more prone to potential fouling by small numbers in the observed record.

$$MSRE = 1/n \sum_{i=1}^{n}\left(\frac{y_i - \hat{y}_i}{y_i} \right)^2 \tag{17}$$

Mean Relative Error (MRE) is a relative metric that is sensitive to the forecasting errors that occur in the lower magnitudes of each dataset. In this case, because the errors are not squared, the evaluation metric is less sensitive to the larger errors that usually occur at higher values.

$$MRE = 1/n \sum_{i=1}^{n}\left(\frac{y_i - \hat{y}_i}{y_i} \right) \tag{18}$$

Mean Absolute Error (MAE) provides no information about underestimation or overestimation. It is not weighted towards higher-magnitude or lower-magnitude events, but instead evaluates all deviations from the observed values, in an equal manner and regardless of sign.

$$MAE = 1/n \sum_{i=1}^{n}|y_i - \hat{y}_i| \tag{19}$$

Mean Absolute Percentage Error (MAPE) is a relative metric that is sensitive to the forecasting errors that occur in the lower magnitudes of each dataset. In this case, because the errors are not squared, the evaluation metric is less sensitive to the larger errors that usually occur at higher magnitudes. It is nevertheless subject to potential "fouling" by small numbers in the observed record.

$$MAPE = 1/n \sum_{i=1}^{n} \frac{|y_i - \hat{y}_i|}{y_i} \tag{20}$$

6. Results

Figure 2 shows the internal correlation within the time series of rainfall and streamflow data for humid, semi-humid, and semi-arid areas with a 5% level of confidence for a lag time of up to 12 h. Figure 2 indicates that the time series of a humid area is mostly stationary, with significant spikes in the streamflow data and rainfall data collected from Longmengsi station. The semi-humid area data has a few periodic events, noticed later in every rainfall time series, but the majority of the time series is stationary, whereas in semi-arid areas, the time series shows seasonality and there is a significant contribution to the variance in the time series from the many significant spikes showing periodicity within the time series. Table 1 indicates that there are shorter delays (2–4 h) in the time series of humid areas, and longer delays (7–8 h) in semi-humid and semi-arid areas.

Table 1. Optimal lag time using Vector Autoregressive and Schwartz Information Criterion (SC).

Humid

	Changhua	Longmengsi	Taohuacun	Shuangshi	Daoshiwu	Lingxia	Yulingguan	Target
SC	4.07	1.99	3.98	4.59	4.1	4.39	4.45	9.25
Lag	2	2	4	3	2	2	4	3

Semi-Humid

	Chenhe	Diaoyutai	Houzhengzi	Maichang	Shaliangzi	Banfangzi	Laoshuimo	Xiaowangjian	Jinjing	Target
SC	2.67	1.31	1.29	1	1.23	1.25	1.09	0.99	4.06	9.11
Lag	5	7	7	7	7	7	7	7	8	6

Semi-Arid

	Yejicha	Wafangzhuang	Huangcaowan	Bachatai	Shunning	Zhifang	Zhidan	Target
SC	2.36	2.06	1.64	2.46	2.02	2.16	4.81	8.32
Lag	2	1	7	7	7	7	1	2

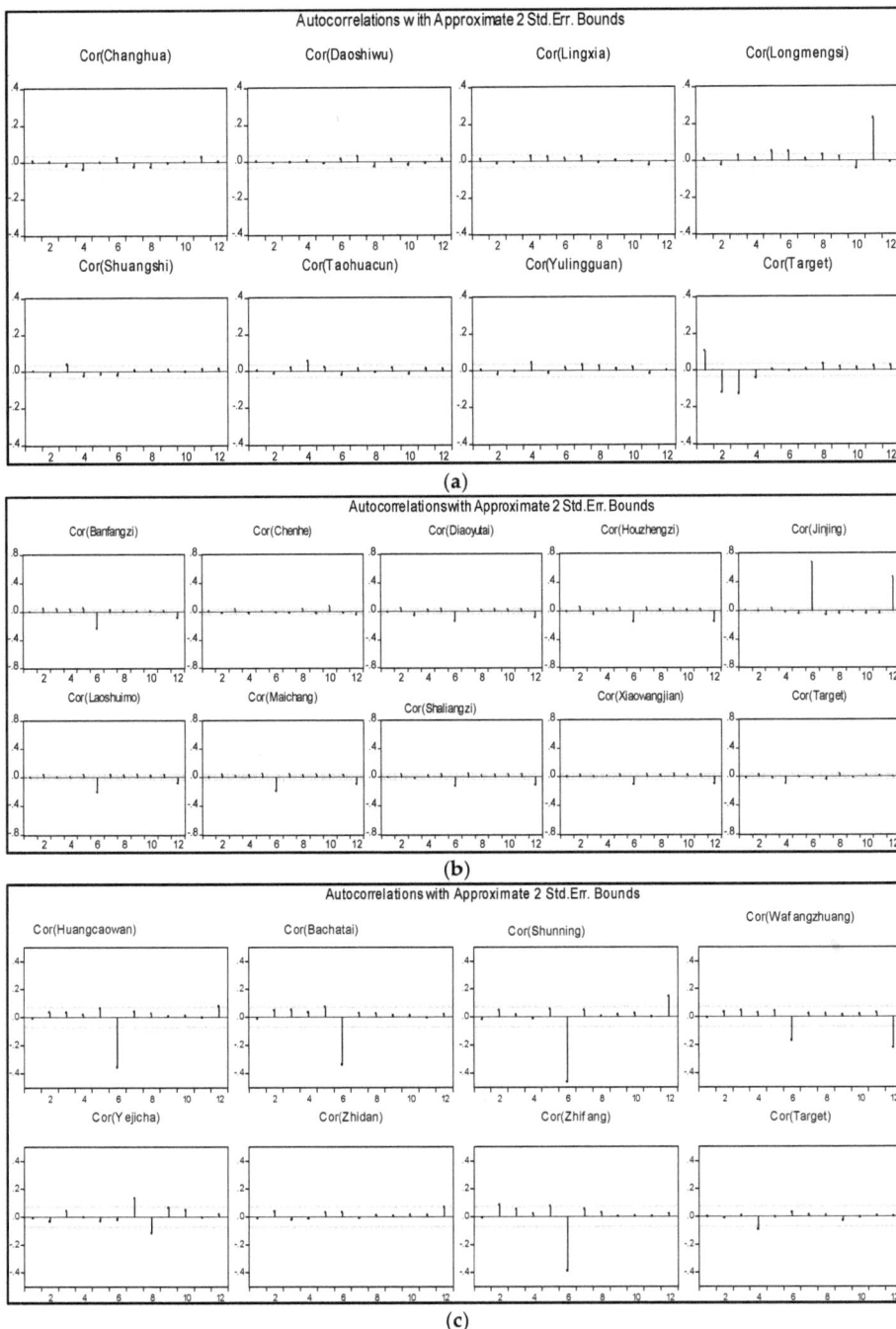

Figure 2. Autocorrelation plots of rainfall and streamflow for (**a**) Humid, (**b**) Semi-Humid, (**c**) Semi-arid catchments.

Selection of significant input variables is an essential step in the development of time series forecasting models to improve the performance of the model by removing irrelevant and redundant variables that add extra noise, which reduces the accuracy and speed of the model [60]. Correlated input variables affect the prediction ability of the model, because they obscure the true relationship that exists between important variables [61]. This study adopted a model-based approach by using a brute force feature selection method to select the significant input variables, trying all possible combinations of attribute selection in an automatic search process that optimized some indicators for model performance. Since models respond differently to input variables, SVM and ANN operate as a subprocess and return a performance vector; then, the brute force operator selects the feature set with the best performance vector Table 2.

Table 2. Selected significant input variables for SVM and ANN models.

SVM Model			ANN Model		
Humid	Semi-Humid	Semi-Arid	Humid	Semi-Humid	Semi-Arid
Longmengsi	Houzhengzi	Yejicha	Taohuacun	Diaoyutai	Yejicha
Taohuacun	Maichang	Wafangzhuang	Yulingguan	Houzhengzi	Wafangzhuang
Shuangshi	Shaliangzi	Bachatai	Shuangshi	Maichang	Bachatai
Daoshiwu	Bafangzi	Shunning	Daoshiwu	Shaliangzi	Shunning
	Xiaowangjian	Zhifang		Laoshuima	Zhifang
				Xiaowangjian	

7. Discussion

Table 3 shows seven statistical measures of performance used to assess the performance of the models for the three catchments. One distinct feature is that the models performed phenomenally during the simulation process of all the catchments. SVM successfully simulated streamflow better than ANN, as indicated by all metrics in Table 3. According to R^2 and NSE, both models accurately predicted the maximum flow for humid and semi-humid regions. However, the value of AME shows that ANN underestimated the minimum streamflow of the humid area. SVM successfully simulated streamflow of the semi-arid area, while ANN poorly simulated the both minimum and maximum flows of the streamflow, as indicated by R^2, NSE, MSRE, and MRE. The results tie in well with those of [22,62]. Due to the high degree of spatial and temporal variability in semi-arid areas, ANN underperformed, because ANN often fails to find global optima in complex and high-dimensional parameter spaces [63].

For the forecasted time in humid areas, SVM successfully forecasted streamflow up to 4 h lead time, and ANN forecasted up to 5 h, according to R^2 and NSE values. This indicated that the models predicted the streamflow very well, though ANN overestimated the low flow events according to MSRE and MAE, signifying a high deviation of predicted values from the observed values. This result is in agreement with those of [22,64], in which the authors compared the performances of ANN and SVM for streamflow forecasting. From Table 3, in the semi-humid area, the ANN model obtained the highest R^2 and NSE values for all of the forecasted period, and also obtained a lower RMSE for all periods than the SVM model. However, SVM performed well when using other evaluation metrics.

Table 3. Performance of SVM and ANN models for streamflow simulation and forecasting of all catchments.

	SVM						ANN					
	Simulation	Forecast (1 h)	Forecast (2 h)	Forecast (3 h)	Forecast (4 h)	Forecast (5 h)	Simulation	Forecast (1 h)	Forecast (2 h)	Forecast (3 h)	Forecast (4 h)	Forecast (5 h)
Changhua (Humid)												
R^2	0.99	0.97	0.90	0.81	0.72	0.63	0.99	0.98	0.94	0.82	0.82	0.74
NSE	0.99	0.97	0.90	0.80	0.70	0.59	0.99	0.98	0.93	0.75	0.73	0.72
RMSE (m^3/s)	0.46	48.34	91.16	128.35	159.99	186.79	19.60	23.25	46.15	86.25	77.54	86.97
MAE	0.34	15.12	29.22	42.07	53.94	64.70	9.39	10.73	29.97	80.90	144.55	73.32
MAPE	0.00	0.10	0.20	0.30	0.41	0.55	0.09	0.11	0.46	1.61	2.84	1.25
MSRE	0.00	0.31	0.70	1.16	1.97	4.40	0.31	0.28	4.56	35.26	88.39	28.52
MRE	0.00	0.04	0.08	0.13	0.20	0.30	-0.01	0.00	0.37	1.58	2.83	1.13
Chenhe (Semi-Humid)												
R^2	0.99	0.94	0.78	0.62	0.56	0.58	0.98	0.98	0.96	0.87	0.89	0.83
NSE	0.99	0.93	0.76	0.58	0.50	0.52	0.98	0.97	0.95	0.82	0.87	0.83
RMSE (m^3/s)	1.74	47.56	90.89	119.21	128.67	126.31	24.31	26.36	35.14	61.12	53.00	75.26
MAE	0.30	9.98	19.75	28.99	37.73	45.97	13.50	22.37	25.04	66.06	41.78	50.52
MAPE	0.00	0.07	0.15	0.25	0.36	0.49	0.45	0.44	1.03	1.47	1.62	1.46
MSRE	0.00	0.07	0.59	1.42	2.72	4.65	2.52	1.32	16.87	17.02	33.42	30.89
MRE	0.00	0.02	0.06	0.11	0.18	0.27	0.41	-0.06	0.98	-0.13	1.52	1.09
Zhidan (Semi-Arid)												
R^2	0.99	0.70	0.39	0.19	0.09	0.06	0.60	0.64	0.46	0.56	0.53	0.37
NSE	0.99	0.68	0.26	-0.11	-0.36	-0.49	0.34	0.54	0.23	0.34	0.22	-1.11
RMSE (m^3/s)	1.49	16.00	22.93	26.48	28.00	28.55	9.20	10.06	9.51	5.63	4.04	4.29
MAE	0.20	4.70	8.00	10.37	12.14	13.52	13.24	9.08	13.43	10.63	16.41	40.11
MAPE	0.26	2.18	5.54	9.80	13.46	15.94	8.97	5.45	9.07	9.05	20.15	59.35
MSRE	0.28	636.73	2758.4	5963.1	8800.2	9720.0	291.7	447.94	401.44	501.11	1599.4	12735
MRE	0.25	2.03	5.29	9.46	13.04	15.46	-7.47	-0.61	-6.69	8.85	20.01	59.30

Regarding relative evaluation metrics such as MRE, MSRE and MAE, ANN did not perform well, for 1 h and 3 h forecast time, especially, the ANN model underestimated the minimum flow, as indicated by the MRE values, which were −0.06 and −0.13, respectively. ANN was applied for hydrological modeling, the author emphasized that ANN models in hydrology tend to perform very well according to statistical metrics sensitive to errors occurring at higher magnitudes (R^2, NSE, RMSE), but perform poorly when estimating low flows because of relative metrics, which are more critical for errors occurring in the lower magnitudes (MRE, MAPE, MSRE) [65]. Ref. [65] used integrated GA to overcome the ANN problem of failing to estimate minimum flows, and also to improve the overall performance of ANN in streamflow simulation. As for semi-arid catchments, both models failed to forecast streamflow, with only the SVM model closely predicting streamflow in the results for the 1-hour-ahead prediction, as indicated by R^2, RMSE, MAE, MAPE and MRE. All metrics critically penalize ANN for 1 h lead time. SVM is penalized more by R^2 than ANN as forecasting time increases, whereas MSRE and NSE severely penalize both models with increasing lead times. Regular ANN was compared with wavelet-ANN (WA-ANN) for 1–3-day lead time forecasting, and as indicated by R^2, ANN and WA-ANN obtained 0.62 and 0.78 for a 1 day lead time, and 0.4 and 0.42 for a 3 day lead time, respectively. These results are in agreement with the findings of this paper regarding the decreasing value of R^2 obtained with increasing lead times [66]. NSE is used to assess the predictive power of hydrological models. The threshold values indicating a model's degree of sufficiency are suggested to be between 0.5 < NSE < 0.65. Therefore, the models performed poorly on semi-arid catchments, and only predicted the one-hour lead time, which is still not satisfactory.

The results from Figure 3 are in agreement with the results in Table 3, that SVM outperformed ANN in streamflow simulation of all catchments; nonetheless, both models successfully simulated streamflow, except for ANN in semi-arid areas, as confirmed by all metric values in Table 3. Points are distributed along the regression line during the first 3 h of lead time for the SVM and ANN models in humid and semi-humid areas, and then spread wide from the line of perfect agreement as the lead time increases. However, the wide distribution of points from the regression line is more significantly noticeable in SVM than in ANN. The notable feature is that the correlation coefficient between the observed discharge and the forecasted discharge also diminishes with the increase in forecast time; Figure 3 is in agreement with Table 3 that the linear regression relationship behavior between observed and estimated streamflow shows that the performance of the models decreases from humid regions to drier regions.

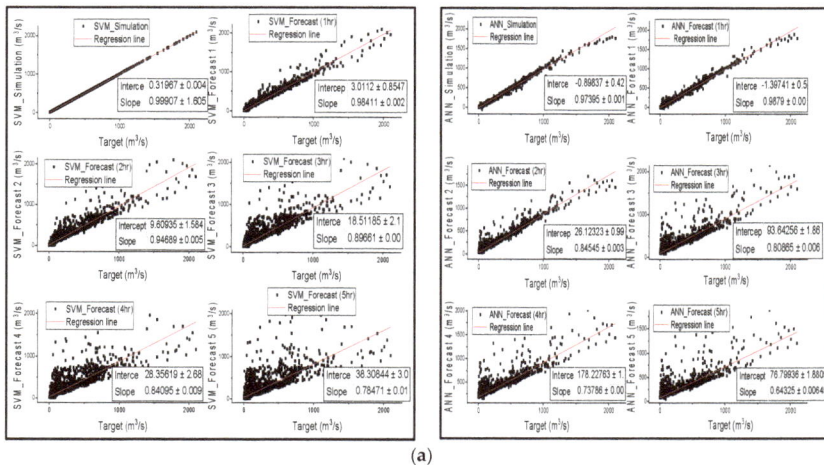

(a)

Figure 3. *Cont.*

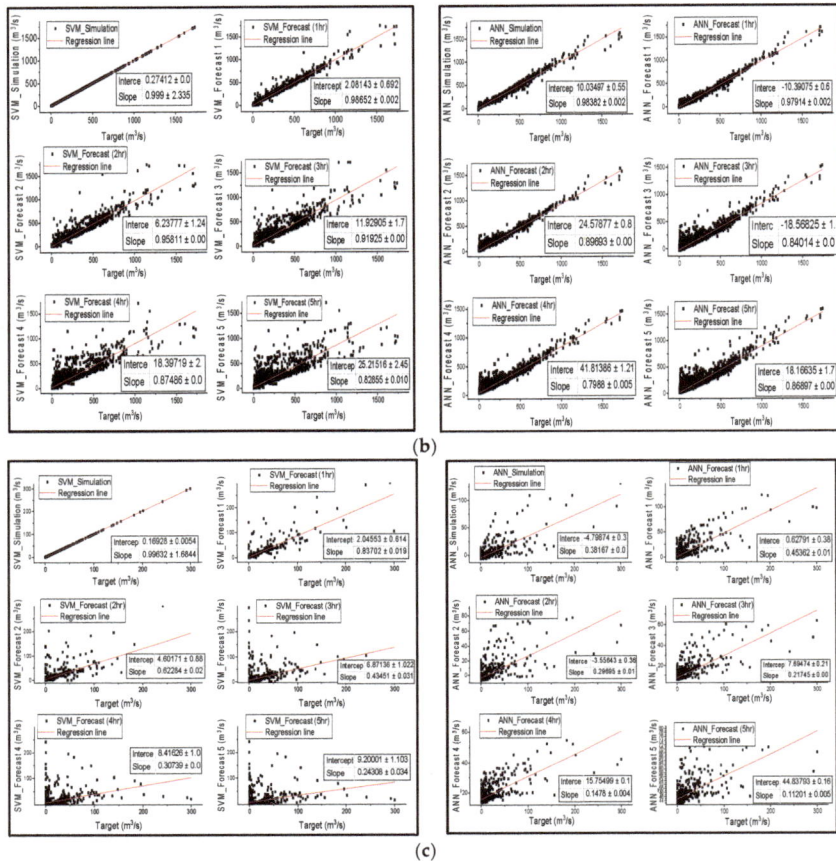

Figure 3. Scatter plots of the target (measure streamflow) versus simulated and forecasted streamflow from 1 h lead time to 5 h lead time for Changhua basin, Chenhe basin and Zhidan basin (**a**), (**b**) and (**c**) respectively for both SVM and ANN models.

Figure 4 gives a clear graphical representation of how the ANN and SVM model has simulated, and forecasted streamflow for all different catchments [67] stated that SVM could be able to prevent the influence of non-SV over the model during training by optimizing SV, and [68] mentioned that SVMs are suitable for nonlinear regression than ANN as they can identify optimal global solution. The SVM model managed to predict the shape of the hydrograph very well for simulation and all forecasted results. Most importantly SVM successfully predicted the lows and peaks of the time series of all catchments. Furthermore, SVM accurately simulated the streamflow of all catchments as indicated in Table 3 and Figure 3. SVM was used for streamflow forecasting, and the model accurately simulated the streamflow of Lang Yang river basin [16]. ANN model also performed very well in humid and semi-humid catchments. Figure 4 clearly shows that SVM outperformed ANN in streamflow simulation of all catchments.

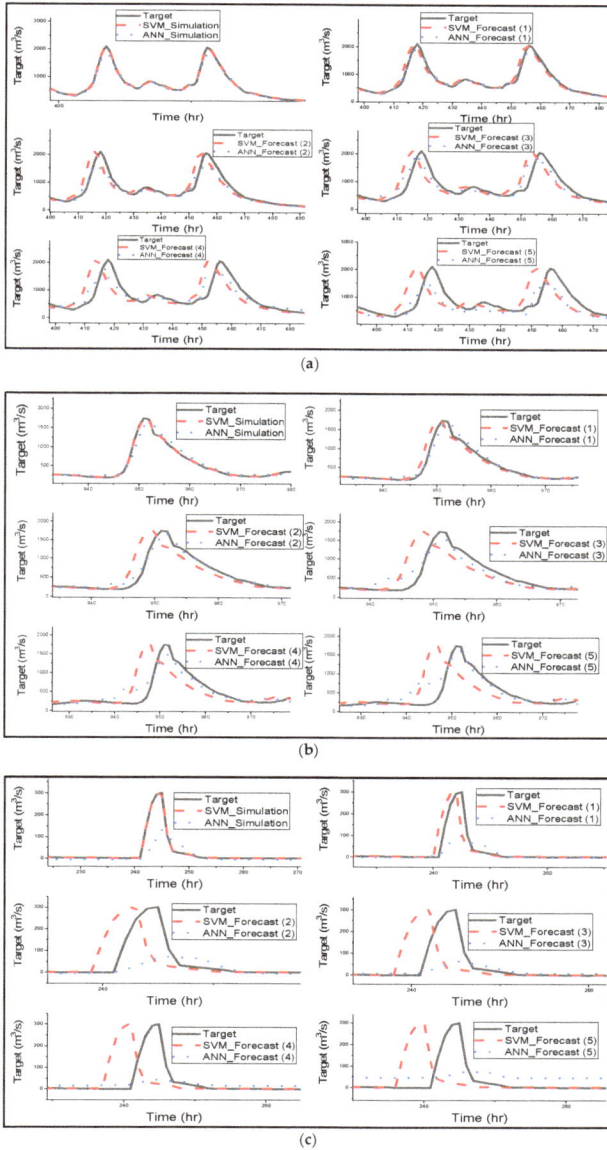

Figure 4. Observed streamflow versus simulated and forecasted streamflow from 1 h lead time to 5 h lead time for Changhua basin, Chenhe basin and Zhidan basin—(**a**), (**b**) and (**c**), respectively—for both SVM and ANN models.

The notable performance from Figure 4 is that as forecast time increases, there is an increase in the lag phase between the predicted hydrograph by SVM and the observed hydrograph. The lag is noticeable in the forecasted period of 3 of Changhua streamflow and increases as forecast time increases. Meanwhile, in the Chenhe catchments, the lag is noticeable within 3 h forecast time. Lastly, for Zhidan, the streamflow in Figure 4 is in agreement with the scatter plots for the performance of the model, as the lag is visible at 1 h lead time. Figure 4 explains the results in Table 3, indicating that all of the metrics that measure overall performance are more sensitive to hydrograph lags than the peaks, and also removes the impression given by Figure 3 that SVM overestimated the maximum flows, as this is due to the lags in the predicted hydrographs. Ref. [69] used SVR for flood stage forecasting, and the model successfully forecasted flood stage, although the results were slightly weaker than the simulation results. SVR effectively forecasted the flood stage with 1 h to 6 h lead time, and the time lag is visible for 5 h to 6 h lead times, but the phase lag is insignificant when compared to the SVM results in this study. The authors suggest that the phase lags could be due to the sensitivity of the model with respect to the lag of the input variables.

Figure 4 shows that ANN forecasted streamflow very well for humid and semi-humid catchments, but the model slightly underestimated the peak flows, and there was a drop in estimated peak flows as forecast time increased; a significant decline in estimated peak is visible with a 5 h lead time. Furthermore, the noticeable characteristic of ANN is that as the lead time increases, the model fails to predict the trend or shape of the observed time series, especially the lower and moderate flows. [65] applied ANN for streamflow forecasting, and the results are quite similar to the results of this study. The authors trained ANN using BP and GA, and the results indicate that ANN models trained with the BP algorithm tend to overestimate the minimum streamflow; therefore, Srinivasule and Jain applied GA to solve the problem. However, the ANN model trained with BP also overestimated the peak flows, whereas in this study, ANN has the problem of underestimating the peak flows. Finally, the ANN model failed to forecast semi-arid streamflow; the model completely underestimated the peak of the hydrograph for all forecasted times. This could be due to the effect of low rainfall being overestimated by the model [63].

This study applied different metrics that are critical to errors occurring at low and at peak flow, as well as those that measure the overall performance of the model. This illustrates that every statistical index has its weaknesses and limitations, as observed in Table 3, in which NSE, R^2, and RMSE heavily penalized SVM, but not MSRE, MRE and MAPE; while metrics like MAE, MSRE, MRE, and MAPE punished ANN more heavily than overall measures of performance. Therefore, consideration of other analysis tools such as graphical representation is prudent before accepting or rejecting a model based on the values of the metrics without acknowledging the flaws.

Figure 5 was considered for further analysis in the performance of the ε-SVM and ANN models. The box plots were formed by determining the median of the data set, then the median in the lower and upper quartiles of the data set, and finally the lower and upper extremes of the data set, which are connected by the whisker to the box showing the minimum and the maximum of the data set. Figure 5 shows that the SVM model accurately predicted the observed time series of all catchments, as the predicted results have the same mean, median, minimum and interquartile range. SVM slightly overestimated the peak flows with 4 h and 5 h lead times for humid areas, whereas ANN underestimated the peaks with 1–3 h lead times, and with 3–5 h lead times, the performance of the model declined, as the mean, median and range were significantly different from the observed data. Furthermore, the model overestimated the minimum flows, as indicated in Table 3. Figure 5c clearly shows that the results predicted by SVM are similar to the observed values. This confirms that for semi-arid catchments, the metrics were sensitive to the lags of the predicted hydrographs. Meanwhile, ANN did not perform well in semi-arid regions; the nonlinearity and variability of the basin could have affected the prediction accuracy of the model, because of overparameterization effects and the optimization algorithm failing to reach global optima in complex and high-dimensional spaces [63]. Figure 5 clearly shows that SVM simulated and forecasted the streamflow of all catchments better than ANN. SVM and ANN were applied for streamflow forecasting, and their results concur with the results of this study, indicating that both models performed well in predictions of streamflow, especially in the humid and semi-humid areas [26,64,70]. The results of this study are in agreement with the results of other studies, suggesting that SVM performs better than ANN. This is because SVMs are capable of evaluating more relevant information conveniently [71]. Furthermore, its quality of abiding by a structural risk minimization principle helps SVM to maximize the margin; thus, its generalizability does not decrease [44].

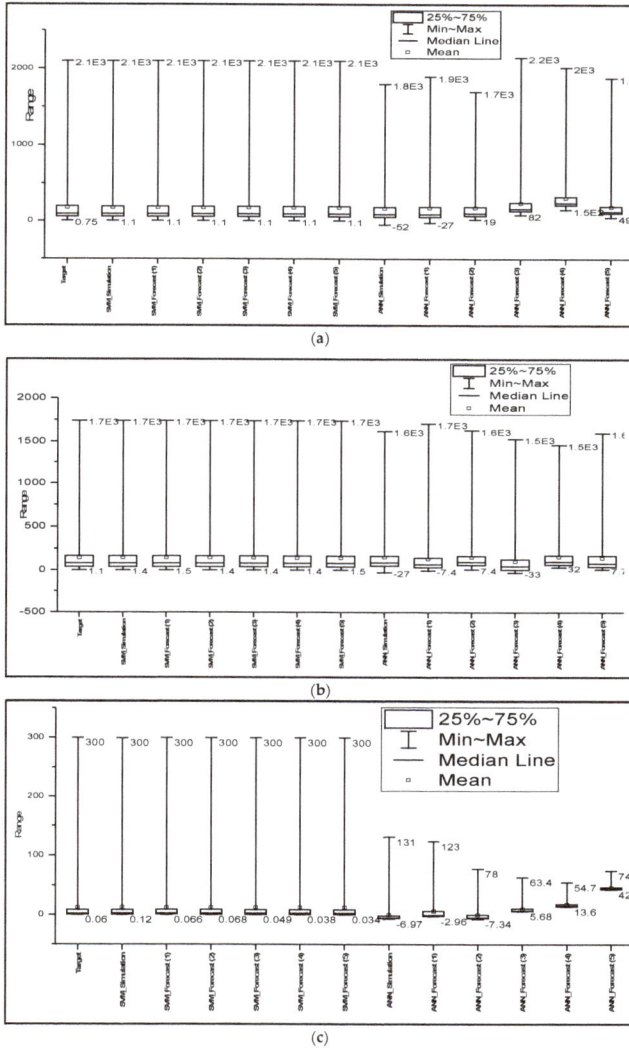

(a)

(b)

(c)

Figure 5. Box plots for forecasted time for (**a**) Changhua, (**b**) Chenhe, and (**c**) Zhidan catchments.

8. Conclusions

This study developed ANN and SVM models for flood simulation and forecasting in humid, semi-humid and semi-arid catchments using input antecedent hourly rainfall date and output antecedent hourly streamflow data. Then, the Brute force method was applied for the selection of the significant input variables of every model, and the ES algorithm was employed for the optimization of model parameters. The models were compared for a 1–5 h lead time for all catchments. The results showed that the ANN model successfully established accurate and reliable streamflow forecasting of humid and semi-humid catchments, although the model had the problem of underestimating the peak flow. Meanwhile, the SVM model successfully simulated and forecasted the streamflow of all basins, and the SVM model was able to maintain excellent accuracy for the minimum and maximum values of all basins and forecast times. The only significant drawback affecting the prediction accuracy of SVM was the presence of lags, and the lag phase increased with the forecast lead time.

Performance of SVM could be improved by removing the lags in the forecasted time series, especially of semi-arid areas, because lags were observed in 1 h lead time predictions in comparison to other areas. Although delays are inevitable when forecasting time series, ANN was found to be efficient in the elimination of lags. However, ANN performed poorly in semi-arid areas, as it overestimated minimum flow and underestimated peak flows. The possible reason for which ANN and SVM performed well in humid and semi-arid areas could be that the rainfall-runoff relationship is not complicated or dynamic, because water storage is near saturation. Whereas, in semi-arid areas, the performance is poor because of the complex and dynamic rainfall-relationship. To improve the performance of the ANN and SVM models in forecasting the streamflow of semi-arid areas, other methods for determining significant input variables should be exploited, such as evolutionary algorithms, or additional model parameter optimization. The ensemble of the models could also help to improve prediction and eliminate the lags in the forecasted time series.

Author Contributions: T.M.B.: Designed, developed the models, and the conceptual framework and analysed the data, wrote the manuscript; Z.L.: Devised and supervised the project, and the findings of this work, verfified analytical methods. All authors discussed the results and contributed to the final manuscript.

Funding: This research was supported by Natural Science Foundation of China (51679061), National Key R&D Program of China (Grant No. 2016YFC0402705) and by the Research Funds for the Central Universities 2018B11014.

Acknowledgments: First of all I would like to deeply thank Almighty God for granting me good health, strength, and knowledge to undertake this research study and complete it soon enough. I woud also like to thank my supervisor Prof Zhijia Li for his exceptional support, guidance, and supervision throughout this research. I exprss my gratitude to Wembu Huo for helping with data collection and his support. Lastly I would like to immensely thank my family and friends for their breathtaking love, support, motivation and encouragement at all times ensuring that the fire keeps burning.

Conflicts of Interest: Authors declare no conflicts of interest.

References

1. Solomatine, D.P. Data-Driven Modeling and Computational Intelligence Methods in Hydrology. In *Encyclopedia of Hydrological Sciences*; John Wiley & Sons, Ltd.: Chichester, UK, 2005. [CrossRef]
2. Solomatine, D.P.; Ostfeld, A. Data-driven modelling: Some past experiences and new approaches. *J. Hydroinform.* **2008**, *10*, 3. [CrossRef]
3. Solomatine, D.P.; Price, R.K. Innovative approaches to flood forecasting using data driven and hybrid modelling. *Education* **2004**, 1–8. [CrossRef]
4. Anderson, M.; McDonnell, J. Encyclopedia of Hydrological Sciences. 2005. Available online: http://www.citeulike.org/group/1428/article/764778 (accessed on 28 July 2018).
5. Mosavi, A.; Ozturk, P. Flood Prediction Using Machine Learning, Literature Review. *Water* **2018**, *10*, 1536. [CrossRef]
6. Jin, H.; Liang, R.; Wang, Y.; Tumula, P. Flood-runoff in semi-arid and sub-humid regions, a case study: A simulation of Jianghe watershed in northern China. *Water* **2015**, *7*, 5155–5172. [CrossRef]
7. Kan, G.; He, X.; Ding, L.; Li, J.; Liang, K.; Hong, Y. Study on applicability of conceptual hydrological models for flood forecasting in humid, semi-humid semi-arid and arid basins in China. *Water* **2017**, *9*. [CrossRef]
8. Wang, L.; Li, Z.; Bao, H. Application of developed grid-ga distributed hydrologic model in semi-humid and semi-arid basin. *Trans. Tianjin Univ.* **2010**, *16*, 209–215. [CrossRef]
9. Pilgrim, D.H.; Chapman, T.G.; Doran, D.G. Problems of rainfall-runoff modelling in arid and semiarid regions. *Hydrol. Sci. J.* **1988**, *33*, 379–400. [CrossRef]
10. Hao, G.; Li, J.; Song, L.; Li, H.; Li, Z. Comparison between the TOPMODEL and the Xin'anjiang model and their application to rainfall runoff simulation in semi-humid regions. *Environ. Earth Sci.* **2018**, *77*. [CrossRef]

11. Lin, X. Flash Floods in Arid and Semi-Arid Zones. *International Hydrological Programme—UNESCO*. 1999. Available online: http://bases.bireme.br/cgi-bin/wxislind.exe/iah/online/?IsisScript=iah/iah.xis&src=google&base=REPIDISCA&lang=p&nextAction=lnk&exprSearch=92304&indexSearch=ID (accessed on 20 October 2018).
12. Dunne, T.; Black, R.D. Partial Area Contributions to Storm Runoff in a Small New England Watershed. *Water Resour. Res.* **1970**, *6*, 1296–1311. [CrossRef]
13. Marchi, L.; Borga, M.; Preciso, E.; Gaume, E. Characterisation of selected extreme flash floods in Europe and implications for flood risk management. *J. Hydrol.* **2010**, *394*, 118–133. [CrossRef]
14. Ragettli, S.; Zhou, J.; Wang, H.; Liu, C.; Guo, L. Modeling flash floods in ungauged mountain catchments of China: A decision tree learning approach for parameter regionalization. *J. Hydrol.* **2017**, *555*, 330–346. [CrossRef]
15. Huo, W.; Li, Z.; Wang, J.; Yao, C.; Zhang, K.; Huang, Y. Multiple hydrological models comparison and an improved Bayesian model averaging approach for ensemble prediction over semi-humid regions. *Stoch. Environ. Res. Risk Assess.* **2018**. [CrossRef]
16. Yu, P.S.; Chen, S.T.; Chang, I.F. Support vector regression for real-time flood stage forecasting. *J. Hydrol.* **2006**. [CrossRef]
17. Li, X.-L.; Lü, H.; Horton, R.; An, T.; Yu, Z. Real-time flood forecast using the coupling support vector machine and data assimilation method. *J. Hydroinform.* **2014**, *16*, 973. [CrossRef]
18. Dibike, Y.B.; Velickov, S.; Solomatine, D.; Abbott, M.B. Model induction with support vector machines: Introduction and applications. *J. Comput. Civ. Eng.* **2001**, *15*, 208–216. [CrossRef]
19. Tongal, H.; Booij, M.J. Simulation and forecasting of streamflows using machine learning models coupled with base flow separation. *J. Hydrol.* **2018**, *564*, 266–282. [CrossRef]
20. Adnan, R.M.; Yuan, X.; Kisi, O.; Adnan, M.; Mehmood, A. Stream Flow Forecasting of Poorly Gauged Mountainous Watershed by Least Square Support Vector Machine, Fuzzy Genetic Algorithm and M5 Model Tree Using Climatic Data from Nearby Station. *Water Resour.* **2018**, *32*, 4469–4486. [CrossRef]
21. Sharifi Garmdareh, E.; Vafakhah, M.; Eslamian, S.S. Regional flood frequency analysis using support vector regression in arid and semi-arid regions of Iran. *Hydrol. Sci. J.* **2018**, *63*, 426–440. [CrossRef]
22. Adnan, R.; Yuan, X.; Kisi, O.; Yuan, Y. streamflow forecasting using artificial neural network and support vector machine model. *Am. Sci. Res. J. Eng. Technol. Sci.* **2017**, *29*, 286–294. [CrossRef]
23. Yu, P.-S.; Yang, T.-C.; Chen, S.-Y.; Kuo, C.-M.; Tseng, H.-W. Comparison of random forests and support vector machine for real-time radar-derived rainfall forecasting. *J. Hydrol.* **2017**, *552*, 92–104. [CrossRef]
24. Su, J.; Wang, X.; Liang, Y.; Chen, B. GA-Based Support Vector Machine Model for the Prediction of Monthly Reservoir Storage. *J. Hydrol. Eng.* **2014**, *19*, 1430–1437. [CrossRef]
25. Yin, Z.; Wen, X.; Feng, Q.; He, Z.; Zou, S.; Yang, L. Integrating genetic algorithm and support vector machine for modeling daily reference evapotranspiration in a semi-arid mountain area. *Hydrol. Res.* **2017**, *48*, 1177–1191. [CrossRef]
26. Aichouri, I.; Hani, A.; Bougherira, N.; Djabri, L.; Chaffai, H.; Lallahem, S. River Flow Model Using Artificial Neural Networks. *Energy Procedia* **2015**, *74*, 1007–1014. [CrossRef]
27. Dounia, M.; Sabri, D.; Yassine, D. Rainfall—Rain off Modeling Using Artificial Neural Network. *APCBEE Procedia* **2014**, *10*, 251–256. [CrossRef]
28. He, Z.; Wen, X.; Liu, H.; Du, J. A comparative study of artificial neural network, adaptive neuro fuzzy inference system and support vector machine for forecasting river flow in the semiarid mountain region. *J. Hydrol.* **2014**. [CrossRef]
29. Raghavendra, S.; Deka, P.C. Support vector machine applications in the field of hydrology: A review. *Appl. Soft Comput. J.* **2014**. [CrossRef]
30. Perera, E.D.P.; Lahat, L. Fuzzy logic based flood forecasting model for the Kelantan River basin, Malaysia. *J. Hydro-Environ. Res.* **2015**. [CrossRef]
31. El-Shafie, A.; Taha, M.R.; Noureldin, A. A neuro-fuzzy model for inflow forecasting of the Nile river at Aswan high dam. *Water Resour. Manag.* **2007**. [CrossRef]
32. Mukerji, A.; Chatterjee, C.; Raghuwanshi, N.S. Flood Forecasting Using ANN, Neuro-Fuzzy, and Neuro-GA Models. *J. Hydrol. Eng.* **2009**. [CrossRef]

33. Al-Abadi, A.M. Modeling of stage–discharge relationship for Gharraf River, southern Iraq using backpropagation artificial neural networks, M5 decision trees, and Takagi–Sugeno inference system technique: A comparative study. *Appl. Water Sci.* **2016**, *6*, 407–420. [CrossRef]

34. Khan, M.Y.A.; Hasan, F.; Panwar, S.; Chakrapani, G.J. Neural network model for discharge and water-level prediction for Ramganga River catchment of Ganga Basin, India. *Hydrol. Sci. J.* **2016**, *61*, 2084–2095. [CrossRef]

35. Wang, L.; Zeng, Y.; Chen, T. Back propagation neural network with adaptive differential evolution algorithm for time series forecasting. *Expert Syst. Appl.* **2015**, *42*, 855–863. [CrossRef]

36. Veintimilla-Reyes, J.; Cisneros, F.; Vanegas, P. Artificial Neural Networks Applied to Flow Prediction: A Use Case for the Tomebamba River. *Procedia Eng.* **2016**, *162*, 153–161. [CrossRef]

37. Nourani, V. An Emotional ANN (EANN) approach to modeling rainfall-runoff process. *J. Hydrol.* **2017**, *544*, 267–277. [CrossRef]

38. Kişi, Ö. River flow forecasting and estimation using different artificial neural network techniques. *Hydrol. Res.* **2008**, *39*, 27–40. [CrossRef]

39. Campolo, M.; Soldati, A.; Andreussi, P. Artificial neural network approach to flood forecasting in the River Arno. *Hydrol. Sci. J.* **2003**, *48*, 381–398. [CrossRef]

40. Jayawardena, A.W.; Fernando, T. River flow prediction: An artificial neural network approach. In *Regional Management of Water Resources*; Iahs Publication: Wallingford, UK, 2001; pp. 239–246.

41. Boser, J.A. Microcomputer Needs Assessment of American Evaluation Association Members. *Am. J. Eval.* **1992**, *13*, 92–93. [CrossRef]

42. Vapnik, V. The Support Vector Method of Function Estimation. In *Nonlinear Modeling*; Springer: Boston, MA, USA, 1998; pp. 55–85. [CrossRef]

43. Wu, M.C.; Lin, G.F. An hourly streamflow forecasting model coupled with an enforced learning strategy. *Water* **2015**, *7*, 5876–5895. [CrossRef]

44. Smola, A. Regression Estimation with Support Vector Learning Machines. Master's Thesis, Technische Universitat Munchen, Munich, Germany, 1996; pp. 1–78. [CrossRef]

45. Wang, W.; Xu, D.; Chau, K.; Chen, S. Improved annual rainfall-runoff forecasting using PSO–SVM model based on EEMD. *J. Hydroinform.* **2013**, *15*, 1377–1390. [CrossRef]

46. Thissen, U.; van Brakel, R.; de Weijer, A.; Melssen, W.; Buydens, L.M. Using support vector machines for time series prediction. *Chemometr. Intell. Lab. Syst.* **2003**, *69*, 35–49. [CrossRef]

47. Simon, D. *Evolutionary Optimization Algorithms*; John Wiley & Sons: Hoboken, NJ, USA, 2013. [CrossRef]

48. Beyer, H.-G.; Schwefel, H.-P. Evolution strategies–A comprehensive introduction. *Nat. Comput.* **2002**, *1*, 3–52. [CrossRef]

49. Costa, L.; Oliveira, P. Evolutionary algorithms approach to the solution of mixed integer non-linear programming problems. *Comput. Chem. Eng.* **2001**, *25*, 257–266. [CrossRef]

50. Richter, J.N. On Mutation and Crossover in the Theory of Evolutionary Algorithms. *ProQuest Dissertations and Theses*. 2010. Available online: https://search.proquest.com/docview/305202204?accountid=11664 (accessed on 15 September 2018).

51. Yin, X.; Zhang, J.; Wang, X. Sequential injection analysis system for the determination of arsenic by hydride generation atomic absorption spectrometry. *Fenxi Huaxue.* **2004**, *32*, 1365–1367. [CrossRef]

52. Hansen, N.; Ostermeier, A. Completely Derandomized Self-Adaptation in Evolution Strategies. *Evol. Comput.* **2001**, *9*, 159–195. [CrossRef] [PubMed]

53. Gholamipoor, M.; Ghadimi, P.; Alavidoost, M.H.; Feizi Chekab, M.A. Application of evolution strategy algorithm for optimization of a single-layer sound absorber. *Cogent Eng.* **2014**, *1*. [CrossRef]

54. Chen, Z.; He, Y.; Chu, F.; Huang, J. Evolutionary strategy for classification problems and its application in fault diagnostics. *Eng. Appl. Artif. Intell.* **2003**, *16*, 31–38. [CrossRef]

55. Zivot, E.; Wang, J. Vector Autoregressive Models for Multivariate Time Series. In *Modeling Financial Time Series with S-PLUS®*; Springer: Berlin, Germany, 2006; pp. 385–429. [CrossRef]

56. Ledolter, J. The analysis of multivariate time series applied to problems in hydrology. *J. Hydrol.* **1978**, *36*, 327–352. [CrossRef]

57. Hau, M.C.; Tong, H. A practical method for outlier detection in autoregressive time series modelling. *Stoch. Hydrol. Hydraul.* **1989**, *3*, 241–260. [CrossRef]

58. Nugroho, A.; Hartati, S.; Mustofa, K. Vector Autoregression (Var) Model for Rainfall Forecast and Isohyet Mapping in Semarang–Central Java–Indonesia. *J. Adv. Comput. Sci. Appl.* **2014**, *5*, 44–49. [CrossRef]
59. Iddrisu, W.A.; Nokoe, K.S.; Akoto, I. Modelling the trend of flows with respect to rainfall variability using vector autoregression. *Int. J. Adv. Res.* **2016**, *4*, 125–140. [CrossRef]
60. Tran, H.D.; Muttil, N.; Perera, B.J.C. Selection of significant input variables for time series forecasting. *Environ. Model. Softw.* **2015**, *64*, 156–163. [CrossRef]
61. Alexandridis, A.; Patrinos, P.; Sarimveis, H.; Tsekouras, G. A two-stage evolutionary algorithm for variable selection in the development of RBF neural network models. *Chemometr. Intell. Lab. Syst.* **2005**, *75*, 149–162. [CrossRef]
62. Gizaw, M.S.; Gan, T.Y. Regional Flood Frequency Analysis using Support Vector Regression under historical and future climate. *J. Hydrol.* **2016**, *538*, 387–398. [CrossRef]
63. De Vos, N.J.; Rientjes, T.H.M. Constraints of artificial neural networks for rainfall-runoff modelling: Trade-offs in hydrological state representation and model evaluation. *Hydrol. Earth Syst. Sci. Discus.* **2005**, *2*, 365–415. [CrossRef]
64. Kalteh, A.M. Monthly river flow forecasting using artificial neural network and support vector regression models coupled with wavelet transform. *Comput. Geosci.* **2013**, *54*, 1–8. [CrossRef]
65. Srinivasulu, S.; Jain, A. Rainfall-Runoff Modelling: Integrating Available Data and Modern Techniques. In *Practical Hydroinformatics: Computational Intelligence and Technological Developments in Water Applications*; Springer: Berlin/Heidelberg, Germany, 2008; pp. 59–70. [CrossRef]
66. Adamowski, J.; Sun, K. Development of a coupled wavelet transform and neural network method for flow forecasting of non-perennial rivers in semi-arid watersheds. *J. Hydrol.* **2010**, *390*, 85–91. [CrossRef]
67. Bhagwat, P.P.; Maity, R. Multistep-ahead River Flow Prediction using LS-SVR at Daily Scale. *J. Water Resour. Prot.* **2012**, *4*, 528–539. [CrossRef]
68. Tehrany, M.S.; Pradhan, B.; Mansor, S.; Ahmad, N. Flood susceptibility assessment using GIS-based support vector machine model with different kernel types. *Catena* **2015**, *125*, 91–101. [CrossRef]
69. Yu, P.S.; Chen, S.T.; Chang, I.F. Flood stage forecasting using support vector machines. *Geophys. Res. Abstr.* **2005**, *7*, 41–76.
70. Suliman, A.; Nazri, N.; Othman, M.; Abdul, M.; Ku-mahamud, K.R. Artificial Neuaral Network and Support Vector Machine in Flood Forecasting: A Review. *J. Hydroinform.* **2013**, *15*, 327–332. [CrossRef]
71. Cristianini, N.; Shawe-Taylor, J. *An Introduction to Support Vector Machines and Other Kernel-Based Methods*. 2000. Available online: https://books.google.com.br/books?hl=en&lr=&id=_PXJn_cxv0AC&oi=fnd&pg=PR9&dq=:+An+Introduction+to+Support+Vector+Machines+and+Other+Kernel-Based+Learning+Methods&ots=xSQi5BXq3e&sig=e0GieLD8UrBJf8Xf060CumoL0wA (accessed on 21 December 2018).

water MDPI

Article

Statistical Analysis of Extreme Events in Precipitation, Stream Discharge, and Groundwater Head Fluctuation: Distribution, Memory, and Correlation

Shawn Dawley [1], Yong Zhang [1,*], Xiaoting Liu [2], Peng Jiang [3], Geoffrey R. Tick [1], HongGuang Sun [3], Chunmiao Zheng [4] and Li Chen [5]

[1] Department of Geological Sciences, University of Alabama, Tuscaloosa, AL 35487, USA; sfdawley@crimson.ua.edu (S.D.); gtick@ua.edu (G.R.T.)
[2] College of Mechanics and Materials, Hohai University, Nanjing 210098, China; lxt5572918@foxmail.com
[3] State Key Laboratory of Hydrology-Water Resources and Hydraulic Engineering, Hohai University, Nanjing 210098, China; peng.jiang.j@gmail.com (P.J.); shg@hhu.edu.cn (H.S.)
[4] Guangdong Provincial Key Laboratory of Soil and Groundwater Pollution Control, School of Environmental Science & Engineering, Southern University of Science and Technology, Shenzhen 518055, Guangdong, China; zhengcm@sustc.edu.cn
[5] Division of Hydrologic Sciences, Desert Research Institute, Las Vegas, NV 89119, USA; Li.Chen@dri.edu
*   Correspondence: yzhang264@ua.edu

Received: 16 February 2019; Accepted: 26 March 2019; Published: 5 April 2019

Abstract: Hydrological extremes in the water cycle can significantly affect surface water engineering design, and represents the high-impact response of surface water and groundwater systems to climate change. Statistical analysis of these extreme events provides a convenient way to interpret the nature of, and interaction between, components of the water cycle. This study applies three probability density functions (PDFs), Gumbel, stable, and stretched Gaussian distributions, to capture the distribution of extremes and the full-time series of storm properties (storm duration, intensity, total precipitation, and inter-storm period), stream discharge, lake stage, and groundwater head values observed in the Lake Tuscaloosa watershed, Alabama, USA. To quantify the potentially non-stationary statistics of hydrological extremes, the time-scale local Hurst exponent (TSLHE) was also calculated for the time series data recording both the surface and subsurface hydrological processes. First, results showed that storm duration was most closely related to groundwater recharge compared to the other storm properties, while intensity also had a close relationship with recharge. These relationships were likely due to the effects of oversaturation and overland flow in extreme total precipitation storms. Second, the surface water and groundwater series were persistent according to the TSLHE values, because they were relatively slow evolving systems, while storm properties were anti-persistent since they were rapidly evolving in time. Third, the stretched Gaussian distribution was the most effective PDF to capture the distribution of surface and subsurface hydrological extremes, since this distribution can capture the broad transition from a Gaussian distribution to a power-law one.

Keywords: statistical analysis; hydrological extremes; stretched Gaussian distribution; Hurst exponent

---

1. Introduction

Low probability and high impact extremes in hydrology, such as storms, play an important role in characterizing the hydrologic system and affecting water infrastructure design [1–4]. Capturing and defining these extremes is still a difficult task in hydrology because of the complexity of natural systems, as well as their variability in both space and time [5,6]. Compared to the prohibitive physical process-based models requiring intensive data that are typically not available for many study sites, statistical analysis of hydrological extremes is an attractive tool to interpret these extreme events within and across systems from simple measurements [7–10].

There are two major challenges when applying basic statistics to analyze hydrologic extremes. First, hydrologic processes in the water cycle are interconnected, while basic statistical analysis of extremes often does not consider multiple systems and tends to oversimplify the complexity of the correlated processes like precipitation and groundwater table fluctuations [11,12]. Understanding how the subtle properties of one system's extreme events can affect the other interconnected systems requires more in-depth analysis and use of advanced statistical techniques. Second, these basic statistical techniques often rely on assumptions or major simplifications of properties for water systems, such as stationarity in both space and time, which may not always be valid for real world dynamics [13–15]. These issues motivated this study.

One example of the assumption/simplification used by basic statistical studies in hydrological extremes is the well-known Gumbel distribution. Statistics of extremes is one of the historical topics in hydrology, including for example probability density functions (PDFs) developed for analyzing the distribution of hydrologic extremes [13]. One of the fundamental distributions used in hydrology is the Gumbel distribution, a case of the generalized extreme value distribution where the shape parameter is 0, proposed by Gumbel to fit the frequency of floods [16]. This distribution was developed before many advances in statistics and computing, and it has been widely used for decades by hydrologists. To apply this theory, data must be assumed to be homogenous, meaning that there should be no change in climate or basin characteristics during both the observation period and any period that predictions are made. This assumption, however, may not be valid considering the intrinsic evolution of the dynamic, natural systems [13,17,18]. One promising way to overcome the assumption of the homogeneous system is non-stationary statistics [19,20].

This study aims to fill two knowledge gaps when analyzing the distribution, memory, and correlation embedded in the hydrological extremes. First, we will identify the PDF that can define the overall distribution pattern of real-world hydrological extremes. Several studies [21,22] found that various random processes in hydrology usually follow a one-sided distribution with a heavy tail. This finding motivated us to test two physically meaningful PDFs, the stretched Gaussian distribution and the α-stable distribution, in capturing the hydrological extremes and comparing them with the classical Gumbel distribution. All three of the distributions allow for a one-sided, heavy tailed PDF, which is a common occurrence in natural water systems [23–25]. Each distribution has different parameters, and they can all be conveniently computed and parameterized for series. Since the extremes of water systems are often needed to determine the infrastructure design and management plans, improving the prediction using the most reasonable distribution as a model can greatly improve water management practices [26].

Second, we will evaluate the non-stationary statistics for hydrological extremes. One example of a non-stationary statistic is the Hurst exponent, first developed to quantify the long-term persistence of water storage of reservoirs by Hurst [27]. Peng et al. [28] introduced detrended fluctuation analysis (DFA) to investigate long-range correlation (also called memory) of a series that contains significant noise, such as DNA nucleotides, financial series, seismic analysis, and hydraulic data [28–32]. Peng et al.'s contribution [28] allowed for a time-scale local Hurst exponent (TSLHE) to be calculated, defining a non-stationary statistic. Zhang and Schilling [32] applied DFA to investigate the scaling behavior of hydraulic head and base flow, which is related to the groundwater recharge and can be used to determine the fractal dimension and Hurst exponent of the series. Zhou et al. [33] used a multi fractal DFA (MF-DFA) method to show that river discharge in the Yangtze basin was non-stationary and had different correlation properties depending on the measurement location in the watershed. Tong et al. [34] used the Hurst exponent to quantify the variation of droughts in both space and time. These successful applications motivated us to apply the TSLHE to explore the evolution of hydrological extreme properties.

Different from most of the previous works, this study tries to evaluate and correlate surface and subsurface hydrological extreme events. We will investigate the effects that extreme storm events, of different properties, have on the fluctuations in surface and subsurface water systems. To the best of our knowledge, these fluctuations have not been compared to different storm properties. With potential changes in climate, the storm properties are expected to evolve in time [35–37]. These fluctuations are correlated to the properties over time and compared across systems and storm properties. The memory of the water series is also investigated using the TSLHE, which is correlated to the storm properties. Finally, we will investigate the distributions that fit each of the different data sets and storm properties.

The rest of this study is organized in four sections. Section 2 briefly introduces the study site and the data sources used for statistical analysis. Methods are then described, including the calculations of storm properties, the Hurst exponent, and the distributions used to fit the time series data. Section 3 presents the results of statistical analysis for both the surface water and groundwater. Section 4 discusses the statistical results, and Section 5 draws the main conclusions.

2. Study Site and Methodologies

2.1. Background of the Study Site and Data Source

The study site was the Lake Tuscaloosa watershed in Tuscaloosa, northern Alabama, USA. The lake has been the primary source of drinking water for the city of Tuscaloosa (200,000 consumers in 2014) since 1970. Lake Tuscaloosa has an approximate volume of 150,000,000 m^3 and a surface area of 23.82 km^2 [38]. The lake is fed by four major streams which have U.S. Geological Survey (USGS) gauge stations—North Creek, Binion Creek, Bush Creek, and Carroll Creek—which represent most of the surface flow into the lake, as well as many smaller streams that do not significantly contribute to the lake. North and Binion Creeks have the highest discharge and the best coverage of measurements, and hence they are used as the streams for this study. The lake sits primarily in the Pottsville Formation, a Pennsylvanian aged sandstone interbedded with shale and siltstone, as well as the lower Coker Formation, a Cretaceous unit with sand and gravel beds. The lake is partially fed by groundwater from these aquifers, as are the streams that flow into it [39].

The study site has a humid subtropical climate, typical of the Deep South weather region of the U.S., with abundant rain (with an average precipitation of 1336.04 mm/year) and rare measurable snowfall. The local climate is affected significantly by the Gulf of Mexico which brings relatively warmer and moist air. This causes precipitation during fall, winter, and spring seasons, when the warmer/moist air from the south interacts with the cooler/drier air from the north of the southeastern U.S. Extreme weather conditions, such as hurricanes, can occur in the spring and fall, especially in April. For example, two tornadoes (EF3 and EF4, where "EF" stands for the Enhanced Fujita scale for tornado intensity/damage) in a span of twelve days hit the city of Tuscaloosa in 15–27 April, 2011 killing more than fifty people and causing considerable infrastructure damage [40]. Therefore, study of extreme hydrological events in this area is particularly important.

The locations of measurement stations are shown in Figure 1. The primary precipitation station is ~13 km southwest of the lake, at the Tuscaloosa Municipal Airport, and is assumed to be consistent with rainfall in the watershed, as is the case for rain dominated systems with little topographic variation [41]. The additional precipitation stations shown in Figure 1 are used as a supplement to the primary station as discussed in the next section. The data were taken from the USGS National Water Information System (NWIS) for terrestrial water and National Centers for Environmental Information (NCEI) from National Oceanic and Atmospheric (NOAA) for precipitation. Data from the USGS were at a daily resolution and from NOAA at an hourly resolution.

Figure 1. Left: The study area and surrounding counties, showing the precipitation stations as red dots with their National Oceanic and Atmospheric ID. **Right**: The study area with gray points denoting surface water stations and the red point showing the position of the groundwater well.

Both the surface and subsurface records were relatively abundant to support reasonable statistical analysis in this study. The data with the longest period of record was the North Creek discharge, with the earliest record from 1938 to the present (i.e., a record of ~80 years with a daily resolution). Precipitation also had a long period of record, starting in 1958 and continuing to 2005. Lake stage and Binion Creek were recorded from 1982 and 1986, respectively. Groundwater data were recorded from 1979 to the present. Groundwater had an average depth from the land surface of approximately 13 m and was generally low in the winter. The vadose zone was comprised mostly of soils, which tended to be loam type soils that are often rich in clay minerals as is typical in the southeast U.S.

2.2. Storm Properties

A complete data series of hourly precipitation was built first, so that there were no missing values. The stations in Perry and Hale counties were first determined to be reasonable analogs by annual statistics. The storm properties were then calculated using the method proposed by Jiang et al. [42]. We briefly review the methodology here, and further details can be found in that reference [42].

Four properties were calculated for each storm, including storm duration, intensity, total precipitation, and inter-storm period. Storm duration was the number of consecutive hours of precipitation for a single storm which ended when there were six consecutive hours of no precipitation. Intensity was the average rainfall per hour in a single storm. Total precipitation was the amount of precipitation that fell during that storm. Inter-storm period was the number of consecutive hours with no precipitation that occurred between storms.

There are different possible values that can be used to define the inter-storm periods that have been used in the literature. For this study, six hours was chosen as the minimum value for an inter-storm period as it was a frequently used minimum in previous work [36,43,44]. This minimum threshold for an inter-storm period also allowed for direct comparison to our previous work since it used the same minimum value for the inter-storm period [42].

To explore the extremes of these properties, the data set was filtered so that the 95th percentile of each property was isolated. These extreme data points were then analyzed against groundwater head fluctuations and fitted with the distributions separately, as well as the entire set of storm properties for further comparison. The distributions used are discussed later in this section.

2.3. Hurst Exponent

The Hurst exponent is a measure of the memory of a time series, meaning how strong the influence of past values is on future values. The Hurst exponent was originally developed to optimize the dam's size in the 1950s when evaluating the reservoir storage [27]. It has since been improved and used in many signal processing applications [28–32]. This study used the method proposed by Habib et al. [45] to calculate a time-scale local Hurst exponent using the following four equations:

$$F(S) \approx S^H \tag{1}$$

$$Y(j) = \sum_{k=1}^{i} [X_k - \langle x \rangle] \tag{2}$$

$$F_k^2(S) = \frac{1}{S} \sum_{j=1}^{j} \left(Y_{j,k} - P_{j,k}^n \right)^2 \tag{3}$$

$$F(S) = \left[\frac{1}{m} \sum_{k=1}^{m} F_k^2(S) \right]^{1/2} \tag{4}$$

Here Equation (1) gives the scaling function ($F(S)$) which is approximately equal to scale (S) raised to the Hurst exponent (H). Equation (2) develops a cumulative sum (Y) where X_k is a specific value and $\langle x \rangle$ is the series mean. Equation (3) determines the variance (F^2) of each section by subtracting a best fit polynomial of order n (P^n). Finally, Equation (4) finds the (square root of the) average variance for all segments which defines the scaling function. Different values of Hurst exponents H represent different properties in a system. In this study, the calculated H is between 0 and 1. The range of $0 < H < 0.5$ represents an anti-persistent series where high values are usually followed by low, and the range of $H > 0.5$ represents persistent series where high values are followed by high. It is also noteworthy that we used a window of 30 samples as our scale (S) when calculating the TSLHE.

2.4. Random Variable Distributions for Hydrological Processes

Here we introduce the three distributions for random hydrological variables. First, the Gaussian distribution, also known as the exponentially modified Gaussian distribution, can be used to capture the distribution for processes with a heavy tail in one direction. This distribution is defined by the following function with a stability index (α), location parameter (D), and scale parameter (T):

$$f(T, S) = \frac{1}{\sqrt{\pi D(S - T)^{.5\alpha}}} \times e^{\frac{-x^2}{4D(S-T)^{.5\alpha}}} \tag{5}$$

This distribution is closely related to the normal (Gaussian distribution) except that it is modified by the stability index α, which controls the tailing behavior of the distribution.

The stable distribution is defined by the following function with a stability index (α), skewness parameter (β), scale parameter (γ), and location parameter (δ):

$$\phi 1(t) \begin{cases} -exp(-\gamma^\alpha |t|^\alpha [1 - i\beta sin(t)(tan\frac{\pi\alpha}{2}) + i\delta t]) & \alpha \neq 1 \\ -exp(-\gamma |t| [1 - i\beta\frac{2}{\pi}sin(t)ln|t| + i\delta t]), & \alpha = 1 \end{cases} \tag{6}$$

Here the stability index α controls the overall shape (i.e., pattern) of the distribution, with $\alpha = 2$ reducing the distribution to a Gaussian distribution. The skewness parameter β controls the skewness of the distribution with a negative β resulting in a skewness to the left (representing extreme minimums), and a positive one causing a skewness to the right (representing extreme maximums). The other parameters do not affect the overall shape of the distribution, except for the overall expansion (by the scale parameter γ) and shift (by the location parameter δ).

The widely used Gumbel distribution is also used here as a control and comparison. The following equation gives the PDF of the Gumbel distribution with a location parameter (μ) and scale parameter (β) as the only two parameters for the distribution:

$$f(x) = \frac{1}{\beta}e^{-(z-e^{-z})} \tag{7}$$

where

$$z = \frac{x - \mu}{\beta} \tag{8}$$

Another commonly used extreme value distribution is the Log-Pearson Type 3 distribution. This distribution is commonly used to fit a frequency distribution data, often when determining flood occurrence [46]. The distribution is based on three parameters, which are location (μ), scale (β), and skewness (γ). We introduced this function on the measurement data series since we did not find a single most effective distribution among the first three listed above. The distribution is defined by the following PDF:

$$f(x) = \frac{\beta}{\Gamma(x)}(x - \mu)^{\gamma-1}e^{-\beta(x-\mu)} \tag{9}$$

where $\Gamma(x)$ represents the Gamma function.

The distributions mentioned above were parameterized in MATLAB or R using convenient optimization toolboxes that estimated the best fit parameters from the data. The Gumbel distribution was parameterized using the "evfit" function which estimated a maximum likelihood estimate of the parameters of the type one extreme value distribution (Gumbel) within a 95% confidence interval. The stretched Gaussian distribution was parameterized using the "exgauss_fit" function, which also used a maximum likelihood method but was bounded by a simple algorithm. The stable distribution was parameterized using the "stblfit" function which used Koutrouvelis' method, an iterative, regression method which used an initial estimate of parameters and repeated using weighted regression runs until convergence criteria was met. The Log-Pearson distribution was parameterized using the "fisdist" function in R, which also used a maximum likelihood estimator method. We used the root mean square error (RSME) as a quantitative metric to determine goodness of fit and compare across different distributions.

3. Results

3.1. Temporal Variation of Precipitation Properties and Their Extremes

First, the seasonal distribution of storm properties was investigated. Box plots for each property are depicted for each month to show subtle temporal variations (Figure 2). Duration and total precipitation had similar trends with highs coming in the winter season and lows in the summer. Particularly, February was on average the wettest month (Figure 2), which was consistent with the local weather conditions. These two properties both tended to have their extremes come during winter months (especially February) when the average values were higher. Intensity had the opposite trend with the most intense precipitation coming in spring/fall and less intense storms in the winter. Intensity extremes had a random pattern with most extremes events occurring in spring and fall, such as April and September, which are the typical seasons for land-falling hurricanes. Inter-storm period did not have a significant trend annually for either average or extreme values, but did have its three largest values all occurring in January, the winter season.

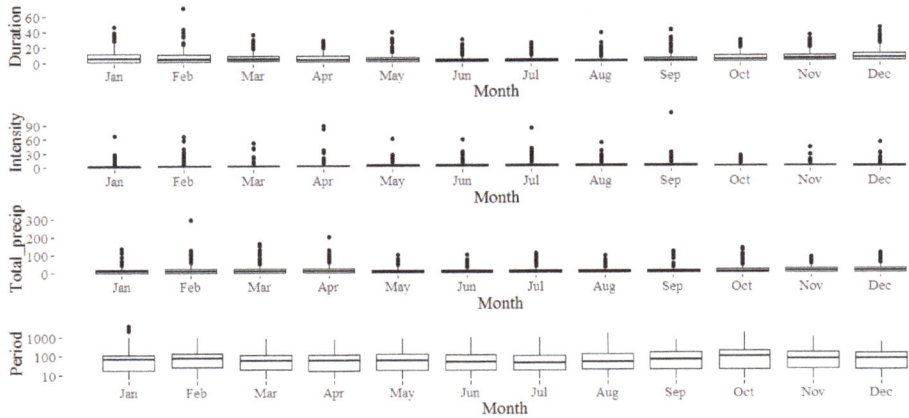

Figure 2. Box and whisker plots for each of the storm properties annual variation showing duration (**row 1**), intensity (**row 2**), total precipitation (**row 3**), and inter-storm period (**row 4**) values using a one-month bin.

Evolution of the distribution for all storm properties was then evaluated to show when the extremes of precipitation occurred in the area (Figure 3). Many of the duration extremes occurred in the start of the study period and their occurrence rate slowly declined with time. Inter-storm period had the opposite behavior with many of the extreme values coming in the most recent portion of the study period. Total precipitation and intensity had less obvious changes in the occurrence of their extremes. Extreme events of intensity became more frequent over the study period, with the plot of intensity extremes occurrence vs. time increasing slightly with time (Figure 3).

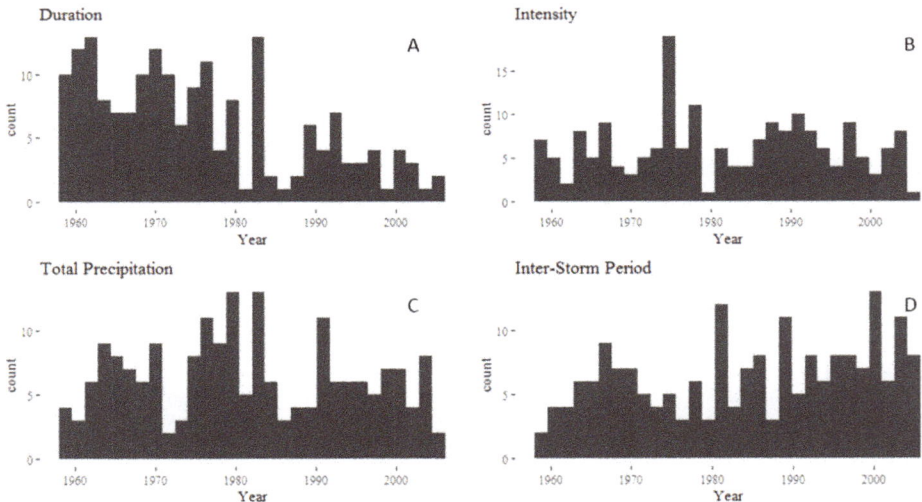

Figure 3. Histogram of different extreme events and their occurrences over time to show the frequency of occurrence for extreme events during the study period. Each sub figure shows this for each storm property: Duration (**A**), intensity (**B**), total precipitation (**C**), and inter-storm period (**D**).

3.2. Correlation between Storm Properties and River/Groundwater

Figure 4 shows the correlation between the extreme values of precipitation characteristics (duration, intensity, and total precipitation), which are taken as the 95th percentile of each unique property compared to the fluctuation of groundwater head. Since groundwater fluctuation was measured using depth to groundwater surface, a decrease in depth to water corresponded to an increase in water storage in the aquifer since the water table was closer to the surface. The strongest correlation of all the studied storm properties was with intensity, with a maximum correlation coefficient of −0.6 at 12 days after the end of a rainfall event. Duration shows a relatively weaker correlation, but it is more consistent with peaks at −0.5 at 10 days. Total precipitation was weakly correlated to increased groundwater storage, peaking much after the other two storm properties mentioned above, at −0.25 at 21 days.

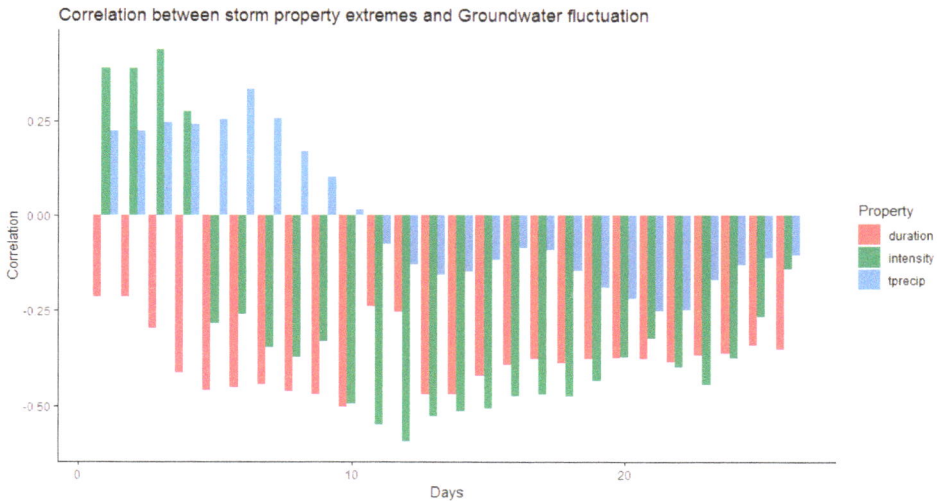

Figure 4. Correlation between the storm property extremes and depth to groundwater surface at different number of days lags, showing the slow evolving process. In the legend, "tprecip" represents "total precipitation".

Due to the high frequency of precipitation at the study site (the average number of days with rainfall ≥0.25 mm is 111.3 days per year at the study site), the inter-storm period was often too short to effectively capture the effects of the interval arrival period between storms. Most of the inter-storm periods were usually only a few days long (13 days was the 95th percentile of inter-storm periods) and only 17 dry periods longer than 31 days were recorded over the entire precipitation record (~60 years).

Since surface water recharging to groundwater is a slow acting process (involving relatively slow infiltration through the ~13 m thick vadose zone), with other results showing ~13 days for peak influence for a storm to occur, these short periods only capture the effects of the previous precipitation event on groundwater head.

Stream discharge, however, cannot be correlated to storm properties in this study, because river response (within minutes to hours) to precipitation was faster than our measurement resolution (daily). We could not find a strong correlation at any time period measurable within the resolution of our data. This was likely due to the issue of storms ending in the middle of the day. This makes it difficult to correlate to fluctuations in river stages that occur within the first hours after a storm ends, since the next river measurements after the storm may occur as much as 23 h after the storm actually ends.

Due to these issues, the small sample size of usable inter-storm periods, and the rapid fluctuation in stream discharge, the results were not presented here or found to be significant at the site. Further investigations for inter-storm periods in more arid areas may provide meaningful and interesting insights. High resolution (ideally hourly) stream discharge data are needed to better correlate these fluctuations.

3.3. Time-Scale Local Hurst Exponent

Time-scale local Hurst exponents for all of the surface water data sets were also calculated and compared to those for extreme values. Figure 5 shows the distribution of TSLHE for the four precipitation properties. The modes of each data set are listed in Table 1. The Hurst exponent revealed the memory present in each system. Streams tended to have a mode of H slightly above 0.5, indicating that they have memory and that high values (in stream discharge) likely follow high values. Hence, streamflow discharge time series have at least some impact of memory in their behavior. Groundwater head fluctuation had a mode of H very close to 1 (~0.963), indicating a highly persistent system with strong memory. This follows the expected trend of groundwater being a slowly evolving system that cannot rapidly transition from high to low values and vice versa.

The storm properties however exhibited different behaviors. The storm properties are also a time series that can be analyzed using DFA techniques in the TSLHE. All of the storm properties showed strongly anti-persistent TSLHE values with modes between $H = 0.09$ and $H = 0.25$ with maximum values never reaching 0.5 or the minimum threshold for some persistence. The TSLHE values for terrestrial water systems showed no significant correlation to storm properties for either the extreme values or the entire data sets.

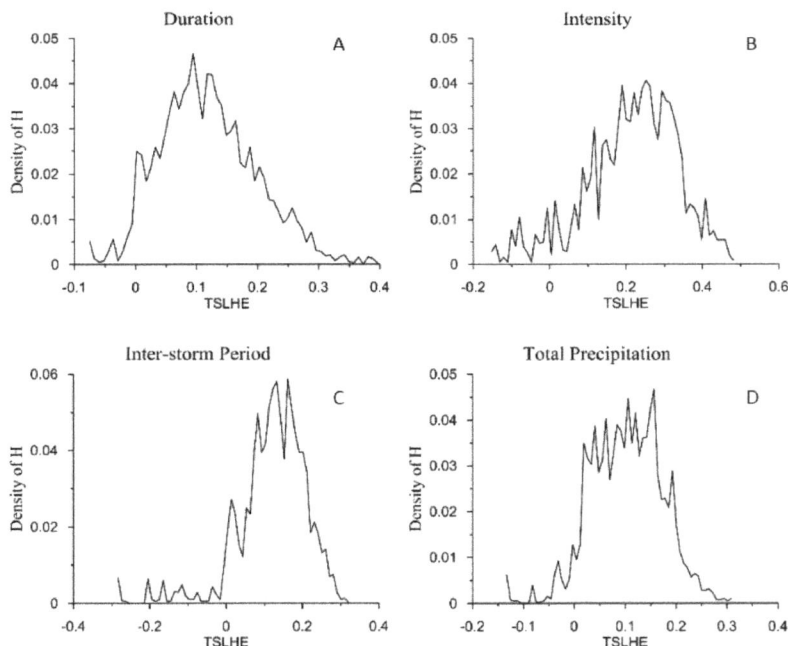

Figure 5. Probability density functions (PDF) of the time-scale local Hurst exponent (TSLHE) calculated for each of the storm properties using the full data sets with duration (**A**), intensity (**B**), inter-storm period (**C**), and total precipitation (**D**).

Table 1. The mode of the TSLHEs for each of the distributions. The North and Binion Creek values represent stream discharge series.

	Groundwater		Surface Water		Storm Properties			
	Depth to Water	Lake Stage	North Creek	Binion Creek	Intensity	Inter-Storm Period	Total precipitation	Duration
Mode TSLHE	0.963	0.792	0.530	0.396	0.252	0.161	0.156	0.095

3.4. Distribution Fittings

Below are the results from the distribution fittings using the three PDFs introduced in Section 2.4. Figure 6 shows the graphical representation of the three distributions which fitted the actual data points for the extreme values of precipitation properties. Figure 7 shows the graphs of the fittings of the full storm property value data sets to the three distributions. Figure 8 shows fitting results for lake stage, groundwater head fluctuation, and stream discharge using the Gumbel, Stretched Gaussian and Stable distributions. Figure 9 shows these same series but fits the log-Pearson type 3 distribution. Tables 2 and 3 show the root mean square error (RMSE) for each distribution compared to the storm properties and surface water/groundwater systems. Tables 4 and 5 show the distribution parameters for the different direct measurements, the storm property full series, and the extremes.

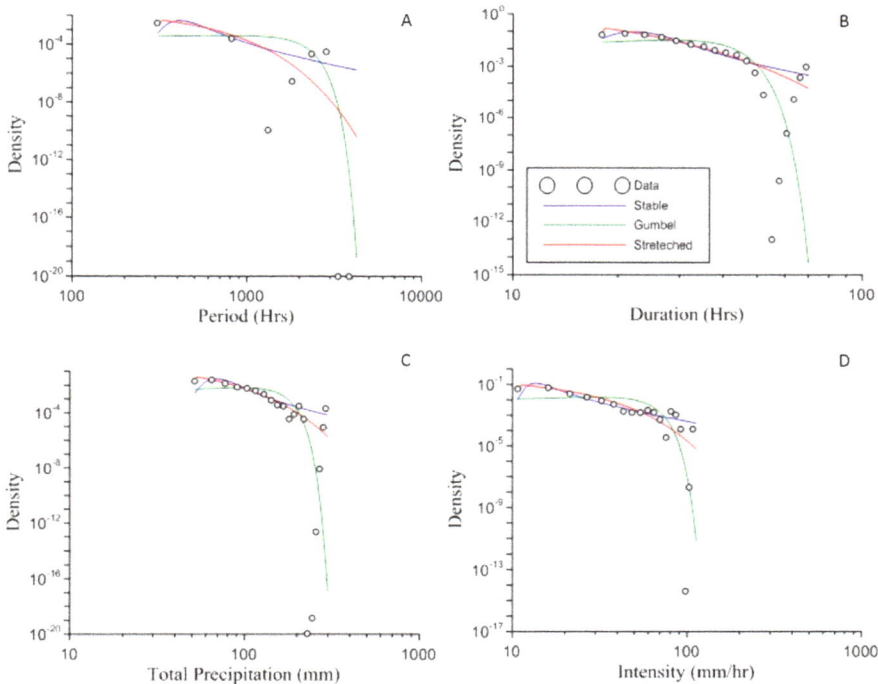

Figure 6. The best-fit PDF distribution for the extreme value set of each storm property: Inter-storm periods (**A**), duration (**B**), total precipitation (**C**), and intensity (**D**), using the three proposed distributions: Stable (the blue line), Gumbel (green line), and stretched Gaussian (red line) compared to the actual data (open circles).

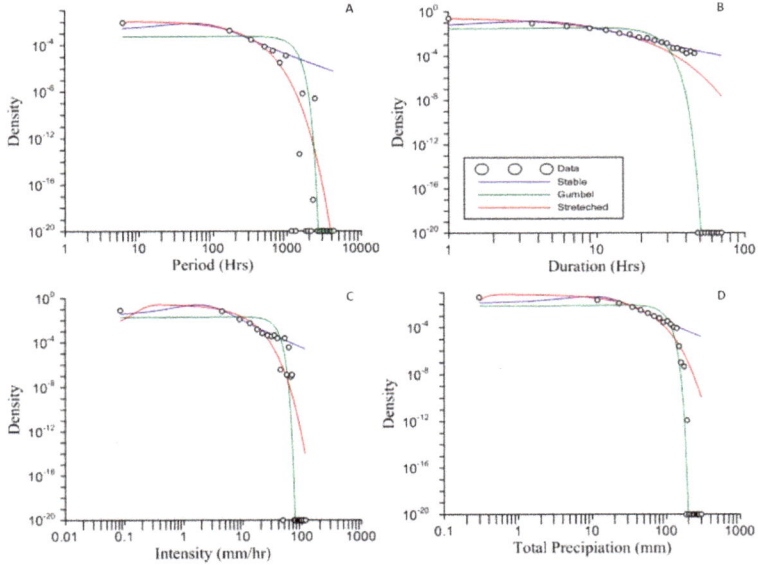

Figure 7. The best-fit PDF distribution for the full data set of each storm property: Inter-storm periods (**A**), duration (**B**), intensity (**C**), and total precipitation (**D**) using the three proposed distributions: Stable (the blue line), Gumbel (green line), and stretched Gaussian (red line) compared to the actual data (open circles).

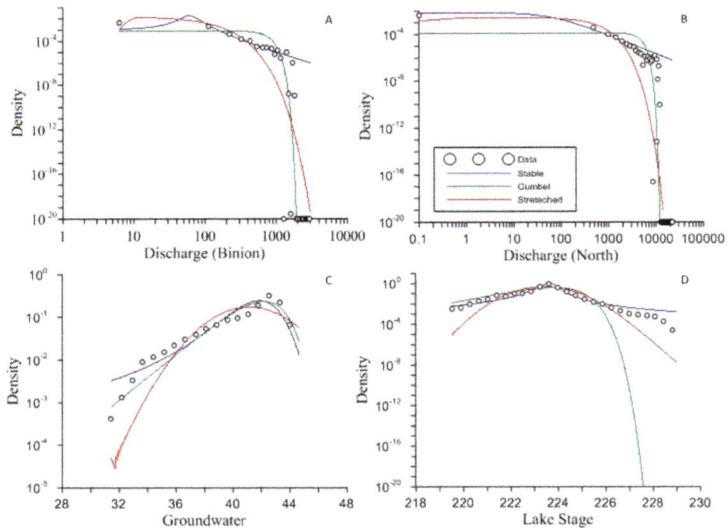

Figure 8. PDF distribution fittings for measurements of: Binion Creek discharge (**A**), North Creek discharge (**B**), depth to groundwater (**C**), and lake stage (**D**) using each of the proposed distributions: Stable (blue), Gumbel (green), and stretched Gaussian (red) compared to the actual data (white circles w/black outline).

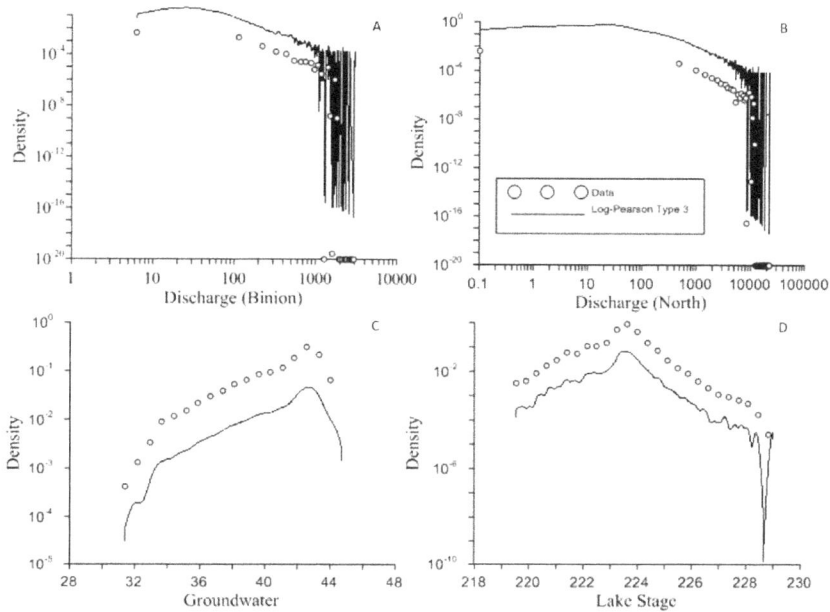

Figure 9. PDF distribution fittings for measurements of: Binion Creek discharge (**A**), North Creek discharge (**B**), depth to groundwater (**C**), and lake stage (**D**) using only the Log-Pearson Type 3 distribution (black lines) compared to the actual data (white circles w/black outline).

Table 2. Root mean square error (RMSE) of each of the calculated distributions for the four storm properties, including both extreme values and full sets. The lowest RMSE is highlighted by bold font for each distribution.

PDF	Extreme Events Only				Full Series			
	Total Precipitation	Duration	Intensity	Period	Total Precipitation	Duration	Intensity	Period
Gumbel	0.0053	0.0183	0.0159	6.31×10^{-4}	0.0092	0.0273	0.0091	9.08×10^{-4}
Stretched	0.002	0.0118	0.0034	1.49×10^{-4}	0.0027	0.0102	0.0026	1.64×10^{-4}
Stable	0.0039	**0.0072**	0.0078	3.12×10^{-4}	0.0068	0.0203	0.0052	4.46×10^{-4}

Table 3. The RMSE of each of the calculated distributions for the different water systems. Numbers shown for the creeks are RMSE for fitting the stream discharge. The lowest RMSE is highlighted by bold font for each distribution.

Distribution	Groundwater	Lake Stage	North Creek	Binion Creek
Gumbel	0.0331	0.1305	2.99×10^{-4}	0.0015
Stretched	0.0539	0.1148	1.49×10^{-4}	$\mathbf{4.18 \times 10^{-4}}$
Stable	0.428	0.0390	$\mathbf{7.42 \times 10^{-5}}$	0.0014
Log-Pearson III	**0.0172**	**0.0172**	0.0563	0.0470

Table 4. The distribution parameters for the water measurement distribution fittings.

		Lake Stage	Groundwater	North Creek	Binion Creek
Stable	α	1.15	1.47	0.57	0.96
	β	−0.31	−1.00	1.00	0.33
	γ	0.30	1.16	51.86	18.15
	δ	223.21	40.38	−43.75	−34.43
Stretched Gaussian	μ	223.20	40.64	0.10	8.74
	D	0.82	2.28	3.82×10^{-8}	1.81
	T	0.22	0.53	370.19	74.65
Gumbel	μ	223.81	42.17	979.53	187.12
	β	0.96	1.61	2962.88	463.87

Table 5. The distribution parameters for storm property distribution fittings.

		Full Series				Extreme Events Only			
		Period	Duration	Intensity	Total Precipitation	Total Precipitation	Duration	Intensity	Period
Stable	α	1.21	1.37	1.18	1.32	1.23	1.42	0.81	1.15
	β	1.00	1.00	1.00	1.00	1.00	1.00	0.84	1.00
	γ	37.43	2.18	1.07	6.39	9.63	2.99	2.66	65.94
	δ	172.67	7.44	5.76	22.79	97.97	26.62	7.32	699.67
Stretched Gaussian	μ	6.00	1.00	0.22	0.34	51.80	18.00	10.67	308.00
	D	1.09×10^{-10}	1.49×10^{-12}	0.072	0.085	5.21×10^{-13}	2.31×10^{-14}	1.16×10^{-12}	3.06×10^{-11}
	T	93.22	4.24	3.63	14.73	24.68	6.49	10.58	210.14
Gumbel	μ	220.67	8.65	7.89	26.44	94.33	28.55	30.39	794.30
	β	641.93	10.89	17.75	45.17	55.99	11.75	25.28	934.64

4. Discussion

4.1. Surface Water

The distribution for all storm properties were compared to the time in the study period. Occurrence of extreme events over time were plotted to show the changes in precipitation behavior in the area. Many of the duration extremes occurred in the start of the study period and slowly declined with time. Inter-storm period had the opposite behavior, with many of the extreme values coming in the most recent portion of the study period. This would indicate that the region was experiencing shorter storms with longer dry spells in between, but does not necessarily indicate a dryer climate, since intensity is increasing, and total precipitation extremes are generally varying without a stable pattern. Changing climate was likely the driver for these changes, creating more intense, shorter storms, with longer dry periods in between. Furthermore, more intensive climate modeling is needed to fully investigate trends in storm properties. Current research suggests that these properties are changing but there is not a scientific consensus for a projection, and projections are variable in both space and time.

4.2. Storm Propertires Correlated with Groundwater Head Fluctutations

For the fluctuations of groundwater table, the strongest average correlation of any storm property was with intensity, with a maximum correlation coefficient of 0.5 at 10 days after the end of a rainfall event. There are two likely causes for this high correlation between storm duration and groundwater head fluctuation. First, a longer storm allows more time for a higher percentage of precipitation to infiltrate into the aquifer, instead of flowing away as overland flow/surface runoff and eventually discharging into streams and lakes, or evaporating. Second, the extreme values of storm duration occur more frequently in the winter months when there is lower evapotranspiration (ET) and withdrawal from the aquifer is lower during this time since there is less water used for irrigation.

Intensity had the next strongest correlation with groundwater head fluctuation. The highest correlation coefficient for intensity was actually higher than for storm duration; however, across the period where fluctuations were analyzed, there was generally a weaker correlation, and for the first four days the correlation was positive indicating a decrease in water supply. This transition time showed that there was a lag between the precipitation event and the first portion of precipitation to infiltrate to the water table. The peak correlation was at 12 days after the precipitation event ended and the correlation was −0.595, representing the point where the groundwater table reached its highest level after a precipitation event. Intensity tended to be higher in the summer months, so this explains the initial positive decrease in water level. This is because there is higher ET in the summer and there is some water withdrawal for irrigation in the area. After the peak correlation, the correlation slowly decreased, representing the return from the peak increase at 12 days.

Interestingly, total precipitation had the weakest correlation with groundwater head fluctuation of the three properties analyzed with a peak correlation coefficient of −0.253 at 20 days after precipitation events. There is a positive correlation (indicating a decrease in groundwater storage) for the first 10 days, peaking at six days after precipitation, transitioning to a negative correlation (increased GW storage) after day 10. The high total precipitation likely exceeded the infiltration capacity of the soils and a significant portion of the precipitation was discharged into surface water systems or as overland flow. Extreme values of total precipitation are less temporally dependent than the other variables, and generally the winter months have higher total precipitation.

4.3. Time-Scale Local Hurst Exponents and System Memory

The TSLHE values for groundwater head fluctuation, lake stage, and river discharge all showed that they were persistent or semi-persistent series since their modes were above $H = 0.5$. Groundwater was the slowest evolving system so that likely caused it to have the highest value, followed by lake stage. The lake stage was faster evolving than groundwater, but slower than river discharge, leading to the increase in total correlation. The TSLHE also showed that precipitation properties do not have memory acting on them and are actually anti-persistent across all values, and reverse more frequently than white noise would. This may be because a major storm would use much of the available atmospheric moisture and result in following storms being weaker and not as intense or long lasting.

4.4. Distribution Evaluation

For the distribution fittings of storm properties, the stretched Gaussian distribution performed the best across all of the data series except for the extreme values of duration where the tempered stable distribution was found to be the more effective. The Gumbel distribution was consistently the least effective distribution for capturing storm properties, often overestimating low values and underestimating the extremes of precipitation. The stretched Gaussian produced a RMSE of approximately half the RMSE of the stable distribution for the values where it was most effective. The stretched Gaussian is effective in these data sets because it mixes the properties of the standard Gaussian distribution, which captures the small, more frequent values, and the power law distribution, which is able to capture the extreme values. The stable distribution generally overestimates the lower, more frequent values and underestimates the extreme values, but remains a viable option for fitting these data. The different data sets all had differing parameters for the distribution, however, there was no major trend across all sets. Generally, the extreme events and full data sets for storm properties have similar parameters for each distribution.

For the measured data from lake stage, groundwater head fluctuations, and stream discharge there were varying results in distribution effectiveness. First the Gumbel, stable and stretched Gaussian distributions were tested. The Gumbel distribution was the best of these distributions to fit the values of depth to groundwater. The depth to groundwater surface had the narrowest range of all the systems, and hence the Gumbel distribution was able to effectively capture it. The stable distribution was the most effective of these three distributions for lake stage. Lake stage had a negative tail and very sharp peak value, likely due to its quick response to storm events. This peak and light negative tail allowed the stable distribution to fit it most effectively. The two creeks had different results. The larger creek, North Creek, was best fit by the stable distribution as it had a weaker tail compared to the smaller creek. Binion Creek had a heavier tail of extreme values and was captured most effectively by the stretched Gaussian distribution.

Since there was not a single distribution that proved most effective in capturing these different values, we also tested the Log-Pearson Type 3 distribution to try to find a single most effective distribution. The Log-Pearson gave the best RSME value for both lake stage and for the depth to groundwater measurements. Both of these measurement series had a much more "normal" shape (i.e., closer to standard Gaussian behavior, and weaker tailing behavior), suggesting that the Log-Pearson Type 3 distribution may perform better on series without large tails. The Log-Pearson also exhibited erratic behavior at the extreme values of river discharge since there were values that did not occur in our data sets (i.e., a density of 0) which caused the density curve to change values rapidly. Since these series were from different parts of the water cycle, and controlled by different processes, there may not be a single distribution that will most effectively capture all of them. There are a variety of different extreme value distributions that could possibly be used to effectively capture these series more effectively, such as the generalized extreme value, two-component extreme value, or generalized Pareto distributions, which will be evaluated for details in a future study.

5. Conclusions

This study conducted statistical analysis to reveal the distribution, memory, and correlation in surface and subsurface water observed in the Lake Tuscaloosa watershed in Tuscaloosa, AL. Three main conclusions are obtained.

First, statistical analysis shows that precipitation properties can be correlated to groundwater head fluctuations and different properties can have an influence on the motion of water. Duration was more closely related to the fluctuation of groundwater head than any of the other studied properties, and had a peak correlation at 10 days after the end of the precipitation event for the ~13 m deep well. Intensity had a stronger peak correlation to groundwater head fluctuation at 12 days but was more variable and had a worse average correlation than duration. Total precipitation was weakly correlated compared to the other two properties indicating that it had the least control over recharge.

Second, the surface and groundwater series are found to be persistent from the TSLHE values. Groundwater followed by lake stage was the most persistent. The TSLHE for stream discharge was very close to the values for white noise. Precipitation properties were anti-persistent, with the values entirely constrained in the anti-persistent range of TSLHE for all properties and hourly values. This was the expected result since the most persistent series were the slowest evolving with more volatile systems being heavily anti-persistent. TSLHE itself was not distributed with any meaningful relationship to storm properties and was likely driven by annual processes.

Third, the stretched Gaussian distribution was found to be the most effective distribution in capturing the storm properties for both the extreme values and the entire data sets. This distribution was likely most effective due to its mix of properties of the Gaussian and power law distributions. In other words, it allows to capture the peak frequency values as well as the lower frequency extreme values that are vital to understanding a system like the water cycle. For the surface and groundwater distributions, however, there was not a clear best distribution, since the different systems had different best RSME values, and the stream discharge best distributions differed between the two streams. Further analysis with more data is needed to resolve the best distribution, which may not be any of those tested here, for these different systems.

Author Contributions: Conceptualization, S.D. and Y.Z.; methodology, S.D., Y.Z., X.L, P.J. and L.C.; software, S.D., X.L, P.J. and L.C.; validation, S.D.; formal analysis, S.D., Y.Z., X.L, and P.J.; investigation, S.D., Y.Z., X.L, and P.J.; resources, S.D.; data curation, S.D.; writing—original draft preparation, S.D.; writing—review and editing, Y.Z. and G.R.T.; visualization, S.D.; supervision, Y.Z.; project administration, Y.Z.; funding acquisition, Y.Z., H.S. and C.Z.

Funding: This research was partially funded by the National Natural Science Foundation of China (under grants 41330632, 41628202, and 11572112). This paper does not necessarily reflect the views of the funding agency.

Conflicts of Interest: The authors declare no conflicts of interest. Any use of trade, firm, or product names was for descriptive purposes only and does not imply endorsement by the U.S. Government.

References

1. Yuan, J.; Emura, K.; Farnham, C.; Alam, M.A. Frequency Analysis of annual maximum hourly precipitation and determination of best fit probability distribution for regions in Japan. *Urban Clim.* **2018**, *24*, 276–286. [CrossRef]
2. Hailegeorgis, T.T.; Alfredsen, K. Analyses of extreme precipitation and rain events including uncertainties and reliability in design and management of urban water infrastructure. *J. Hydrol.* **2017**, *544*, 290–305. [CrossRef]
3. Hui, R.; Herman, J.; Lund, J.; Madani, K. Adaptive water infrastructure planning of nonstationary hydrology. *Adv. Water Resour.* **2018**, *118*, 83–94. [CrossRef]
4. Jakob, D. Nonstationarity in extremes and engineering design. *Extremes Changing Clim.* **2012**, *65*, 363–417.
5. Evin, G.; Farve, A.C.; Hingray, B. Stochastic generation of multi-site daily precipitation focusing on extreme events. *Hydrol. Earth Syst. Sci.* **2018**, *22*, 655–672. [CrossRef]
6. Kundzewicz, W.; Dennis, P.; Milly, D.; Betancourt, P.; Falkenmark, M.; Hirsch, R.M.; Kundzewicz, Z.W.; Lettenmaier, D.P.; Stouffer, R.J. Stationarity is dead: whither water management? *Science* **2008**, *319*, 573–574.
7. Farzad, F.; Yaseen, Z.; Ahemd, E.S. Application of soft computing based hybrid models in hydrological variables modeling: a comprehensive review. *Theor. Appl. Climatol.* **2017**, *128*, 875–903.
8. Bresciani, E.; Gleeson, T.; Goderniaux, P.; de Dreuzy, J.R.; Werner, A.D.; Worman, A.; Zijl, W.; Batelann, O. Groundwater flow systems theory: research challenges beyond the specified-head top boundary condition. *Hydrogeol. J.* **2016**, *24*, 1087–1090. [CrossRef]
9. Baynes, E.R.C.; van de Lageweg, W.I.; McLelland, D.R.; Aberle, J.; Dijkstra, J.; Henry, P.Y.; Rice, S.P.; Thom, M.; Moulin, F. Beyond equilibrium: Re-evaluating physical modeling of fluvial systems to represent climate changes. *Earth Sci. Rev.* **2018**, *181*, 82–97. [CrossRef]
10. Fatichi, S.; Vivoni, E.R.; Ogden, F.L.; Ivanov, Y.Y.; Mirusm, B.; Gochis, D.; Downer, C.W.; Camporese, M.; Davison, J.H.; Ebel, B.; et al. An overview of current applications, challenges, and future trends in distributed process-based models in hydrology. *J. Hydrol.* **2018**, *537*, 45–60. [CrossRef]
11. Guadagnini, A.; Riva, M.; Neuman, S.P. Recent advances in scalable non-Gaussian geostatistics: The generalized sub-Gaussian model. *J. Hydrol.* **2018**, *562*, 685–691. [CrossRef]
12. Jensen, A.; Hamaker, H.C.; Cramer, H.; Stene, E. A characteristic application of statistics in hydrology. *Rev. Int. Stat. Inst.* **1970**, *38*, 42–48. [CrossRef]
13. Katz, R.W.; Parlange, M.B.; Naveau, P. Statistics of extremes in hydrology. *Adv. Water Resour.* **2002**, *25*, 1287–1305. [CrossRef]
14. Rawat, K.S.; Singh, S.K.; Jacintha, T.G.A.; Nemcic-Jurec, J. Appraisal of long term groundwater quality of peninsular India using water quality index and fractal dimension. *J. Earth Syst. Sci.* **2017**, *126*, 122–144. [CrossRef]
15. Jiang, C.; Xiong, L.; Guo, S.; Xia, J.; Xu, C.Y. A process-based insight into nonstationarity of the probability distribution of annual runoff. *Water Resour. Res.* **2017**, *53*, 4214–4235. [CrossRef]
16. Gumbel, E.J. The Return Period of Flood Flows. *Ann. Math. Stat.* **1941**, *12*, 163–190. [CrossRef]
17. Serago, J.M.; Vogel, R.M. Parsimonious nonstationary flood frequency analysis. *Adv. Water Resour.* **2018**, *112*, 1–16. [CrossRef]
18. Call, B.C.; Belmont, P.; Schmidt, J.C.; Wilcock, P.R. Changes in floodplain inundation under nonstationary hydrology for an adjustable alluvial river channel. *Water Resour. Res.* **2017**, *53*, 3811–3834. [CrossRef]
19. Yu, Z.; Miller, S.; Montalto, F.; Lall, U. The bridge between precipitation and temperature-pressure change events: modeling future non-stationary precipitation. *J. Hydrol.* **2018**, *562*, 346–357. [CrossRef]
20. Jagtap, R.S.; Gedam, V.K.; Kale, M.M. Generalized extreme value model with cyclic covariate structure for analysis of non-stationary hydrometeorological extremes. *J. Earth Syst. Sci.* **2019**, *128*, 14. [CrossRef]
21. Cvetkovic, V. The tempered one-sided stable density: A universal model for hydrological transport? *Environ. Res. Lett.* **2011**, *6*, 034008. [CrossRef]

22. Zhang, Y.; Meerschaert, M.M.; Baeumer, B.; LaBolle, E.M. Modeling mixed retention and early arrivals in multidimensional heterogeneous media using an explicit Lagrangian scheme. *Water Resour. Res.* **2015**, *51*, 6311–6337. [CrossRef]

23. Tapiero, C.S.; Vallois, P. Randomness and fractional stable distributions. *Statistical Mech. Appl.* **2018**, *511*, 54–60. [CrossRef]

24. Golubev, A. Exponentially modified Gaussian (EMG) relevance to distributions related to cell proliferation and differentiation. *J. Theor. Biol.* **2010**, *262*, 257–266. [CrossRef] [PubMed]

25. Gomez, Y.M.; Bolfarine, H.; Gomez, H. Gumbel distribution with heavy tail and applications to environmental data. *Math. Comput. Simulat.* **2019**, *157*, 115–129. [CrossRef]

26. Ye, L.; Hanson, L.S.; Ding, P.; Wang, D.; Vogel, R.M. The probability distribution of daily precipitation at the point and catchment scales in US. *Hydrol. Earth Syst. Sci.* **2018**, *22*, 6519–6531. [CrossRef]

27. Hurst, H.E. Long-term storage capacity of reservoirs. *Tran. Am. Soc. Civ. Eng.* **1951**, *116*, 770–799.

28. Peng, C.K.; Buldyrev, S.V.; Havlin, S.; Simon, M.; Stanley, H.E.; Goldberger, A.L. Mosaic Organization of DNA nucleotides. *Phys. Rev.* **1994**, *49*, 1685–1689. [CrossRef]

29. Qian, B.; Rasheed, K. Foreign Exchange Market Prediction with Multiple Classifiers. *J. Forecasting* **2010**, *29*, 271–284. [CrossRef]

30. Shadkhoo, S.; Jagari, G.R. Multifractal Detrended Cross-Correlation Analysis of Temporal and Spatial Seismic Data. *Eur. Phys. J.* **2009**, *72*, 679–683. [CrossRef]

31. Nath, S.K.; Dewangan, P. Detection of Seismic Reflections from Seismic Attributes through Fractal Analysis. *Geophys. Prospect.* **2002**, *50*, 341–360. [CrossRef]

32. Zhang, Y.K.; Schilling, K. Temporal Scaling of Hydraulic Head and Base flow and its Implication for Groundwater Recharge. *Water Resour. Res.* **2004**, *40*, 9. [CrossRef]

33. Zhou, Y.; Zhang, Q.; Singh, V. Fractal-based Evolution of the Effect of Water Reservoirs on Hydrological Process: the Dams in the Yangtze River as a Case Study. *Stoch. Environ. Res. Risk Assess.* **2014**, *28*, 263–279. [CrossRef]

34. Tong, S.; Lai, Q.; Zhang, J.; Bao, Y.; Lusi, A.; Ma, Q.; Li, X.; Zhang, F. Spatiotemporal drought variability on the Mongolian Plateau from 1980–2014 based on the SPEI-PM, intensity analysis and Hurst exponent. *Sci. Total Environ.* **2018**, *615*, 1557–1565. [CrossRef] [PubMed]

35. Jiang, P.; Yu, Z.; Gautam, M.R.; Yuan, F.; Acharya, K. Changes of Storm Properties in the United States: Observations and Multimodel Ensemble Projection. *Global Planet Change* **2016**, *142*, 41–52. [CrossRef]

36. Yu, Z.; Jiang, P.; Gautam, M.R.; Zhang, Y.; Acharya, K. Changes in Seasonal Storm Properties in California and Nevada from an Ensemble of Climate Projections. *J. Geophys. Res.-Atmos.* **2016**, *120*, 2676–2688. [CrossRef]

37. Alexander, L.V.; Zhang, X.; Peterson, T.C.; Caesar, J.; Gleason, B.; Klein Tank, A.M.G.; Haylock, M.; Collins, D.; Trewin, B.; Rahimzadeh, F.; et al. Global observed changes in daily climate extremes of temperature and precipitation. *J. Geophys. Res.-Atmos.* **2006**, *111*, 1042–1063. [CrossRef]

38. City of Tuscaloosa: Lakes Division. Available online: https://www.tuscaloosa.com/city-services/water/lakes (accessed on 24 October 2017).

39. Slack, L.J.; Pritchett, J.L. *Sedimentation in Lake Tuscaloosa, Al, 1982-86*; Water-Resources Investigation Report; USGS: Denver, CO, USA, 1988.

40. Doswell, C.A.; Carbin, G.W.; Brooks, H.E. The tornadoes of spring 2011 in the USA: an historical perspective. *Weather* **2012**, *47*, 88–94. [CrossRef]

41. Larson, L.W.; Peck, E.L. Accuracy of precipitation measurements for hydrologic modeling. *Water Resour. Res.* **1974**, *10*, 857–863. [CrossRef]

42. Jiang, P.; Dawley, S.; Lu, B.; Zhang, Y.; Tick, G.R.; Sun, H.; Zheng, C. Precipitation Storm Property Distributions with Heavy Tails Follow Tempered Stable Density Relationships. *J. Phys. Conf. Ser.* **2018**, *1053*, 012119. [CrossRef]

43. Samuel, J.M.; Sivapalan, M. A comparative modeling analysis of multiscale temporal variability of rainfall in Australia. *Water Resour. Res.* **2008**, *44*, W07401.

44. Robinson, J.S.; Sivapalan, M. Temporal scales and hydrological regimes: Implications for flood frequency scaling. *Water Resour. Res.* **1997**, *33*, 2981–2999. [CrossRef]

45. Habib, A.; Sorensen, J.P.R.; Bloomfield, J.P.; Muchan, K.; Newell, A.J.; Butler, A.P. Temporal Scaling Phenomena in Groundwater-Floodplain systems using robust Detrended Fluctuation Analysis. *J. Hydrol.* **2017**, *549*, 715–730. [CrossRef]

46. Liang, Z.; Hu, Y.; Li, B.; Yu, Z. A modified weighted function method for parameter estimation of Pearson type three distribution. *Water Resour. Res.* **2014**, *50*, 3216–3228. [CrossRef]

![water logo] *water*

MDPI

Article

Third-Order Polynomial Normal Transform Applied to Multivariate Hydrologic Extremes

Yeou-Koung Tung [1,*], Lingwan You [1] and Chulsang Yoo [2]

[1] Disaster Prevention and Water Environment Research Center, National Chiao Tung University, Hsinchu 300, Taiwan; lucie751111@gmail.com
[2] Department of Civil, Environment, and Architecture, Korea University, Seoul 02841, Korea; envchul@korea.ac.kr
*   Correspondence: yk2013tung@gmail.com

Received: 26 December 2018; Accepted: 4 March 2019; Published: 8 March 2019

Abstract: Hydro-infrastructural systems (e.g., flood control dams, stormwater detention basins, and seawalls) are designed to protect the public against the adverse impacts of various hydrologic extremes (e.g., floods, droughts, and storm surges). In their design and safety evaluation, the characteristics of concerned hydrologic extremes affecting the hydrosystem performance often are described by several interrelated random variables—not just one—that need to be considered simultaneously. These multiple random variables, in practical problems, have a mixture of non-normal distributions of which the joint distribution function is difficult to establish. To tackle problems involving multivariate non-normal variables, one frequently adopted approach is to transform non-normal variables from their original domain to multivariate normal space under which a large wealth of established theories can be utilized. This study presents a framework for practical normal transform based on the third-order polynomial in the context of a multivariate setting. Especially, the study focuses on multivariate third-order polynomial normal transform (TPNT) with explicit consideration of sampling errors in sample L-moments and correlation coefficients. For illustration, the modeling framework is applied to establish an at-site rainfall intensity–duration-frequency (IDF) relationship. Annual maximum rainfall data analyzed contain seven durations (1–72 h) with 27 years of useable records. Numerical application shows that the proposed modeling framework can produce reasonable rainfall IDF relationships by simultaneously treating several correlated rainfall data series and is a viable tool in dealing with multivariate data with a mixture of non-normal distributions.

Keywords: polynomial normal transform; multivariate modeling; sampling errors; non-normality; extreme rainfall analysis

---

1. Introduction

In hydrosystem design, performance evaluation, and simulation, the problems often involve multiple random variables that are correlated with a mixture of non-normal marginal distributions. Under this condition, it is generally difficult, if not impossible, to establish an analytical joint probability distribution for these variables. In comparison with univariate distributions, there are relatively few analytical multivariate distribution functions under special combinations of parametric marginal distributions, and most of them are of the same type, which can be found in [1,2]. Examples of using analytical multivariate distributions in hydrology are bivariate Gamma distribution [3] and bivariate generalized extreme distribution [4]. Their use is somewhat limited to many practical problems because of different marginal distributions.

Due to the difficulty in establishing a truly multivariate joint distribution model for problems involving mixtures of several correlated, non-normal variables, approximated approaches, such as copula or normal transform, are often used by preserving marginal distributions or moments, including the correlation features among the variables. However, one should realize that, unlike using a true multivariate joint distribution function, preservation of the marginal distributions and dependence structure represents the retention of partial information of the concerned multivariate random variables in the analysis [5].

The concept of copula is one type of approximated multivariate approaches that has recently received tremendous attention and applications by researchers in various disciplines, including in hydrology [6]. Some examples of applying copula in multivariate hydrologic modeling can be found in analyzing floods [7,8], droughts [9–12], dam safety [13], and extreme rainfalls [14,15]. Most of the copula-based applications deal with bivariate problems and some trivariate problems under some restrictive conditions on correlation structures [8]. Applications of copula to higher dimension multivariate problems are rare primarily because there are only a few copula families that are rather restrictive in describing the dependence structure. Recently, the introduction of vine copulas has shown the advantage of overcoming the limitation of currently used copulas in multivariate analysis [16–19]. A copula-based approach is parametric by nature in that analytical marginal distribution models for the involved variables are specified.

Alternatively, another viable scheme in treating multivariate problems involving correlated non-normal random variables is to apply a NORTA (normal-to-anything) algorithm [20]. By a NORTA algorithm, normal transformation of an individual non-normal variable is made by preserving its marginal probability content in the normal variable domain as $\Phi(z) = F_X(x)$ with $\Phi(\cdot)$ and $F_X(\cdot)$, respectively, being the cumulative distribution functions (CDFs) of the standard normal variable Z and the original variable X. In addition, a relationship must be established to allow the determination of an equivalent correlation coefficient, ρ_{z_j,z_k}, of a pair of normal transformed variables, Z_j and Z_k, from the correlation coefficient, ρ_{x_j,x_k}, of the corresponding random variables, X_j and X_k, in the original space. Once the correlation matrix of standard normal variables Z's, $\left\{\rho_{z_j,z_k}\right\}$, is obtained from that of the non-normal variables X's, $\left\{\rho_{x_j,x_k}\right\}$, appropriate orthogonal transformation can be implemented to transform the original correlated variables into uncorrelated standard normal space for analysis.

The determination of ρ_{z_j, z_k} from ρ_{x_j, x_k} is made through the Nataf transform [21], which requires solving an implicit non-linear equation in the form of a double integration involving marginal distributions of a pair of random variables, X_j and X_k, under consideration:

$$\rho_{x_j,x_k} = \int_{-\infty}^{\infty}\int_{-\infty}^{\infty} \left(\frac{x_j - \mu_j}{\sigma_j}\right)\left(\frac{x_k - \mu_k}{\sigma_k}\right) \phi_{jk}\left(z_j, z_k \middle| \rho_{z_j,z_k}\right) dz_j\, dz_k \tag{1}$$

where $x_j = F_j^{-1}[\Phi(z_j)]$, and $\phi_{jk}(\cdot)$ = bivariate standard normal joint probability density function (PDF). Lebrun and Dutfoy [22] provide an insightful analysis of Nataf transform and uncover that it is a special modeling of dependence structure using Gaussian copula. To facilitate practical engineering applications, a set of empirical equations for 10 commonly used distribution functions has been established to relate ρ_{z_j,z_k} to ρ_{x_j,x_k} and their distribution properties [23]. Such empirical relations were applied to reliability analysis of engineering systems [5,24]. Later, computationally more efficient methods based on root finding and linear search [25], the false position method [26], and the artificial neural network method [27] were proposed to solve Equation (1) for ρ_{z_j,z_k} from the known ρ_{x_j,x_k} and marginal PDFs of X_j and X_k.

The above mentioned schemes (i.e., copula, NORTA, and Nataf transform) all require the stipulation of marginal PDFs. The stipulation of a distribution function implies knowing the complete statistical information of the random variable, including its moments of all orders. This ideal situation is attainable only when one has a large amount of data, which generally is not the case in practice. Therefore, to relax the information requirement without having to specify the distribution functions, third-order polynomial normal transform (TPNT) can be used. By TPNT, each individual non-normal random variable is related to a 3rd-order polynomial function of the corresponding standard normal variable [28]. The polynomial coefficients are determined by matching the statistical moments or quantiles of the individual random variables. The multivariate version of TPNT was first proposed by Vale and Maurelli [29] to simultaneously consider statistical moments and correlation coefficients. A multivariate TPNT procedure has been applied to different fields including, but not limited to, Monte Carlo simulation for generating multivariate random variates [24,30–33], wind power modeling [34], load computation in power network planning [35], and reliability analysis [36,37].

It should be noted that the great majority of multivariate TPNT applications are done under the assumption of known marginal statistical moments (i.e., product-moments and L-moments) and correlation coefficients. However, in real-life hydrologic applications, the amount of available data generally is not sufficiently large to reliably ascertain the true marginal probability distribution functions, statistical moments, and correlation coefficients. Therefore, the sample statistical moments and correlation coefficients used could be subject to sampling errors. In this study, a procedure is proposed to (1) optimally estimate multivariate TPNT coefficients by explicitly incorporating sampling errors associated with the sample moments and correlation coefficients, and (2) comply with a one-to-one monotonicity increasing relation between quantiles of the original and normal transformed variables. The procedure is illustrated by analyzing annual maximum rainfall data series involving seven different durations to establish at-site rainfall intensity–duration–frequency (IDF) and depth–duration–frequency (DDF) relationships.

2. Methods

2.1. Third-Order Polynomial Normal Transform (TPNT)

2.1.1. Univariate TPNT

By TPNT, a univariate non-normal random variable, X, is approximated by the standard normal variable, Z, in the form of a 3rd-order polynomial functional relation as [28]

$$X = \text{TPNT}(Z \mid a_0, a_1, a_2, a_3) = a_0 + a_1 Z + a_2 Z^2 + a_3 Z^3 \tag{2}$$

where $\text{TPNT}(Z \mid a_0, a_1, a_2, a_3)$ denotes the 3rd-order polynomial transform with a_0, a_1, a_2, and a_3 being the transformation coefficients. The TPNT coefficients can be determined by several methods of varying mathematical complexity. By preserving the first four product-moments, the TPNT coefficients are related to the first four product moments of the standardized variable, $X' = (X - \mu_x)/\sigma_x$, as [28]

$$0 = a_0 + a_2 \tag{3}$$

$$1 = a_1^2 + 6a_1 a_3 + 2a_2^2 + 15a_3^2 \tag{4}$$

$$\gamma_x = 2a_2 \left(a_1^2 + 24a_1 a_3 + 105a_3^2 + 2 \right) \tag{5}$$

$$\kappa_x = 3 + 24 \left[a_1 a_3 + a_2^2 \left(1 + a_1^2 + 28a_1 a_3 \right) + a_3^2 \left(12 + 48a_1 a_3 + 141a_2^2 + 225a_3^2 \right) \right] \tag{6}$$

in which γ_x = skew coefficient; κ_x = kurtosis of the original random variable X. Alternatively, the TPNT coefficients in Equation (2) can also be related to the first four L-moments as [38]

$$a_0 + a_2 = \lambda_1 \tag{7}$$

$$0.5642a_1 + 1.4104a_3 = \lambda_2 \tag{8}$$

$$0.5513a_2 \ \lambda_3 \tag{9}$$

$$0.0692a_1 + 0.8078a_3 = \lambda_4 \tag{10}$$

in which λ_m = the mth order L-moment [39] of the original non-normal random variable, X. Other than the above two moment-matching methods, TPNT coefficients can also be determined by the quantile-based least square method and the Fisher–Cornish asymptotic expansion (FC) method [40]. Chen and Tung [41] investigated the performance of different methods in determining the TPNT coefficients with regard to their accuracy and robustness in capturing the probabilistic features of the random variable X under the condition that the population distribution is known. It was found that, among the various methods for estimating TPNT coefficients, the L-moment based method is computational simplistic and can yield a satisfactory performance under a wide range of distribution conditions. The product-moment method can also yield a satisfactory normal transformation provided that accurate estimations of skew coefficient and kurtosis in Equations (3)–(6) can be made. However, when the statistical moments are to be estimated from finite data, the sample L-moments have been proven to be more stable and robust than those of product-moments [42], especially when the sample size is not large.

By referring to Equations (3)–(6), one also realizes that determining TPNT coefficients based on the product-moments requires solving a system of non-linear equations. It is expected that solving Equations (3)–(6) would be more difficult than solving L-moments based on Equations (7)–(10), which is linear. Sometimes, the solution to the system of non-linear equation may not be attainable. According to Equations (7)–(10), TPNT coefficients can be easily obtained in terms of L-moments as

$$a_0 = \lambda_1 - 1.8138\lambda_3 \tag{11}$$

$$a_1 = 2.2552\lambda_2 - 3.9376\lambda_4 \tag{12}$$

$$a_2 = 1.8138\lambda_3 \tag{13}$$

$$a_3 = -0.1931\lambda_2 + 1.5751\lambda_4 \tag{14}$$

In the transformation process, it is necessary to preserve probability content in both original space and standard normal space, i.e., $F_X(x_p) = \Phi(z_p) = p$. This implies that quantiles of the two variables should satisfy the following relationship:

$$x_p = a_0 + a_1 z_p + a_2 z_p^2 + a_3 z_p^3 \tag{15}$$

where x_p and z_p = pth-order quantiles of random variable X and standard normal random variable, Z, respectively, that is, $x_p = F^{-1}(p)$ and $z_p = \Phi^{-1}(p)$. Furthermore, inherently embedded in Equation (15) is a requirement of one-to-one monotonically increasing relations between x_p and z_p. This, then, requires that TPNT coefficients must comply with the following conditions:

$$a_3 > 0 \text{ and } a_2{}^2 - 3a_1 a_3 < 0. \tag{16}$$

It should be noted that the TPNT coefficients obtained from solving Equations (3)–(6), Equations (7)–(10), or other methods mentioned above do not guarantee the compliance of the monotonicity condition stipulated in Equation (16). This is especially a major concern when sample statistics are used in determining TPNT coefficients.

2.1.2. Multivariate TPNT

The TPNT coefficients can be determined by preserving the statistical moments of individual random variables. Specifically, L-moments are used herein to determine the multivariate TPNT coefficients due to simple, linear functional relationships between the TPNT coefficients and the L-moments as shown in Equations (11)–(14). Furthermore, sample L-moments have several desirable sampling properties over the product-moments as proven by Hosking [42]. In the context of fitting the first four L-moments of a total of N correlated variables, Equations (7)–(10) can be re-written as

$$a_{0j} + a_{2j} = \lambda_{1j} \tag{17}$$

$$0.5642a_{1j} + 1.4104a_{3j} = \lambda_{2j} \tag{18}$$

$$0.5513a_{2j} \, \lambda_{3j} \tag{19}$$

$$0.0692a_{1j} + 0.8078a_{3j} = \lambda_{4j} \tag{20}$$

in which λ_{mj} = the mth order L-moment of the jth random variable X_j for $j = 1, 2, \ldots, N$.

In addition to preserving marginal statistical moments of involved variables, multivariate TPNT must also simultaneously preserve the statistical dependence between random variables in the transformation. The correlation coefficient of any two correlated random variables, X_j and X_k, is imbedded in their 2nd-order cross-product moment of which Vale and Maurelli [29] had shown the explicit expressions in terms of TPNT coefficients as

$$CP_{j,k}\left(a_j, \, a_k; \rho_{z_j, \, z_k}\right) = E\left[X_j X_k\right] = \mu_j \mu_k - \rho_{x_j, \, x_k} \sigma_j \sigma_k$$
$$= \left(6a_{3j}a_{3k}\right) \rho^3_{z_j, \, z_k} + \left(2a_{2j}a_{2k}\right) \rho^2_{z_j, \, z_k} + \left[\left(a_{1j} + 3a_{3j}\right)\left(a_{1k} + 3a_{3k}\right)\right]\rho_{z_j, \, z_k} + \left[\left(a_{0j} + a_{2j}\right)\left(a_{0k} + a_{2k}\right)\right] \tag{21}$$

in which $\rho_{x_j, \, x_k}$, $\rho_{z_j, \, z_k}$ = correlation coefficient of random variables $\left(X_j, X_k\right)$ and its equivalent $\left(Z_j, Z_k\right)$ in normal space; μ_j and σ_j = mean and standard deviation of random variable X_j, respectively. The correlation coefficient in the original scale, $\rho_{x_j, \, x_k}$, is related to its counterpart in the normal space, $\rho_{z_j, \, z_k}$, in a 3rd-order polynomial relationship through TPNT coefficients.

Upon the determination of TPNT coefficients for the two concerned random variables, the correlation coefficient in the normal space, $\rho_{z_j, \, z_k}$, corresponding to that in the original space, $\rho_{x_j, \, x_k}$, can be obtained by finding the real root of Equation (21). The mathematical relations between the two correlation coefficients are [20]

$$\rho_{x_j, \, x_k} \times \rho_{z_j, \, z_k} > 0; \; \left|\rho_{x_j, \, x_k}\right| \leq \left|\rho_{z_j, \, z_k}\right| \tag{22}$$

Equation (21) is used repeatedly to solve for $\rho_{z_j, \, z_k}$ for all pairs of correlated random variables to establish the correlation matrix in multivariate normal space.

2.2. Optimization Framework for Determining Multivariate TPNT Coefficients

2.2.1. Objective Function

To determine the multivariate TPNT coefficients that best preserve the known values of L-moments, the least-square criterion is used in the study by which the objective function can be expressed as

$$\text{Minimize} \sum_{m=1}^{4} \sum_{j=1}^{N} \delta^2_{mj} \tag{23}$$

where δ_{mj} = a decision variable defining the deviation between the mth-order TPNT-based L-moments computed by the left-hand side of Equations (17)–(20) and the known values, λ_{mj}, of the jth random variable, X_j. Of course, other forms of objective function, such as minimizing the sum of absolute deviations, can be used.

2.2.2. Constraints

Several constraints are essential to make sure that multivariate TPNT coefficients obtained are able to preserve the known statistical features and mathematical relationships of concerned random variables.

(a) Preservation of L-moments for the individual variable X_j:

The deviation δ_{mj} in the objective function defining the degree of preserving the known values of the first-four L-moments of individual variable X_j can be written, according to Equations (17)–(20), as

$$a_{0j} + a_{2j} + \delta_{1j} = \lambda_{1j} \tag{24}$$

$$0.5642a_{1j} + 1.4104a_{3j} + \delta_{2j} = \lambda_{2j} \tag{25}$$

$$0.5513a_{2j} + \delta_{3j} = \lambda_{3j} \tag{26}$$

$$0.0692a_{1j} + 0.8078a_{3j} + \delta_{4j} = \lambda_{4j}. \tag{27}$$

Note that the value of δ_{mj} is unrestricted-in-sign, meaning that its value can be negative, zero, and positive, depending on the relative magnitudes of TPNT-based L-moments and those of the known values.

In reality, statistical properties of a random variable are estimated from a finite number of sample data. Consequently, sample L-moments of random variables, X_j, are subject to uncertainty. In practice, two approaches are used to estimate sample L-moments: plotting position-based estimators and unbiased estimators. This study adopts the latter by which the first four unbiased estimators of L-moments, $\{\lambda_1, \lambda_2, \lambda_3, \lambda_4\}$, can be computed respectively as [42]

$$\ell_1 = \binom{n}{1}^{-1} \sum_{i=1}^{n} x_{i:n} \tag{28}$$

$$\ell_2 = \frac{1}{2!} \binom{n}{2}^{-1} \sum_{i=1}^{n} \left\{ \binom{i-1}{1} - \binom{n-i}{1} \right\} x_{i:n} \tag{29}$$

$$\ell_3 = \frac{1}{3!} \binom{n}{3}^{-1} \sum_{i=1}^{n} \left\{ \binom{i-1}{2} - 2\binom{i-1}{1}\binom{n-i}{1} + \binom{n-i}{2} \right\} x_{i:n} \tag{30}$$

$$\ell_4 = \frac{1}{4!} \binom{n}{4}^{-1} \sum_{i=1}^{n} \left\{ \binom{i-1}{3} - 3\binom{i-1}{2}\binom{n-i}{1} + 3\binom{i-1}{1}\binom{n-i}{2} - \binom{n-i}{3} \right\} x_{i:n} \tag{31}$$

in which ℓ_m = sample estimator of the mth-order L-moment, λ_m; $x_{i:n}$ = the ith ranked sample (ascending order) in a data of size n.

Suppose that the sampling distributions of sample L-moments are derived or approximated. Proper bounds can then be incorporated into Equations (24)–(27) for determining the suitable and probabilistically plausible TPNT coefficients for all N random variables $X_1, X_2, \ldots, X_N$. Assuming that the lower and upper bounds of the L-moments can be determined from their corresponding sampling distributions, constraint Equations (24)–(27) then can be modified as

$$\ell_{1j}^{(L)} \leq a_{0j} + a_{2j} + \delta_{1j} \leq \ell_{1j}^{(U)} \tag{32}$$

$$\ell_{2j}^{(L)} \leq 0.5642a_{1j} + 1.4104a_{3j} + \delta_{2j} \leq \ell_{2j}^{(U)} \tag{33}$$

$$\ell_{3j}^{(L)} \leq 0.5513a_{2j} + \delta_{3j} \leq \ell_{3j}^{(U)} \tag{34}$$

$$\ell_{4j}^{(L)} \le 0.0692a_{1j} + 0.8078a_{3j} + \delta_{4j} \le \ell_{4j}^{(U)} \tag{35}$$

for $j = 1, 2, \ldots, N$. In Equations (32)–(35), $\ell_{mj}^{(U)}$ and $\ell_{mj}^{(L)}$ are, respectively, the upper and lower bounds containing the unknown population mth-order L-moment of random variable X_j, λ_{mj}. The derivation of bounds for unknown population L-moments is described in Section 2.3.1.

(b) Preservation of the monotonic probability–quantile relationship for individual variable X_j:

$$a_{3j} > 0; \; a_{2j}^2 - 3a_{1j}a_{2j} < 0, \; \text{for } j = 1, 2, \ldots, N \tag{36}$$

(c) Preservation of the correlation between all pairs of different variables, X_j and X_k:

Based on Equation (21), any pair of two correlated random variables X_j and X_k must satisfy the following equation.

$$CP_{j,k}\left(a_j, a_k; \rho_{z_j, z_k}\right) - \left(\mu_j \mu_k - \rho_{x_j, x_k}\sigma_j\sigma_k\right) = 0, \; \text{for all variable pairs } j \ne k \tag{37}$$

where $CP_{j,k}\left(a_j, a_{j'}; \rho_{z_j, z_k}\right) = E(X_j, X_k)$ = cross-product moment of variables, X_j and X_k, defined in Equation (21), which are functions of the corresponding TPNT coefficients $a_j = (a_{0j}, a_{1j}, a_{2j}, a_{3j})$ and $a_k = (a_{0k}, a_{1k}, a_{2k}, a_{3k})$; ρ_{x_j, x_k} and ρ_{z_j, z_k} = correlation coefficients of random variables, X_j and X_k, and their normal equivalents, Z_j and Z_k, respectively; μ_j, σ_j = mean and standard deviation of random variable X_j, respectively.

Similarly, constraint Equation (37) on correlation coefficients can be modified as

$$r_{x_j, x_k}^{(L)} \le \frac{CP_{j,k}\left(a_j, a_k; r_{z_j, z_k}\right) - m_j(a_j)m_k(a_k)}{s_j(a_j) \times s_k(a_k)} \le r_{x_j, x_k}^{(U)}, \; \text{for all variable pairs } j \ne k \tag{38}$$

in which $r_{x_j, x_k}^{(L)}$, $r_{x_j, x_k}^{(U)}$ = lower and upper bounds, respectively, of the unknown population correlation coefficient, ρ_{x_j, x_k}, between the random variables X_j and X_k (see Section 2.3.2); r_{z_j, z_k} = equivalent correlation coefficient of the random variables in the standard normal domain; $m_j(a_j), s_j(a_j)$ = TPNT-based estimation of mean and standard deviation of random variables X_j which can be computed according to Equations (3) and (4) as

$$m_j(a_j) = a_{0j} + a_{2j} \tag{39}$$

$$s_j^2(a_j) = a_{1j}^2 + 6a_{1j}a_{3j} + 2a_{2j}^2 + 15a_{3j}^2. \tag{40}$$

Equation (38) can alternatively be expressed as

$$CP_{j,k}\left(a_j, a_k; r_{z_j, z_k}\right) - m_j(a_j)\, m_k(a_k) - r_{x_j, x_k}^{(U)}\, s_j(a_j)\, s_k(a_k) \le 0 \tag{41}$$

$$CP_{j,k}\left(a_j, a_k; r_{z_j, z_k}\right) - m_j(a_j)\, m_k(a_k) - r_{x_j, x_k}^{(L)}\, s_j(a_j)\, s_k(a_k) \ge 0 \tag{42}$$

In summary, by considering sampling errors of sample L-moments and correlation coefficients, the optimization model to determine the most plausible TPNT coefficients for establishing multivariate relationships can be summarized as follows:

The objective function is expressed in Equation (23) or its variations, which is subject to the following constraints:

- Equations (32)–(35) for preserving plausible L-moments ($8 \times N$ constraints);
- Equation (36) for complying with a probability–quantile monotonic relationship ($2 \times N$ constraints);
- Equations (41) and (42) for preserving plausible correlation coefficient ($N \times (N - 1)$ constraints); and
- unrestrictive-in-sign of polynomial coefficients ($a_{0j}, a_{1j}, a_{2j}, a_{3j}$) and deviations δ_{mj}.

2.3. Determination of Bounds for L-Moments and Correlation Coefficients

2.3.1. Bounds for L-Moments

To determine the bounds for L-moments, the sampling distributions corresponding to the sample L-moments are needed. For independent random samples of size n from a distribution function $F_X(x)$ having the mth-order population L-moment λ_m, Hosking [39] showed that the statistic $n^{1/2}(\ell_m - \lambda_m)$, with ℓ_m being the sample L-moment of order m, is unbiased, having a sampling distribution asymptotically converge to the normal distribution with the mean zero and variance Λ_{mm}. Therefore, the variance of the mth-order sample L-moment, ℓ_m, has the variance of $\sigma^2(\ell_m) = \Lambda_{mm}/n$. For the first four orders of sample L-moment, the value of Λ_{mm} can be computed by

$$\Lambda_{11} = \iint_{x<y} (y-x)^2 du\, dv \tag{43}$$

$$\Lambda_{22} = \iint_{x<y} [(3-4v)(1-4u)](y-x)^2 du\, dv \tag{44}$$

$$\Lambda_{33} = \iint_{x<y} \left[\left(7-24v+18v^2\right)\left(1-12u+18u^2\right)\right](y-x)^2 du\, dv \tag{45}$$

$$\Lambda_{44} = \iint_{x<y} \left[\left(-13+84v-150v^2+80v^3\right)\left(-1+24u-90u^2+80u^3\right)\right](y-x)^2 du\, dv \tag{46}$$

in which $u = F_X(x)$ and $v = F_X(y)$. To estimate the values of Λ_{mm} based on the ranked sample observations, the double integration stated in Equations (43)–(46) can be carried out numerically as

$$\hat{\Lambda}_{11} = \frac{1}{n(n-1)} \sum_{i=1}^{n-1} \sum_{k=i+1}^{n} \left[x_{(k)} - x_{(i)}\right]^2 \tag{47}$$

$$\hat{\Lambda}_{22} = \frac{1}{n(n-1)} \sum_{i=1}^{n-1} \sum_{k=i+1}^{n} \left\{(3-4p_{k:n})(1-4p_{i:n})\left[x_{(k)} - x_{(i)}\right]^2\right\} \tag{48}$$

$$\hat{\Lambda}_{33} = \frac{1}{n(n-1)} \sum_{i=1}^{n-1} \sum_{k=i+1}^{n} \left\{\left(7-24p_{k:n}+18p_{k:n}^2\right)\left(1-12p_{i:n}+18p_{i:n}^2\right)\left[x_{(k)} - x_{(i)}\right]^2\right\} \tag{49}$$

$$\hat{\Lambda}_{44} = \frac{1}{n(n-1)} \sum_{i=1}^{n-1} \sum_{k=i+1}^{n} \left\{\left(-13+84p_{k:n}-150p_{k:n}^2+80p_{k:n}^3\right)\left(-1+24p_{i:n}-90p_{i:n}^2+80p_{i:n}^3\right)\left[x_{(k)} - x_{(i)}\right]^2\right\} \tag{50}$$

in which $x_{(i)}$ = the ith ranked sample in ascending order, i.e., $x_{(1)} < x_{(2)} < \cdots < x_{(i)} < \cdots < x_{(k)} < \cdots < x_{(n)}$; $p_{i:n}$ = estimated cumulative probability for the ith ranked sample, i.e., $\Pr\left[X \leq x_{(i)}\right]$, by using the well-known Weibull plotting position formula, $p_{i:n} = i/(n+1)$. Makkonen [43] has shown that the Weibull plotting position formula [44] provides the best estimate for the underlying non-exceedance probability. The superiority of the Weibull formula gets more pronounced with a decreasing sample size. By adopting the normality distribution assumption, the α-confidence interval for the unknown population λ_k can be obtained as

$$\left(\ell_m^{(L)}, \ell_m^{(U)}\right) = \left(\ell_m - z_{\alpha/2}\sqrt{\frac{\Lambda_{mm}}{n}}, \ell_m + z_{\alpha/2}\sqrt{\frac{\Lambda_{mm}}{n}}\right) \tag{51}$$

in which $z_{\alpha/2} = \Phi^{-1}(1-\alpha/2)$, a standard normal quantile with an exceedance probability of $\alpha/2$, with $\Phi^{-1}(\cdot)$ being the inverse standard normal CDF.

2.3.2. Bounds for Correlation Coefficients

To quantify the lower and upper bounds of a correlation coefficient, Fisher transform is often used by which the sampling distribution of the inverse hyperbolic tangent function of sample correlation approximately follows a normal distribution as [45,46]

$$tanh^{-1}(r) = \frac{1}{2} \ln\left(\frac{1+r}{1-r}\right) \sim Normal\left(\frac{1}{2} \ln\left(\frac{1+\rho}{1-\rho}\right), \frac{1}{n-3}\right) \tag{52}$$

in which r, ρ = sample and population correlation coefficients, respectively; n = number of sample pairs. With a specified confidence level α, the corresponding lower and upper bounds for the unknown population coefficient ρ can be obtained as

$$\left[r^{(L)}, r^{(U)}\right] = \left[tanh\left(tanh^{-1}(r) - \frac{z_{\alpha/2}}{\sqrt{n-3}}\right), tanh\left(tanh^{-1}(r) + \frac{z_{\alpha/2}}{\sqrt{n-3}}\right)\right] \tag{53}$$

where the hyperbolic tangent function is defined as $tanh(\theta) = (e^{2\theta} - 1)/(e^{2\theta} + 1)$.

2.4. Solution Algorithm

A recursive procedure is proposed to solve the above optimization models for determining multivariate TPNT coefficients. The procedure consists of four steps of initialization, optimization, validation, and updating. Solution algorithm for determining multivariate TPNT coefficients considering sampling errors of L-moments and correlation coefficients is detailed below and outlined in Figure 1.

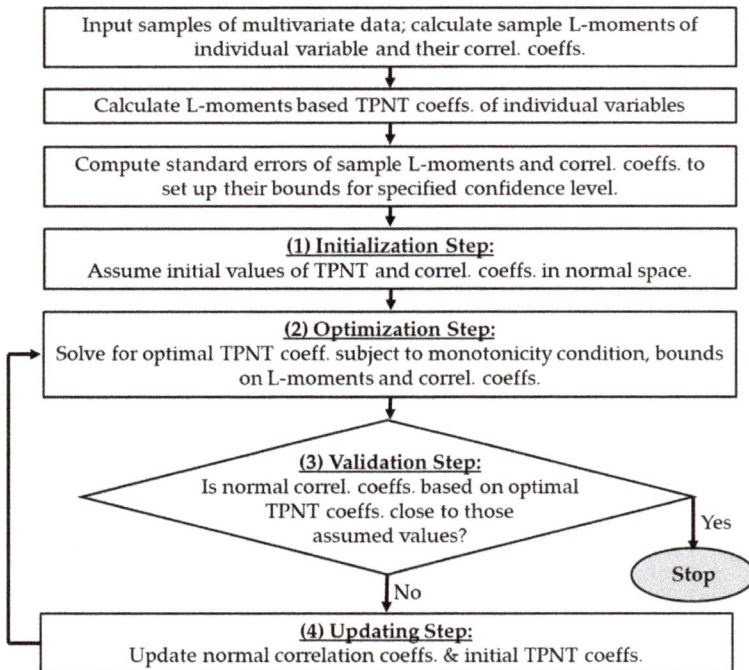

Figure 1. Flow diagram showing the procedure of multivariate TPNT modeling considering sampling errors of sample statistics.

Step (1): Initialization—Since the problem involves a nonlinear optimization model, an initial solution would be needed. One straightforward and sound initial solutions for the TPNT coefficients, $a_j^{(0)}$, are those provided by Equations (11)–(14) for all variables $j = 1, 2, \ldots, N$. The initial TPNT coefficients obtained this way will automatically satisfy the constraint Equations (24)–(27). However, they do not necessarily comply with the monotonicity constraint Equation (36). As for the initial normal correlation coefficients, rather than arbitrarily choosing a set of initial correlation coefficients, let $\left\{ \rho_{z_j, z_k}^{(0)} \right\} = \left\{ r_{x_j, x_k} \right\}$ in which r_{x_j, x_k} is the sample correlation coefficient between random variables X_j and X_k. Alternatively, obtain a feasible set of initial $\left\{ \rho_{z_j, z_k}^{(0)} \right\}$ by solving the 3rd-order polynomial function of ρ_{z_j, z_k} in Equation (21) according to the initially assumed TPNT coefficients, $a_j^{(0)}$.

Step (2): Optimization—Based on the initially adopted TPNT coefficients, $a_j^{(0)}$, and the normal transformed correlation coefficients $\left\{ \rho_{z_j, z_k}^{(0)} \right\}$, solve the optimization model with objective function Equation (23) and constraint Equations (32)–(35), (36), and (41)–(42) for the optimal TPNT coefficients $a_j^* = \left(a_{0j}^*, a_{1j}^*, a_{2j}^*, a_{3j}^* \right)$ for random variable X_j, $j = 1, 2, \ldots, N$.

Step (3): Validation—From the optimal feasible TPNT coefficients $a_j^* = \left(a_{0j}^*, a_{1j}^*, a_{2j}^*, a_{3j}^* \right)$ for $j = 1, 2, \ldots, N$ obtained from Step (2), determine the equivalent normal variates corresponding to the sample data by solving the 3rd-order polynomial function:

$$a_{0j}^* + a_{1j}^* z_{j,i} + a_{2j}^* z_{j,i}^2 + a_{3j}^* z_{j,i}^3 = x_{j,i}, \text{ for all } j = 1 \sim N;\ i = 1 \sim n \tag{54}$$

where $z_{j,i}$ = unknown normal variate corresponding to the ith observation of the jth random variable $x_{j,i}$ under the optimal set of TNPT coefficients $a_j^* = \left(a_{0j}^*, a_{1j}^*, a_{2j}^*, a_{3j}^* \right)$. From the normal-transformed data series of two different variables, $z_j = (z_{j,1}, z_{j,2}, \ldots, z_{j,n})$ and $z_k = (z_{k,1}, z_{k,2}, \ldots, z_{k,n})$, the corresponding correlation coefficient, ρ_{z_j, z_k}^*, in the normal space is calculated.

Step (4): Updating—Compare the discrepancies between the initialized $\rho_{z_j, z_k}^{(0)}$ and validated ρ_{z_j, z_k}^* for all different pairs of concerned random variables. If the discrepancy in any pair of durations is judged to be significant, update the initial normal correlation as $\rho_{z_j, z_k}^{(0)} = \rho_{z_j, z_k}^*$ and TPNT coefficients $a_j^{(0)} = a_j^*$, and the process from Steps (2)–(4) is repeated. Otherwise, the optimal solutions are obtained and the iteration stops.

With regard to the optimization step presented in Step (2), the sequential quadratic programming (SQP) algorithm is implemented [47]. The SQP tackles a nonlinear optimization problem by successively finding the approximated optimum solution to the quadratic programming (QP) representation of the original problem. The approximated solution is improved iteratively by solving the QP problem. Boggs and Tolle [48] elaborated some useful properties of the SQP algorithm. The subroutine "sqp.m" in Matlab is used in this study to solve the optimization model.

3. Numerical Example

In this section, at-site rainfall intensity–duration–frequency (IDF) and depth–duration–frequency (DDF) relations are established to demonstrate the proposed multivariate TPNT method and examine its general performance. Rainfall IDF relations are widely used in the planning, design, and management of hydrosystem infrastructures, such as stormwater sewer systems and detention basins [49,50]. Such relations at a given location involves at-site frequency analysis of annual maximum rainfall intensity (or depth) data of several selected durations. The conventional approach in rainfall frequency analysis chooses a proper parametric probability distribution model to individually fit the observed annual maximum rainfall data of different durations. The choice of a distribution model for the rainfall intensity–frequency relations is largely statistical without much physical justification [51].

By the conventional approach, resulting rainfall intensity–frequency curves of different durations could sometimes intersect within the probability range of practical application. The crossover phenomenon often occurs when data record length is relatively short. According to the physical reality, rainfall intensity–frequency curves of different durations should not crossover or intersect. Porras and Porras [52] attributed the occurrence of crossover of rainfall IDF curves to short record data of questionable representation in which a significant amount of sampling errors existed in the estimated rainfall quantiles by frequency analysis. One other plausible reason for the possible crossover of IDF curves is that frequency analysis of rainfall data is performed separately for each duration without considering the inter-correlations that are intrinsically embedded in rainfall data of different durations. Haktanir [53] earlier pointed out that rainfall frequency analysis of different durations in the process of establishing IDF relationships should not be performed independently of each other, but did not propose a mechanism to handle the correlation directly. Recently, Gräler et al. [54] applied D-vine copula, along with the generalized extreme value distributions, to derive rainfall IDF relationships based on rainfalls of five durations. You and Tung [55], under the TPNT framework, developed a constrained least square model to simultaneously considering rainfall data of seven durations for establishing at-site rainfall IDF relations. However, their model does not explicitly take into account the correlation among rainfall data of different durations.

The multivariate TPNT-based model presented above was applied to establish at-site rainfall IDF relations using annual maximum hourly rainfall data of various durations at a raingauge in Zhongli City of Taoyuan County, Taiwan. Annual maximum rainfall intensity data cover the record period of 1988–2015, but the year 1992 was excluded from the analysis due to long periods of registers with technical issues. Hence, only 27-year data ($n = 27$) with seven ($N = 7$) durations (i.e., 1, 2, 6, 12, 24, 48, and 72 h) are used in this illustration (see data in Table 1). The sample values of the mean, standard deviation, and first-four L-moments of rainfall data of different durations are tabulated in Table 2. Furthermore, the standard error values corresponding to the first four sample L-moments, according to Equations (47)–(50), are listed in Table 3. The sample correlation coefficients of all rainfall intensity pairs of different durations in the original and normal-transformed domains are shown in Tables 3 and 4, respectively. Based on the information given in Tables 2 and 4, one is able to define the lower and upper bounds for the L-moments and correlation coefficients according to the desired confidence level, α, by Equations (51) and (53), respectively. Table 5 lists the values of correlation coefficients in normal-transformed space, r_{z_j, z_k}, provided by the solution to constraint Equations (41) and (42) in the optimization model.

Under different constraint types and confidence levels for the L-moments and correlation coefficients, the corresponding optimal multivariate TPNT coefficients can vary. With the confidence level of $\alpha = 90\%$ for both L-moments and correlation coefficients, Table 6a–d list the optimal TPNT coefficients under four different constraint types, including "LM" for L-moments by Equations (32)–(35), "Mono" for monotonicity by Equation (36), "Corr" for correlation by Equations (41) and (42), and "NC" for no-crossover by Equation (56). Once the optimal TPNT coefficients associated with each rainfall duration are obtained from solving the multivariate TPNT model, the rainfall IDF relations, according to Equation (15), can be established as

$$i^*_{t,T} = a^*_{0,t} + a^*_{1,t}z_T + a^*_{2,t}z_T^2 + a^*_{3,t}z_T^3 \tag{55}$$

where $i^*_{t,T}$ = estimated *t*-h, *T*-year rainfall intensity; $a^*_{0,t}, a^*_{1,t}, a^*_{2,t}$, and $a^*_{3,t}$ = optimum TPNT coefficients corresponding to rainfall of duration *t* (h); z_T is the standard normal quantile corresponding to return period *T*-year having an annual exceedance probability of $1 - \Phi(z_T) = 1/T$.

Table 1. Annual maximum rainfall intensities (mm/h) at Zhongli Station, Taiwan.

Year	1 h	2 h	6 h	12 h	24 h	48 h	72 h
1988	48.0	41.0	18.1	9.0	4.6	2.9	2.2
1989	83.5	69.5	38.2	19.9	10.0	6.1	4.1
1990	89.0	51.3	25.1	14.0	8.1	5.7	3.8
1991	52.5	36.3	13.1	6.8	4.4	3.9	2.9
1993	73.0	50.5	25.8	15.5	8.1	4.3	3.0
1994	44.5	31.0	19.1	11.7	8.7	5.0	3.3
1995	87.5	59.0	27.1	13.7	7.0	4.7	3.5
1996	80.5	40.3	14.0	10.3	8.0	4.6	3.1
1997	35.5	22.8	16.4	11.0	6.3	3.7	2.7
1998	70.5	35.3	21.2	11.1	7.8	4.5	3.0
1999	58.5	29.3	12.2	7.7	5.5	2.9	2.0
2000	43.5	25.8	19.8	14.0	11.2	8.2	5.8
2001	49.5	43.5	28.0	21.5	14.3	12.4	8.7
2002	35.0	28.3	12.4	6.9	5.0	2.7	1.8
2003	30.5	20.5	9.2	7.1	4.7	2.6	1.7
2004	51.5	35.3	19.4	13.3	9.6	6.1	4.1
2005	38.0	30.3	14.8	9.3	7.6	4.6	3.7
2006	48.5	32.3	19.7	13.8	7.7	5.3	4.6
2007	65.5	60.5	24.9	12.6	8.4	6.5	5.3
2008	42.0	32.3	13.5	9.5	7.1	5.5	4.8
2009	33.0	29.3	14.9	10.7	6.2	3.2	2.6
2010	47.5	35.0	13.6	9.3	7.0	3.6	2.4
2011	71.5	48.8	21.0	12.6	7.1	3.7	2.6
2012	72.0	47.8	43.4	34.3	17.6	8.8	5.9
2013	52.0	46.0	25.8	14.5	7.6	6.7	4.5
2014	41.0	26.3	15.0	11.3	7.9	4.9	3.6
2015	35.5	25.5	15.6	10.3	5.6	3.2	2.1

Table 2. Sample moments (in mm/h) of rainfall intensity data.

Moments	1 h	2 h	6 h	12 h	24 h	48 h	72 h
$\hat{\mu}$	54.80	38.26	20.04	12.65	7.89	5.05	3.63
$\hat{\sigma}$	17.8	12.4	7.9	5.6	2.9	2.2	1.5
ℓ_1	54.80	38.26	20.04	12.65	7.89	5.05	3.63
ℓ_2	10.21	7.02	4.25	2.67	1.46	1.13	0.83
ℓ_3	1.673	1.393	1.110	0.879	0.389	0.317	0.200
ℓ_4	0.322	0.715	0.706	0.863	0.460	0.246	0.148

Table 3. Standard errors (in mm/h) of sample L-moments of rainfall intensity data.

Std. Error	1 h	2 h	6 h	12 h	24 h	48 h	72 h
$se(\ell_1)$	3.423	2.386	1.519	1.071	0.554	0.415	0.296
$se(\ell_2)$	1.110	0.952	0.750	0.682	0.324	0.226	0.148
$se(\ell_3)$	1.005	0.660	0.447	0.389	0.159	0.127	0.089
$se(\ell_4)$	0.713	0.404	0.209	0.182	0.088	0.066	0.050

Table 4. Sample correlation coefficients between rainfall intensity of different durations.

Duration	1 h	2 h	6 h	12 h	24 h	48 h	72 h
1 h	1.000						
2 h	0.826	1.000					
6 h	0.624	0.760	1.000				
12 h	0.421	0.496	0.910	1.000			
24 h	0.306	0.335	0.762	0.915	1.000		
48 h	0.210	0.338	0.635	0.732	0.869	1.000	
72 h	0.161	0.324	0.576	0.671	0.812	0.976	1.000

Table 5. Correlation coefficients of normal-transformed rainfall intensity of different durations.

Duration	1 h	2 h	6 h	12 h	24 h	48 h	72 h
1 h	1.000						
2 h	0.849	1.000					
6 h	0.654	0.768	1.000				
12 h	0.511	0.569	0.918	1.000			
24 h	0.430	0.440	0.736	0.870	1.000		
48 h	0.396	0.462	0.713	0.796	0.903	1.000	
72 h	0.323	0.438	0.670	0.755	0.854	0.979	1.000

Table 6. Multivariate TPNT coefficients obtained under different constraints with $\alpha = 0.90$.

(a) Constraints: L-moments only (LM)							
TPNT Coefficients	1 h	2 h	6 h	12 h	24 h	48 h	72 h
a_0	51.76	35.73	18.03	11.06	7.19	4.47	3.27
a_1	21.76	13.02	6.80	2.62	1.47	1.58	1.29
a_2	3.03	2.53	2.01	1.59	0.70	0.58	0.36
a_3	−1.465	−0.230	0.292	0.843	0.444	0.170	0.072
$a_3 > 0$	−1.46 *	−0.23 *	0.29	0.84	0.44	0.17	0.07
$a_2^2 - 3a_1a_2 < 0$	104.8 *	15.4 *	−1.90	−4.09	−1.46	−0.47	−0.15

(b) Constraints: L-moments and Monotonicity (LM/Mono)							
	1 h	2 h	6 h	12 h	24 h	48 h	72 h
a_0	51.99	35.81	18.03	11.06	7.19	4.47	3.27
a_1	17.52	11.98	6.80	2.62	1.47	1.58	1.29
a_2	2.80	2.45	2.01	1.59	0.70	0.58	0.36
a_3	0.150	0.167	0.292	0.843	0.444	0.170	0.072
$a_3 > 0$	0.15	0.17	0.29	0.84	0.44	0.17	0.07
$a_2^2 - 3a_1a_2 < 0$	−0.01	−0.01	−1.9	−4.1	−1.5	−0.5	−0.1

(c) Constraints: L-moments, Monotonicity, and Correlation (LM/Mono/Corr)							
	1 h	2 h	6 h	12 h	24 h	48 h	72 h
a_0	51.76	35.81	18.03	11.06	7.19	4.47	3.27
a_1	17.14	11.98	6.80	2.62	1.47	1.58	1.29
a_2	3.03	2.45	2.01	1.59	0.70	0.58	0.36
a_3	0.382	0.167	0.292	0.843	0.444	0.170	0.072
$a_3 > 0$	0.38	0.17	0.29	0.84	0.44	0.17	0.07
$a_2^2 - 3a_1a_2 < 0$	−10.4	−0.01	−1.9	−4.1	−1.5	−0.5	−0.1

(d) Constraints: L-moments, Monotonicity, Correlation, and No Crossover (LM/Mono/Corr/NC)							
	1 h	2 h	6 h	12 h	24 h	48 h	72 h
a_0	51.99	35.81	18.01	11.06	7.19	4.48	3.26
a_1	17.52	11.98	6.80	2.62	1.47	1.61	1.25
a_2	2.80	2.45	2.03	1.59	0.71	0.57	0.37
a_3	0.150	0.167	0.292	0.843	0.444	0.159	0.089
$a_3 > 0$	0.15	0.17	0.29	0.84	0.44	0.16	0.09
$a_2^2 - 3a_1a_2 < 0$	−0.01	−0.01	−1.8	−4.1	−1.5	−0.4	−0.2

Note: * indicates a violation of monotonicity condition.

4. Results and Discussions

By varying the value of z_T for different return periods in Equation (55), in conjunction with the optimal TPNT coefficients listed in Table 6a–d, one can establish IDF curves as shown in Figures 2 and 3. Part (a) of Table 6 and Figures 2–4 (denoted by "LM") shows the results from considering only the bounding constraints of L-moments, Equations (32)–(35). In fact, the optimal TPNT coefficients corresponding to each duration can be obtained separately from the exact solutions using sample L-moments in Equations (11)–(14). Note that the TPNT coefficients obtained from each rainfall duration at this stage do not necessarily comply with a one-to-one monotonic increasing relation of rainfall quantile and probability. This can be clearly seen in Table 6a for 1 and 2 h rainfalls for which the two monotonicity constraints are violated (shown by *). Part (b) (denoted by "LM/Mono") shows the results by considering both L-moment constraints, Equations (32)–(35), and the monotonicity constraints, Equation (36), for each rainfall duration. In this case, both results presented in Parts (a) and (b) in Table 6 and Figures 2–4 can be obtained separately by treating rainfall data of different durations without considering their inter-correlations. Results in Part (c), denoted by "LM/Mono/Corr," were obtained by incorporating correlation constraints of rainfall data with different durations, Equations (41) and (42), in determining the multivariate TPNT coefficients.

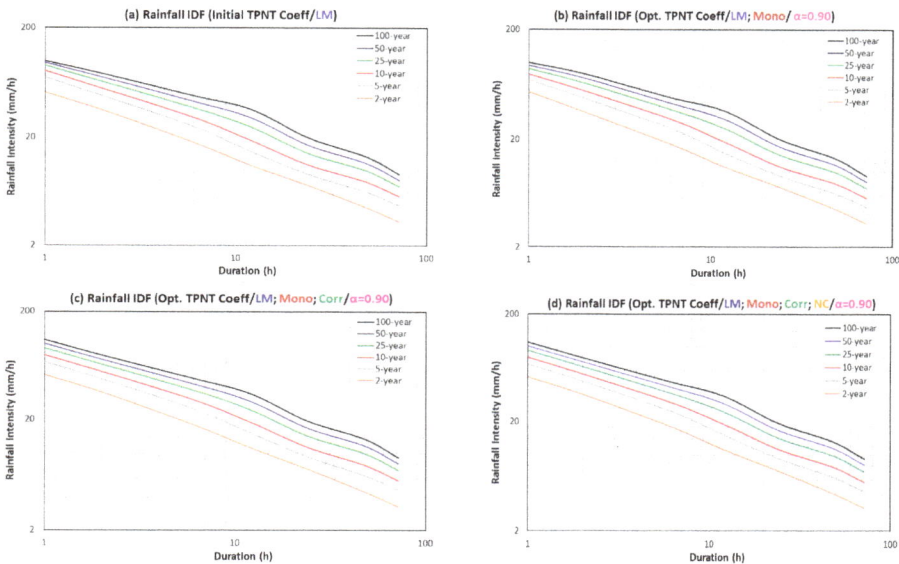

Figure 2. Multivariate TPNT modeling of rainfall intensity–duration relationships of varying return periods under different constraints which consider: (**a**) L-moments only; (**b**) L-moments and monotonicity; (**c**) L-moments, monotonicity and correlation; (**d**) L-moments, monotonicity, correlation and no crossover.

To show the degree of goodness-of-fit of normal transformed rainfall data by the proposed multivariate TPNT procedure, a normal probability plot of 24 h rainfall data (after normal transformation) with the fitted line and 95% confidence band are shown in Figure 5 as an example. The goodness-of-fit test shown in Figure 5 was achieved by the Anderson–Darling test [56] by which the test statistic is 0.535 with a *p*-value of 0.155. Figure 5 represents the worst case among the seven durations considered. The range of *p*-value varies from 0.155 (for 72 h) to 0.933 (for 2 h), which are higher than the generally adopted significance level of 0.05. This indicates that the normal transform by the proposed multivariate TPNT procedure is quite adequate.

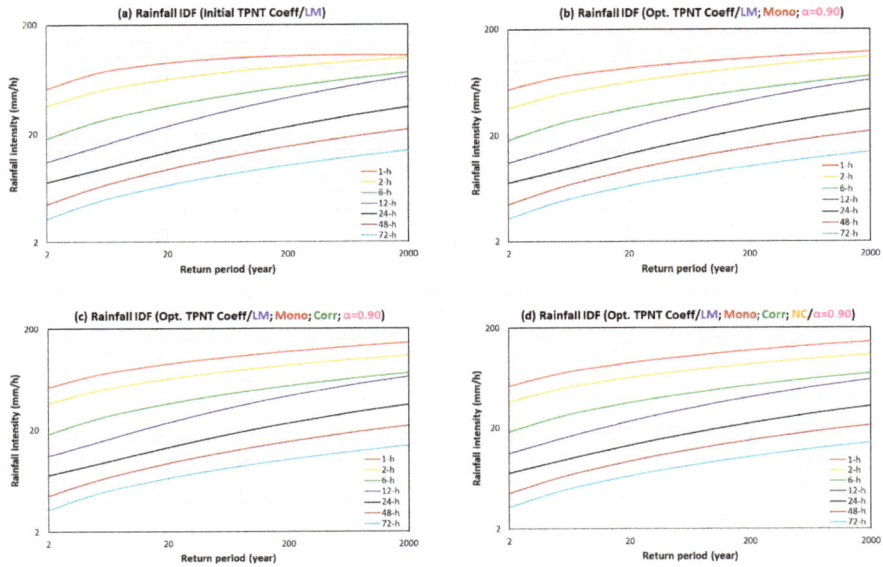

Figure 3. Multivariate TPNT modeling of rainfall intensity–frequency relationships of varying durations under different constraints which consider: (**a**) L-moments only; (**b**) L-moments and monotonicity; (**c**) L-moments, monotonicity and correlation; (**d**) L-moments, monotonicity, correlation and no crossover.

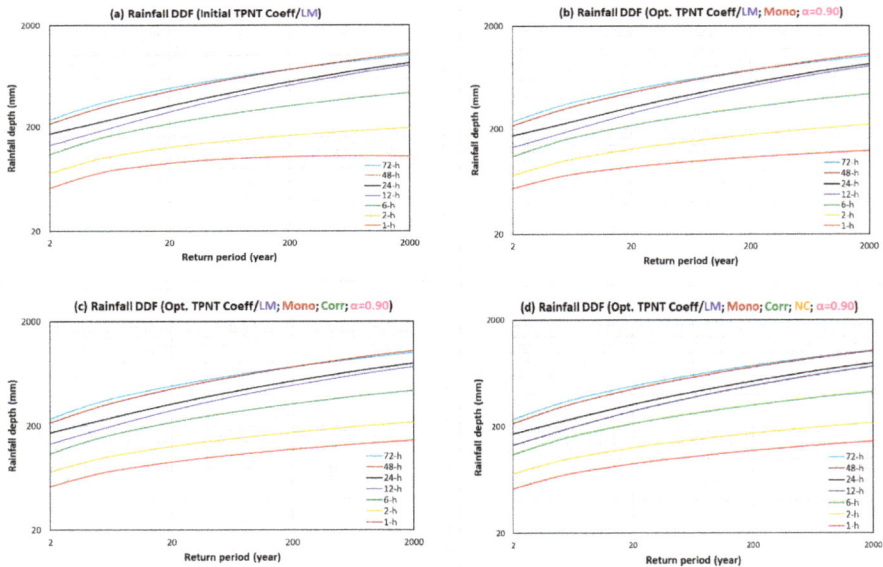

Figure 4. Multivariate TPNT modeling of rainfall depth–frequency relationships of varying durations under different constraints which consider: (**a**) L-moments only; (**b**) L-moments and monotonicity; (**c**) L-moments, monotonicity and correlation; (**d**) L-moments, monotonicity, correlation and no crossover.

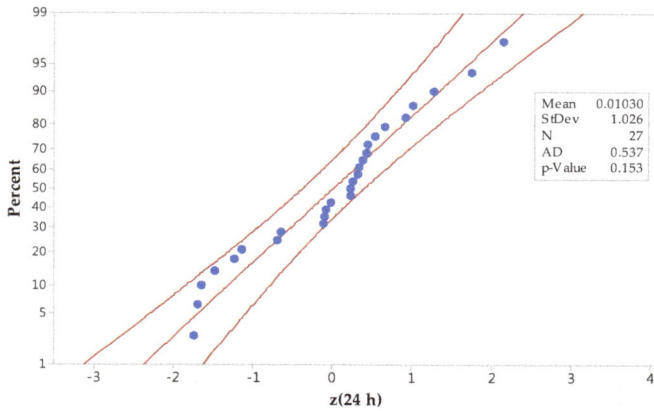

Figure 5. Probability plot of normalized 24 h rainfall data.

Note that the solution obtained up to this stage does not necessarily comply with the physical reality that rainfall intensity (depth) of a given return period is a decreasing (an increasing) function of duration. In other words, rainfall intensity/depth–frequency curves of different durations should not intersect or crossover each other. However, in the process of establishing rainfall IDF/DDF relationships, one does not know in advance if any two resulting two curves would intersect before the statistical model is developed. Therefore, a special set of intersections avoidance constraints are imposed in establishing the IDF curves:

$$\left(a_{0,t_j} - a_{0,t_k}\right) + \left(a_{1,t_j} - a_{1,t_k}\right)z_{T_*} + \left(a_{2,t_j} - a_{2,t_k}\right)z_{T_*}^2 + \left(a_{3,t_j} - a_{3,t_k}\right)z_{T_*}^3 > 0, \text{ for durations } t_j < t_k \tag{56}$$

where T_* = upper limit of selected rainfall return period below which no crossover of IDF curves is permitted to occur; z_{T_*} = standard normal quantile obtainable from $\Phi^{-1}\left(1 - \frac{1}{T_*}\right)$. Hence, additional $N - 1$ no-crossover (NC) constraints are included in the optimization model to solve for multivariate TPNT coefficients. Part (d) results (denoted by "LM/Mono/Corr/NC") show the rainfall IDF relations considering the NC constraints.

Figure 2 shows the rainfall intensity–duration curves corresponding to various frequencies. For this particular data set, by only preserving sample L-moments, Figure 2a reveals two unusual features for those curves when return period is high (say, ≥100 years). They are (1) curves that tend to converge together for rainfall duration in the vicinity of 1 h and (2) the relatively pronounced undulation of curves for medium and long duration. These features are indications of possible anomalies that should not appear in a reasonable rainfall IDF relation. The convergence of rainfall intensity–duration curves in Figure 2a, shown in a different form in intensity–frequency relation as Figure 3a, reveal that the 1 h curve (in red) clearly does not satisfy the monotonicity condition according to Equation (36), which requires a rainfall intensity quantile value to increase continuously with a return period (see also Table 4a). In fact, the 2 h intensity–frequency curve (in gold) also mildly violates the monotonicity requirement as the curve starts to bend down for high return periods. The violation of the monotonicity condition can also be observed in the form of the depth–frequency curve for a 1 h duration (see Figure 4a). In this circumstance, the non-monotonicity of the 1 h rainfall intensity–frequency relation produces a crossover with the 2 h curve shown in Figure 3a.

Interestingly, Figure 3a also reveals that 6 and 12 h rainfall intensity–frequency curves have a strong tendency to intersect as rainfall frequency increases. This tendency to intersect could be attributed to a relatively large undulation of intensity–duration curves in the range of 6–12 h when rainfall frequency increases (see Figure 2a). From 6 to 12 h, the gradient of intensity–duration curves flatten out for larger return periods. The empirical results show some evidence of improvement (in terms of a decrease in undulation, for large frequencies) when more constraints are considered. However, the improvement is not significantly enough to remove undulation. In practical engineering applications, the undulation of rainfall IDF curves such as those shown in Figure 2a–d is removed by fitting the estimated rainfall intensity–duration data by an empirical IDF model, such as Sherman's equation [57].

It is clear that, by considering the monotonicity constraint, Equation (36), the crossover tendency of intensity–duration curves (see Figure 2b) in the vicinity of 1–2 h disappears (see also Table 6b), as does that of the 1 h and 2 h intensity–frequency curves in Figure 3b. Correspondingly, the concave down appearance of the 1 h depth–frequency curve and, to a lesser extent, the 2 h curve is corrected (see Figure 4b).

Notice that joint consideration of complying with L-moments and the monotonicity condition does not truly take into account the inter-correlations of rainfall intensity or depth with different durations. The appearance of undulation in the rainfall intensity–duration curves for medium and long durations ($\geq$6 h), which satisfy the monotonicity condition, is not affected. Hence, the crossover tendency of 6 and 12 h intensity–frequency curves (see Figure 3b) and the actual intersection of 48 and 72 h depth–frequency curves (see Figure 4b) remain unchanged.

With further consideration of inter-correlations of rainfalls of different durations, Equations (41) and (42), the resulting rainfall IDF and DDF curves are shown in Part (c) of Figures 2–4. Figure 2c shows that the rainfall intensity–duration curves in the range of short duration for a high return period completely remove the crossover tendency. Both Figures 3c and 4c show that rainfall intensity–frequency and depth–frequency curves for 1 and 2 h are parallel to each other. Still, the 48 and 72 h rainfall depth–frequency curves intersect (see Figure 4c).

For illustration, this application artificially select $T_* = 5000$-year in Equation (56) as the limiting frequency below which rainfall depth–frequency or intensity–frequency curves of any two durations are not allowed to intersect. The obvious results of imposing no-crossover constraint is that the 48 h rainfall depth–frequency curve in Figure 4d would not intersect with the 72 h curve.

As for the effect of confidence level, numerical results indicate that a feasible solution for TPNT coefficients may not exist when the confidence levels for the unknown true L-moments and correlation coefficients are too low. This is expected because the width of confidence interval shrinks toward the sample L-moments and correlation coefficients as the confidence level reduces. At a certain confidence level, the corresponding width of the confidence band might be too restrictive for the optimization model to find feasible TPNT coefficients that simultaneously satisfy the monotonicity constraints. How low the limiting confidence level is depends on the problem. In this numerical example, the limiting confidence level is about 70%, below which no feasible solution can be found for multivariate TPNT coefficients. On the other hand, a reasonable confidence interval allows one to obtain a suitable set of TPNT coefficients to approximate multivariate relations.

5. Summary and Conclusions

Statistical modeling and data analysis in hydrosystems engineering often encounter multiple correlated random variables following non-normal distributions. Due to the difficulty in establishing a full joint probability density function for the involved variables, most of the methods tackling multivariate problems preserve the marginal statistical properties (e.g., distributions or moments) of individual variables and their correlation structures. In this study, focus is placed on the third-order polynomial transform (TPNT) procedure, which relies on the preservation of marginal L-moments and correlations among variables. In particular, a general framework is presented to optimally determine multivariate TPNT coefficients incorporating the constraints that (1) preserve the statistical L-moments and correlations with explicit consideration of their associated sampling errors; (2) comply with a one-to-one monotonicity increasing relation between quantiles of the original and normal transformed variables. Other than the above basic constraints required to hold the statistical and mathematical validity of the TPNT method, additional constraints that are relevant to the problem at hand can be incorporated into the modeling framework. In the illustrative example of establishing rainfall intensity–duration–frequency (IDF) relations, the no-intersection constraints for rainfall depth–frequency curves of different durations, Equation (56), are introduced in the model formulation to ensure that the resulting IDF relationships comply with the physical reality. The proposed method not only solves for the suitable multivariate TPNT coefficients that satisfy the monotonicity condition for individual variables, but also produces the correlation coefficients between random variables in the normal space. At this stage, the proposed multivariate TPNT procedure has not gone through a formal mathematical testing for its performance under different scenarios of multivariate distributions, correlation structures, and sample sizes. However, the procedure is based on a good logic with sound statistical and mathematical theory. The results from the empirical application to establish at-site rainfall IDF relationships appear to be quite reasonable.

Author Contributions: Conceptualization: Y.-K.T. and L.W.Y.; formal analysis: Y.-K.T., L.W.Y., and C.S.Y.; investigation: Y.-K.T., L.W.Y., and C.S.Y.; writing-original draft preparation: Y.-K.T. and L.W.Y.; writing-review & editing: Y.-K.T., L.W.Y., and C.S.Y.; project administration: Y.-K.T. and C.S.Y.; funding acquisition: Y.-K.T. and C.S.Y.

Funding: This study was primarily supported by the Joint Cooperative Research Program managed by the National Research Foundation of Korea (2016K2A9A1A06922023) and the Ministry of Science & Technology of Taiwan (MOST 105–2923-E-009-004-MY2). Additional funding was received from the General Research Program of the Ministry of Science & Technology of Taiwan (106-2221-E-009 -067).

Acknowledgments: The authors wish to express their gratitude to the two anonymous reviewers for their thorough review and constructive comments.

Conflicts of Interest: The authors declare no conflict of interest.

References

1. Kotz, S.; Balakrishnan, N.; Johnson, N.L. Models and Applications. In *Continuous Multivariate Distributions*, 2nd ed.; Wiley and Sons Inc.: New York, NY, USA, 2005; Volume 1.
2. Hutchinson, T.P.; Lai, C.D. *Continuous Bivariate Distributions—Emphasizing Applications*; Rumsby Scientific Publishing: Adelaide, South Australia, 1990.
3. Yue, S. A bivariate gamma distribution for use in multivariate flood frequency analysis. *Hydrol. Process.* **2001**, *15*, 1033–1045. [CrossRef]
4. Nadarajah, S.; Shiau, J.T. Analysis of extreme flood events for the Pachang River, Taiwan. *Water Resour. Manag.* **2005**, *19*, 363–374. [CrossRef]
5. Der Kiureghian, A.; Liu, P.L. Structural reliability under incomplete probability information. *J. Eng. Mech.* **1986**, *112*, 85–104. [CrossRef]
6. Genest, C.; Chebana, F. Chapter 30: Coupula Modeling in Hydrologic Frequency Analysis. In *Handbook of Applied Hydrology*, 2nd ed.; Singh, V.P., Ed.; McGraw-Hill Book Company: New York, NY, USA, 2017.
7. Favre, A.C.; Adlouni, S.E.; Perreault, L.; Thiemonge, N.; Bobee, B. Multivariate hydrological frequency analysis using copulas. *Water Resour. Res.* **2004**, *40*, W01101. [CrossRef]
8. Ganguli, P.; Reddy, M.J. Probabilistic assessment of flood risks using trivariate copulas. *Theor. Appl. Climatol.* **2013**, *111*, 341–360. [CrossRef]
9. Chen, L.; Singh, V.P.; Guo, S.L.; Mishra, A.K.; Guo, J. Drought analysis using copulas. *J. Hydrol. Eng.* **2013**, *18*, 797–808. [CrossRef]
10. Khedun, C.P.; Chowdhary, H.; Mishra, A.K.; Giardino, J.R.; Singh, V.P. Water deficit duration and severity analysis based on runoff derived from Noah land surface model. *J. Hydrol. Eng.* **2013**, *18*, 817–833. [CrossRef]
11. Sadri, S.; Burn, D.H. Copula-based polled frequency analysis of droughts in the Canadian Prairies. *J. Hydrol. Eng.* **2014**, *19*, 277–289. [CrossRef]
12. Tosunoglu, F.; Kisi, O. Joint modelling of annual maximum drought severity and corresponding duration. *J. Hydrol.* **2016**, *543*, 406–422. [CrossRef]
13. Requena, A.I.; Mediero, L.; Garrote, L. A bivariate return period based on copulas for hydrologic dam design: Accounting for reservoir routing in risk estimation. *Hydrol. Earth Syst. Sci.* **2013**, *17*, 3023–3038. [CrossRef]

14. Li, C.; Singh, V.P.; Mishra, A.K. A bivariate mixed distribution with a heavy tailed component and its application to single site daily rainfall simulation. *Water Resour. Res.* **2013**, *49*, 767–789. [CrossRef]

15. Jun, C.; Qin, X.S.; Gan, T.Y.; Tung, Y.K.; De Michele, C. Bivariate frequency analysis of rainfall intensity and duration for urban stormwater infrastructure design. *J. Hydrol.* **2017**, *553*, 374–383. [CrossRef]

16. Aas, K.; Czado, C.; Frigessi, A.; Bakken, H. Pair-copula constructions of multiple dependence. *Insur. Math. Econ.* **2009**, *44*, 182–198. [CrossRef]

17. Shafaei, M.; Fard, A.F.; Dinpashoh, Y.; Mirabbasi, R.; Michele, D.C. Modeling flood event characteristics using d-vine structures. *Theor. Appl. Climatol.* **2017**, *130*, 713–724. [CrossRef]

18. Vernieuwe, H.; Vandenberghe, S.; De Baets, B.; Verhoest, N.E.C. A continuous rainfall model based on vine copulas. *Hydrol. Earth Syst. Sci.* **2015**, *19*, 2685–2699. [CrossRef]

19. Tosunoglu, F.; Singh, V.P. Multivariate Modeling of Annual Instantaneous Maximum Flows Using Copulas. *J. Hydrol. Eng.* **2018**, *23*, 04018003. [CrossRef]

20. Qing, X. Generating correlated random vector involving discrete variables. *Commun. Stat. Theory Methods* **2017**, *46*, 1594–1605.

21. Nataf, A. Determination des distributions dont les marges sont donnees. *Comptes Rendus l'Academie Sciences* **1962**, *225*, 42–43.

22. Lebrun, R.; Dutfoy, A. An innovating analysis of the Nataf transformation from the copula viewpoint. *Probab. Eng. Mech.* **2009**, *24*, 312–320. [CrossRef]

23. Liu, P.L.; Der Kiureghian, A. Multivariate distribution models with prescribed marginals and covariances. *Probab. Eng. Mech.* **1986**, *1*, 105–112. [CrossRef]

24. Chang, C.H.; Tung, Y.K.; Yang, J.C. Monte Carlo simulation for correlated variables with marginal distributions. *J. Hydraul. Eng.* **1994**, *120*, 313–331. [CrossRef]

25. Chen, H.F. Initialization for NORTA: Generation of random vectors with specified marginal and correlations. *INFORMS J. Comput.* **2001**, *13*, 312–331. [CrossRef]

26. Li, H.S.; Lu, Z.Z.; Yuan, X.K. Nataf transformation based point estimate method. *Chin. Sci. Bull.* **2008**, *53*, 2586–2592. [CrossRef]

27. Niaki, S.T.A.; Abbasi, B. Generating correlation matrices for normal random vectors in NORTA algorithm using artificial neural networks. *J. Uncertain Syst.* **2008**, *2*, 192–201.

28. Fleishman, A.L. A method for simulating non-normal distributions. *Psychometrika* **1978**, *43*, 521–532. [CrossRef]

29. Vale, C.D.; Maurelli, V.A. Simulating multivariate non-normal distributions. *Psychometrika* **1983**, *48*, 465–471. [CrossRef]

30. Headick, T.C.; Sanwilowsky, S.S. Simulating correlated multivariate nonnormal distributions: Extending the Fleishman power method. *Psychometrika* **1999**, *64*, 25–35. [CrossRef]

31. Chen, X.Y.; Tung, Y.K. Applications of TPNT in multivariate Monte Carlo simulation. In *Water Resources Planning and Management*; EWRI/ASCE: Philadelphia, PA, USA, 2003; pp. 23–26.

32. Hodis, F.A. Simulating Univariate and Multivariate Nonnormal Distributions Based on a System of Power Method Distributions. Ph.D. Thesis, Southern Illinois University, Carbondale, IL, USA, 2008.

33. Demirtas, H.; Hedeker, D.; Mermelstein, R.J. Simulation of massive public health data by power polynomials. *Stat Med.* **2012**, *31*, 3337–3346. [CrossRef] [PubMed]

34. Yang, H.; Zou, B. The point estimate method using third order polynomial normal transformation technique to solve probabilistic power flow with correlated wind source and load. In Proceedings of the Asia-Pacific Power and Energy Engineering Conference, Shanghai, China, 27–29 March 2012; pp. 1–4.

35. Cai, D.; Shi, D.; Chen, J. Probabilistic load flow computation with polynomial normal transformation and Latin hypercube sampling. *IET Gen. Trans. Distrib.* **2013**, *7*, 474–482. [CrossRef]

36. Chen, X.Y. Investigating Third-Order Polynomial Normal Transform and Its Applications to Uncertainty and Reliability Analyses. Master's Thesis, Hong Kong University of Science and Technology, Hong Kong, China, 2002.

37. Zhao, Y.G.; Lu, Z.H. Fourth moment standardization for structural reliability assessment. *J. Struct. Eng.* **2007**, *133*, 916–924. [CrossRef]

38. Tung, Y.K. Polynomial normal transformation in uncertainty analysis. *Appl. Probab. Stat.* **1999**, *1*, 167–174.

39. Hosking, J.R.M. *The Theory of Probability Weighted Moments*; IBM Research Report, RC12210; IBM: Yorktown Heights, NY, USA, 1986.

40. Fisher, R.A.; Cornish, E.A. The percentile points of distribution having known cumulants. *Technometrics* **1960**, *2*, 209–225. [CrossRef]
41. Chen, X.Y.; Tung, Y.K. Investigation of polynomial normal transform. *Struct. Saf.* **2003**, *25*, 423–445. [CrossRef]
42. Hosking, J.R.M. L-Moments: Analysis and Estimation of Distributions Using Linear Combinations of Order Statistics. *J. R. Stat. Soc. Ser. B Method* **1990**, *52*, 105–124. [CrossRef]
43. Makkonen, L.; Pajari, M.; Tikanmaki, M. Discussion on "Plotting positions for fitting distributions and extreme value analysis". *Can. J. Civil Eng.* **2013**, *40*, 130–139. [CrossRef]
44. Weibull, W. A statistical theory of the strength of materials. *R. Swed. Inst. Eng. Res. Proc.* **1939**, *151*, 1–45.
45. Fisher, R.A. On the 'probable error' of a coefficient of correlation deduced from a small sample. *Metron.* **1921**, *1*, 1–32.
46. Kennedy, J.B.; Neville, A.M. *Basic Statistical Methods for Engineers and Scientists*, 3rd ed.; Happer and Row Publishing: New York, NY, USA, 1986.
47. Wilson, R.B. A simplicial Method for Convex Programming. Ph.D. Thesis, Harvard University, Boston, MA, USA, 1963.
48. Boggs, P.T.; Tolle, J.W. Sequential quadratic programming. *Acta Numer.* **1995**, *4*, 1–51. [CrossRef]
49. Akan, A.O.; Houghtalen, R.J. *Urban Hydrology, Hydraulics, and Stormwater Quality*; Wiley: Hoboken, NJ, USA, 2003.
50. Sun, S.A.; Djordjevic, S.; Khu, S.T. Decision making in flood risk based storm sewer network design. *Water Sci Technol.* **2010**, *64*, 247–254. [CrossRef]
51. Singh, V.P.; Strupczewski, W.G. On the status of flood frequency analysis. *Hydrol Process.* **2002**, *16*, 3737–3740. [CrossRef]
52. Porras, P.J.S.; Porras, P.J., Jr. New perspective on rainfall frequency curves. *J. Hydrol. Eng.* **2001**, *6*, 82–85. [CrossRef]
53. Haktanir, T. Divergence criteria in extreme rainfall series frequency analyses. *Hydrol. Sci. J.* **2003**, *48*, 917–937. [CrossRef]
54. Gräler, B.; Fischer, S.; Schumann, A. Joint modeling of annual maximum precipitation across different duration levels. In *EGU General Assembly Conference Abstracts*; Discussion Paper SFB 823; EGU: Munich, Germany, 2016.
55. You, L.; Tung, Y.K. Derivation of rainfall IDF relations by third-order polynomial normal transform. *Stoch. Environ. Res. Risk Assess.* **2018**, *32*, 2309–2324. [CrossRef]
56. D'Agostino, R.B.; Stephens, M.A. *Goodness-of-Fit Techniques*; Marcel Dekker: New York, NY, USA, 1986.
57. Sherman, C.W. Frequency and intensity of excessive rainfalls at Boston, Massachusetts. *Transaction* **1931**, *95*, 951–960.

water

MDPI

Article

Spatial Downscaling of Satellite Precipitation Data in Humid Tropics Using a Site-Specific Seasonal Coefficient

Mohd. Rizaludin Mahmud [1,2,*], Mazlan Hashim [1,2], Hiroshi Matsuyama [3], Shinya Numata [3] and Tetsuro Hosaka [4]

[1] Geoscience and Digital Earth Centre, Research Institute of Sustainability and Environment, Universiti Teknologi Malaysia, Skudai, Johor Bharu 81310, Malaysia; mazlanhashim@utm.my

[2] Department of Geoinformation, Faculty of Geoinformation & Real Estate, Universiti Teknologi Malaysia, Skudai, Johor Bharu 81310, Malaysia

[3] Faculty of Urban Environmental Sciences, Tokyo Metropolitan University, 1-1 Minami Osawa, Hachioji, Tokyo 192-0397, Japan; matuyama@tmu.ac.jp (H.M.); nmt@tmu.ac.jp (S.N.)

[4] Graduate School for International Development and Cooperation, Hiroshima University, 1-5-1 Kagamiyama, Higashihiroshima 739-8529, Japan; hosaka3@hiroshima-u.ac.jp

* Correspondence: rizaludin@utm.my

Received: 11 February 2018; Accepted: 26 March 2018; Published: 31 March 2018

Abstract: This paper described the development of a spatial downscaling algorithm to produce finer grid resolution for satellite precipitation data (0.05°) in humid tropics. The grid resolution provided by satellite precipitation data (>0.25°) was unsuitable for practical hydrology and meteorology applications in the high hydrometeorological dynamics of Southeast Asia. Many downscaling algorithms have been developed based on significant seasonal relationships, without vegetation and climate conditions, which were inapplicable in humid, equatorial, and tropical regions. Therefore, we exploited the potential of the low variability of rainfall and monsoon characteristics (period, location, and intensity) on a local scale, as a proxy to downscale the satellite precipitation grid and its corresponding rainfall estimates. This study hypothesized that the ratio between the satellite precipitation and ground rainfall in the low-variance spatial rainfall pattern and seasonality region of humid tropics can be used as a coefficient (constant value) to spatially downscale future satellite precipitation datasets. The spatial downscaling process has two major phases: the first is the derivation of the high-resolution coefficient (0.05°), and the second is applying the coefficient to produce the high-resolution precipitation map. The first phase utilized the long-term bias records (1998–2008) between the high-resolution areal precipitation (0.05°) that was derived from dense network of ground precipitation data and re-gridded satellite precipitation data (0.05°) from the Tropical Rainfall Measuring Mission (TRMM) to produce the site-specific coefficient (SSC) for each individual pixel. The outcome of the spatial downscaling process managed to produce a higher resolution of the TRMM data from 0.25° to 0.05° with a lower bias (average: 18%). The trade-off for the process was a small decline in the correlation between TRMM and ground rainfall. Our results indicate that the SSC downscaled method can be used to spatially downscale satellite precipitation data in humid, tropical regions, where the seasonal rainfall is consistent.

Keywords: rainfall; monsoon; high resolution; TRMM

1. Introduction

Precise information on spatiotemporal rainfall is critical for accurate hydrology predictions and simulations in humid tropical regions. Satellite precipitation data are useful for supporting in-situ measurements, because they provide wide coverage, are publicly available, and are grid-based.

However, their suitability for small basins is hindered by their coarse grid size [1,2]. This is conspicuous for most humid tropical catchments in Southeast Asia, where the region comprises of small land–sea ration area—especially islands and peninsula. Hence, the spatial variability of tropical rainfall variation is rather high [3], and is expected to increase [4]. Although the new satellite precipitation data product from Global Precipitation Mission and GsMAP has higher resolution (0.1°) than its predecessor, the Tropical Rainfall Measuring Mission (TRMM), it is only available from 2015 onwards. Effective climate-hydrologic analysis requires continuous data, especially historical, and therefore it is important to improve those datasets. Due to that conflict, numerous efforts have been made to improve the coarse grids by spatial downscaling.

However, spatial-downscaling algorithms for satellite precipitation data for humid, tropical environments have rarely been reported. Currently, advances in spatial downscaling of satellite precipitation data are centered on using rainfall-related environmental parameters at higher spatial resolutions as predictors. Based on the strong relationship between the rainfall and its site-specific explanatory proxy variables, the rainfall values for a smaller grid were estimated through the regression coefficient. Often, multiple regression analyses are used to assess the relationships between rainfall, vegetation, and elevation [5–10]. Those variables were selected due to their significant relationship at a specific temporal period. In the temperate region, the relationship between seasonal rainfall and vegetation was strong, particularly during late spring and summer, where the photosynthetic rate increased. Meanwhile, topographic variations have significantly influenced regional or local rainfall patterns and distribution, especially in the hilly areas. These orographic effects can be relatively stronger if that region received air masses from the significant seasonal wind flows (e.g., monsoon).

Employing these variables for robust downscaling in humid tropical regions might be less suitable because of the weaker relationships between rainfall, vegetation, and elevation compared to temperate regions. Although applying multivariate regression could be effective in statistically increasing the predictive power of the model, the approach is constrained by several doubts: first is the possibility of a declining relationship between predictors and rainfall from low to finer resolution scales, and second is whether the high predictive power agreed with the physics of the rainfall-environmental perspectives [11–13]. Merging the rain gauge data to downscale the satellite precipitation in the tropics is useful, such as the process done by [14]. However, their method did not improved the spatial resolution of the precipitation. Efforts by [15] in applying the fractal downscaling is effective, but limited by the real-time support of wind and other meteorological data through complex processes. Therefore, an alternative initiative for an effective, operational, and less complex transformation of the satellite spatial downscaling in humid tropics is required.

The proxy variable in humid tropics should be one that influences rainfall patterns and, most importantly, one for which the surrogate data is available at a higher resolution than the satellite precipitation (<0.1°). Anders and Nesbitt [16] highlighted significant variables that influenced the satellite precipitation gradient in the tropics. On a local scale, precipitation was influenced by hydro-meteorological variables, namely prevailing winds, atmospheric moisture, and convective mode. Another important criteria for the spatial downscaling method is the operational aspect. Most of the satellite precipitation spatial downscaling models were developed based on the single or multivariate relationship over specific times and conditions; therefore, downscaling of the future satellite data requires the recalibration or redevelopment of the model, because either the predictor or the rainfall itself might change and influence the predictive power of the regression model (e.g., [8,15]).

Rainfall distribution in the tropics is closely associated with water vapor [17] and monsoons [18]. However, the high resolution data for water vapor is not available regularly, and therefore not suitable to be used as proxy downscaling variable. The Asian monsoon season contributed significantly to the variation in local rainfall in many tropical regions of Southeast Asia [19–21]. The seasonal rainfall pattern is found to be less variable on a local scale, and exhibits specific local zoning [21–23]. Using the ratio product between the satellite precipitation and the corresponding rain gauge to calibrate the satellite precipitation is a well-developed approach in quantitative downscaling, and widely used in merging algorithms [24]. Theoretically, if the rainfall pattern was historically consistent over space and time, the ratio between the satellite and the rain gauge should follow a similar trend. We could expand this concept to developing a spatial downscaling method that is suitable for the humid tropics.

It is our aim to produce high-resolution satellite precipitation data by two process: first, by re-gridding the raw satellite precipitation data; and second, by recalculating the values of each pixel using the historical satellite–rain gauge ratio value. The appropriately high resolution would depend on two main factors: the density of rain gauges and the desired scale. For humid tropics, the challenge is to model atmospheric and hydro-meteorological variables at a mesoscale resolution (2–20 km) or lower [25,26]. We hypothesized that in humid tropical regions of low seasonal rainfall variability, the bias ratio between the previous satellite and ground measurements is consistent, and therefore can be used as a coefficient to estimate the accurate rainfall values of the future satellite precipitation datasets. If the site-specific coefficient were available at a smaller grid, a fine-scale estimation of the satellite precipitation would be achievable.

Based on the above-mentioned concept and hypothesis, we attempted to conduct an experiment. To test this hypothesis, Peninsular Malaysia was selected as an experimental site, because the coefficient of variance (COV) for seasonal precipitation is low [27]. The experiment has two main objectives: (1) Derive the site-specific coefficient (SSC) for each individual pixel, using the average bias ratio between the high-resolution ground rainfall data and re-gridded satellite rainfall data; and (2) validate the SSC to produce high-resolution precipitation maps. The proposed downscaling algorithm can be used to create high-resolution precipitation maps in the highly dynamic hydro-meteorological status quo of humid tropics, with less complex computation and more reliable results.

2. Materials and Methods

2.1. Study Site

Peninsular Malaysia (99.7–104.5° E, 1.3–6.8° N) is located in the western part of Malaysia (Figure 1a). It has a population of 18 million and an area of ~132,000 km^2. The general land cover is agricultural (52%), forest (22%), and built-up areas (26%) [28]. The climate is that of a tropical rainforest, with temperatures ranging from 24 °C to 32 °C and an annual rainfall of 2500 mm. The rainfall distribution pattern over Peninsular Malaysia is strongly influenced by seasonal monsoons, and the area is classified into five local climate regions: northwest, east, west, southwest, and highland (>400 m above sea level) (Figure 2) [27]. There are two distinct wet seasons: one from November to February, during which the northeast monsoon (NEM) produces heavy rainfall in the eastern region (Figure 1b); and the other from May until mid-September, when the southwest monsoon (SWM) affects areas in the west and southwest regions (Figure 1d). The northwest, west, and southwest regions experience two annual wet seasons, from mid-March until May (IM1) (Figure 1c), and from mid-September until August (IM2) [21,27]. Both of these seasons occur during the inter-monsoon periods between the NEM and SWM seasons. Substantial rainfall occurs during the inter-monsoon periods, because of the directional wind change and effects of the local topography.

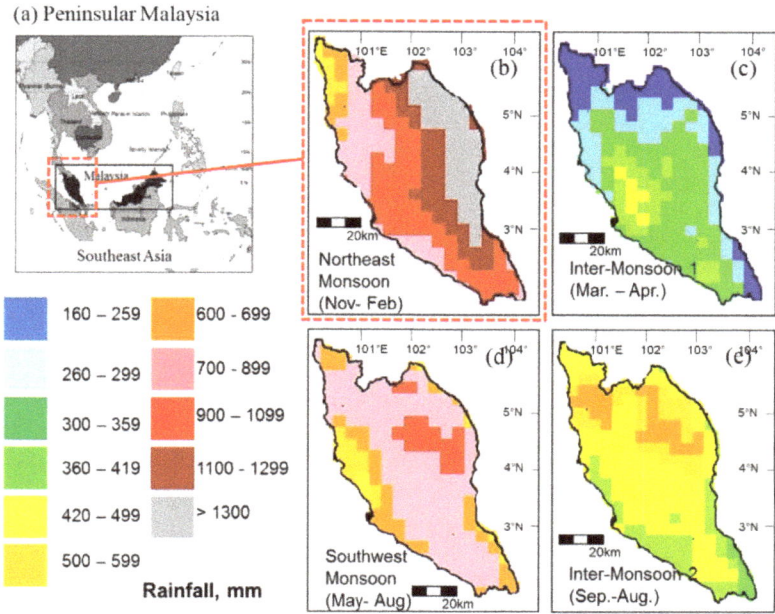

Figure 1. Peninsular Malaysia and its seasonal rainfall variation.

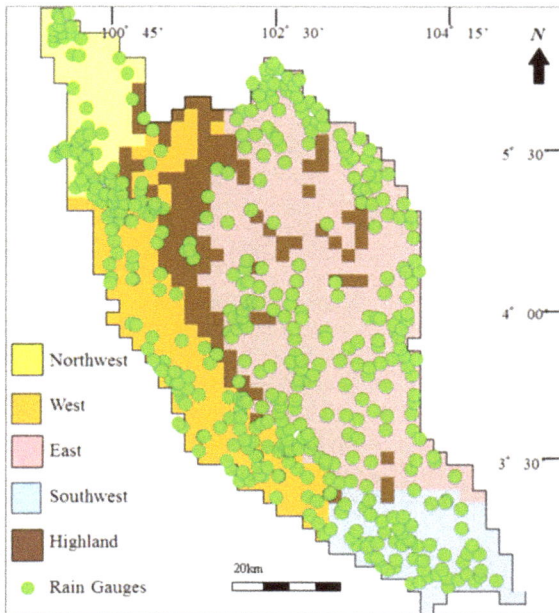

Figure 2. Rain gauge distribution in Peninsula Malaysia and rainfall zones (based on seasonal and intensity).

2.2. Data

2.2.1. Tropical Rainfall Measuring Mission Satellite Data

The Tropical Rainfall Measuring Mission (TRMM) Multi-Satellite Precipitation Analysis (TMPA) data product, which provides rainfall estimates from multiple satellites and other sources, was selected for this study. The TRMM satellite orbits the Earth at an altitude of 402 km, carrying three primary sensors, including precipitation radar (PR), a TRMM microwave imager (TMI), and a visible and infrared scanner (VIRS). The PR sensor is designed to provide detailed vertical distribution of radar reflectivity related to the amount of precipitation inside the system. The TMI sensor measures the vertically-integrated ice and water path, and the VIRS provides information on cloud-top temperatures and reflectance. Using the fundamental concept of precipitation and radar reflectivity, the rain rate is estimated. A general description of the data product, including algorithms and other parameters, can be found in the TRMM instruction manual (2005 and 2011).

Precipitation data products from the TRMM satellite were used because they provide frequent, current, and consistent data (scaling from three hourly to monthly readings) with high spatial resolution (0.25), and because the data are publicly available. The high spatial and temporal resolution satisfies the requirement of primary inputs for hydrological modelling and spatial analysis. The data were downloaded from the official website of the National Aeronautics and Space Administration (NASA), with the collaboration of the Japanese Aerospace Exploration Agency (JAXA). The rainfall data can be accessed from the following link (http://daac.gsfc.nasa.gov/data/datapool/TRMM/01_Data_Products/02_Gridded/index.html). The TMPA data for Peninsular Malaysia were extracted using the corresponding global coordinates for this region.

2.2.2. Rain Gauge Data

A total of 984 rain gauges, covering the entire Malaysia peninsula from 1998 (prior to the availability of the TRMM data) to 2011, were collected from the Malaysian Department of Irrigation and Drainage (Figure 2). Rain gauge measurements were conducted on a daily basis with a 24 h observation period, beginning and ending each day at 8:00 a.m. Then, the daily rainfall measurements were summed over one month to produce monthly rainfall data. After that, these data and their corresponding geographical coordinates were exported into a geographic information system (GIS) in shapefile format.

2.3. Downscaling Tropical Rainfall Measuring Mission Data Using the Seasonal Site-Specific Coefficient

2.3.1. Phase 1: Preparation of the High-Resolution Precipitation Data for Coefficient Derivation

Monthly areal precipitation at 0.05° resolution was generated using the dense rain gauge network and the universal co-kriging interpolation method in the ArcGIS software package (Esri, Redlands, CA, USA). Meanwhile, the satellite precipitation data was re-gridded from the original 0.25° resolution to a resolution of 0.05°, though within this new fine resolution grid (0.05°), the original precipitation value of the TRMM data was retained. For further understanding of this process, an illustration has been provided (Figure 3a). Two types of areal precipitation datasets were produced: the first is from the satellite; while the second is from the ground rain gauges, where both have an identical 0.05° grid. This process was done to the dataset from 1998 to 2008.

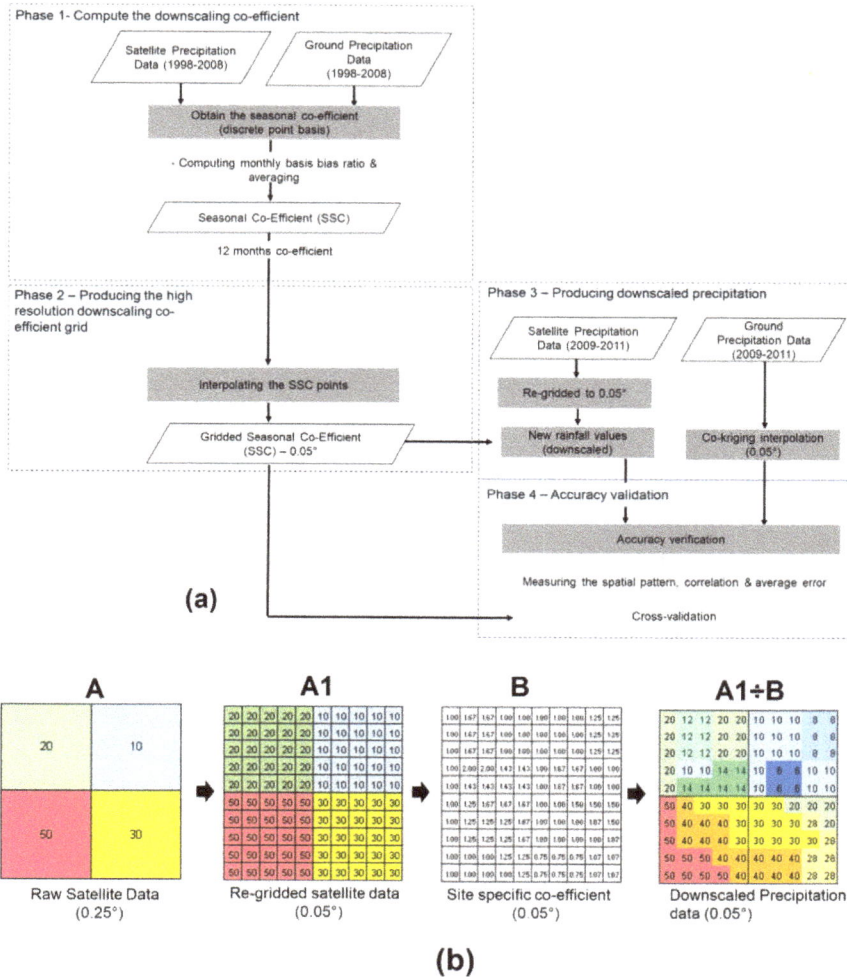

Figure 3. (**a**) Overall methodology. (**b**) Methodological flowchart of the spatial downscaling process—generating the high resolution precipitation data.

2.3.2. Phase 2: Deriving the Seasonal Site-Specific Coefficient

By using the input from phase 1, the SSC was derived through two fundamental steps. First, the re-gridded satellite precipitation data were divided by the ground precipitation. This process was performed at monthly scale. Prior to this process, new images were produced, where each pixel has a specific monthly bias ratio value. This was done for the dataset from 1998–2008. The second step was to calculate the average bias ratio for each pixel. The outcome of this process is known as the seasonal coefficient, or SSC. In total, there were 12 unique coefficient images, each representing the monthly basis downscale coefficient from January to December. Equation (1) shows the SSC-downscaled precipitation calculation, and Figure 3 illustrates the process (Phase 2).

$$HRC_{(i,j)} = \frac{1}{n} \sum \frac{Sat_{(i,j)}}{Rg_{(i,j)}} \tag{1}$$

$$DSat_{(i,j)} = \frac{RSat_{(i,j)}}{HRC_{(i,j)}} \tag{2}$$

where *Sat* is the satellite precipitation data, *Rg* is the areal ground precipitation data, *DSat* and *RSat* are the downscale and raw re-gridded precipitation values, respectively, and *i* and *j* are the pixel coordinates.

2.3.3. Phase 3: Downscaling the Tropical Rainfall Measuring Mission Satellite Data Using the Site-Specific Coefficient

The next step of the downscaling process is to apply the derived SSC to an independent dataset from 2009–2011. The satellite precipitation data was re-gridded from 0.25° to 0.05°. Subsequently, each pixel value for the re-gridded raw TRMM data (0.05°) was divided by the corresponding SSC derived in phase 2. An SSC value of 1.0 represents a perfect condition where no modification occurs. Meanwhile, an SSC value greater than 1.0 indicated an overestimate, and vice versa. Equations (1) and (2), as well as Figure 3 (phase 3), summarize the process of the SSC derivation and downscale process, respectively.

2.3.4. Phase 4: Accuracy Validation

To verify the performance of the SSC-downscale procedures, four indicators were used on their respective products, generated in phase 3. The first two indicators were the bias ratio reduction capacity and root mean square error (RMSE) between the precipitation product from the SSC-downscaled data and the interpolated rain gauges. To determine the quality of the SSC-downscaled products, we first computed the bias ratio reduction, which is the percent difference between the average bias ratios of the direct re-gridded raw TRMM data against that of the downscaled product. High bias ratio reduction capacity (~100%) indicated good quality (low bias), and vice versa. In addition, the coefficient of variance (COV) for the bias ratio was computed to examine whether the bias records were developed under low seasonal variance. Equation (3) shows the calculation for the bias ratio reduction capacity.

$$BR\ Capacity = \left(\frac{Bias\ ratio_{\overline{RSat}} - Bias\ ratio_{\overline{DSat}}}{Bias\ ratio_{\overline{RSat}}} \right) x\ 100\% \tag{3}$$

$$RMSE = \sqrt{\left[\frac{1}{N} \sum_{i=1}^{N} (S_i - G_i)^2 \right]} \tag{4}$$

where *RSat* is the average of the directly re-gridded satellite precipitation data, *DSat* is the average SSC-based downscale precipitation, S_i is the satellite precipitation, and G_i is the rain gauge measurement. RMSE was computed for two measurement pairs: downscale rainfall vs rain gauge, and raw rainfall vs rain gauge.

The third indicator was the bias ratio comparison to other gridded precipitation data products, measured either by satellite or rain gauge interpolation, or also by hybrids that are publicly available. This was carried out to determine the relative performance of the highest resolution of SSC-downscaled precipitation data upon other data products. Our expectation was that the SSC-downscaled should perform better than other products, or at least have comparable performance. Statistically, a small bias ratio was taken to indicate the spatial predictive increment after the downscaling process.

There were six gridded precipitation data products that were taken into account: (1) Global Satellite Mapping of Precipitation (GsMAP), (2) Precipitation Estimation from Remotely Sensed Information using Artificial Neural Networks (PERSIANN), (3) CPC Morphing precipitation product (CMORPH), (4) CPC Unified Gauge-Based Analysis of Daily Precipitation (CPC), (5) CPC Merged

Analysis Precipitation (CMAP) data, and (6) Global Precipitation Climatology Project (GPCP) precipitation data.

The GSMaP Project was sponsored by Japan Science and Technology—Core Research for Evolutional Science and Technology (JST-CREST) and is promoted by the JAXA Precipitation Measuring Mission (PMM) Science Team. The GsMAP products currently provide 0.1° resolution data, which is distributed by the Earth Observation Research Center, Japan Aerospace Exploration Agency [29]. The PERSIANN product, produced by the Center for Hydrometeorology and Remote Sensing (CHRS) at the University of California, uses neural network function classification procedures to compute an estimate of rainfall rate at 0.25° × 0.25° for each pixel of the infrared brightness temperature image provided by geostationary satellites [30]. Gridded precipitation numbers three to five were different types of products produced by the NOAA Climate Prediction Center, using various types of data and processing methods. The CPC product utilized the optimal interpolation (OI) objective analysis technique [31] provided by the NOAA Climate Prediction Center. Meanwhile, the CMORPH product produces high spatial and temporal resolution global precipitation estimates from passive microwave and infrared data. Morphing technique refers to the process of performing a time-weighting interpolation between multi-temporal, microwave-derived precipitation at a given location. Details about the morphing process can be found in [32]. Another product, CMAP, is a global precipitation product that merged precipitation estimates from several satellite-based algorithms and rain gauges. The creators of CMAP used the merging technique of reducing random error [33] and blending [34]. The last product, GPCP, eventually utilized a similar input, but with different merging techniques [24].

The fourth indicator was Moran's I. It was used to determine the qualitative performance of the downscaled result. Moran's I was able to define the rainfall pattern, and was reliable to be used in hydrology (e.g., [35]). The idea is that the pattern of the downscaled precipitation should more closely resemble the ground areal rainfall pattern. Therefore, the difference of the Moran's I values between the downscaled and ground areal rainfall should be small compared to those of the non-downscaled values. We computed the value based on monsoon preferences, because the rainfall patterns were strongly influenced by that factor. The corresponding equations [36] are shown below:

$$I = \frac{n \sum_{i=1}^{n} \sum_{j=1}^{n} w_{i,j} z_i z_j}{S_o \sum_{i=1}^{n} z_i^2} \tag{5}$$

$$S_o = \sum_{i=1}^{n} \sum_{j=1}^{n} w_{i,j} \tag{6}$$

where z_i is the rainfall value deviation from its mean, $w_{i,j}$ is the spatial weight between rainfall at i and j location, n is the total samples, and S_o is the aggregate of all the weights.

2.4. Determining the Effect of Interpolation to the Gridded Areal Ground Rainfall

To evaluate the effects of the interpolation process to the gridded site-specific coefficient, as well as the areal rainfall, k-fold cross-validation analysis was conducted. We applied the holdout method, which is based on separating the data into two sets: one is used for training and the other for testing. Prior to that, the rain gauge data was divided into two datasets. The samples for testing and validation were divided to be 60 and 40%, respectively. This is to ensure that there were a balanced number of samples between testing and validation, and also adequate samples to cover the whole study area [37]. A well-distributed selection was made to ensure the equivalent spatial coverage for both datasets. Two indicators were computed: the mean average error, known as root mean square error (RMSE); and the datasets' corresponding percentage against the average rainfall. We justified that the effect should be small and not affect the entire downscaling quality (<10%) [38].

3. Results

3.1. Performance of the Site-Specific Coefficient Tropical Rainfall Measuring Mission Downscaled Precipitation

(a)   Quantitative Assessment

The three-year average showed that the SSC-TRMM downscaled precipitation had a lower bias ratio compared to the raw TRMM precipitation products over all hydro-climate regions (Table 1). The bias ratio reduction capacity is an indicator quantifying the effectiveness of the downscaling method in reducing bias, and represents high similarity value with the ground reference value, which had an average score of 54%. The greatest improvement was identified in the northwest, with a 94% bias reduction. Meanwhile, continuous performance over time showed that the downscaled precipitation data scored a lower RMSE compared to the raw precipitation data (Figure 4a–e). However, there was a slight decrease of the correlation between the downscaled precipitation data against the ground rainfall data. Nonetheless, it can be justified to be a very minimal effect. Hence, it can be considered as a positive trade-off, because the downscaled precipitation had improved the data's overall quality. The spatial refinement of factor five (from 0.25° to 0.05°) resulted in a remarkable improvement of the rainfall predictions, as indicated by a reduction in bias ratio of 54% and an RMSE of 40%.

Table 1. Bias ratio comparison between raw Tropical Rainfall Measuring Mission (TRMM) data and site-specific coefficient (SSC)-downscaled product.

Region	2009		2010		2011		Average Bias Ratio Reduction Capacity (%)	Average SSC Ratio *
	Raw	SSC	Raw	SSC	Raw	SSC		
Northwest	3.14	2.01	3.59	1.52	2.70	1.30	94	1.6
East	2.70	1.70	1.60	1.03	1.34	0.86	49	1.2
West	1.28	0.83	1.39	0.85	1.35	0.84	53	0.8
Southwest	1.45	1.09	2.16	1.57	1.72	1.31	31	1.3
Highland	1.04	0.97	1.43	0.91	1.35	0.85	41	0.9

* 1.0 is perfect ratio, >1.0 is a satellite overestimate, <1.0 is a satellite underestimate. The numbers in italics represent underestimate cases.

(b)   Qualitative Assessment

From the qualitative perspective, a visual assessment showed that the spatial pattern of the SSC-downscaled precipitation more closely resembled the ground rainfall (Figure 5). This was statistically proven, where the differences between the Moran's I value of the downscaled precipitation and rain gauge-interpolated rainfall surfaces was getting smaller (Table 2). The corresponding average difference was 2%. Meanwhile, the average difference was 6% for the raw TRMM precipitation vs rain gauge-interpolated precipitation. These findings clarified that qualitatively, the SSC-downscaled precipitation was effective in depicting the actual rainfall on the ground compared to the raw version of the TRMM precipitation. Combining the results from quantitative and qualitative assessment showed that the overall SSC-downscaled precipitation results were able to precisely depict the actual ground rainfall over continuous spatial dimension and time, with a trade-off in decreased monthly correlation.

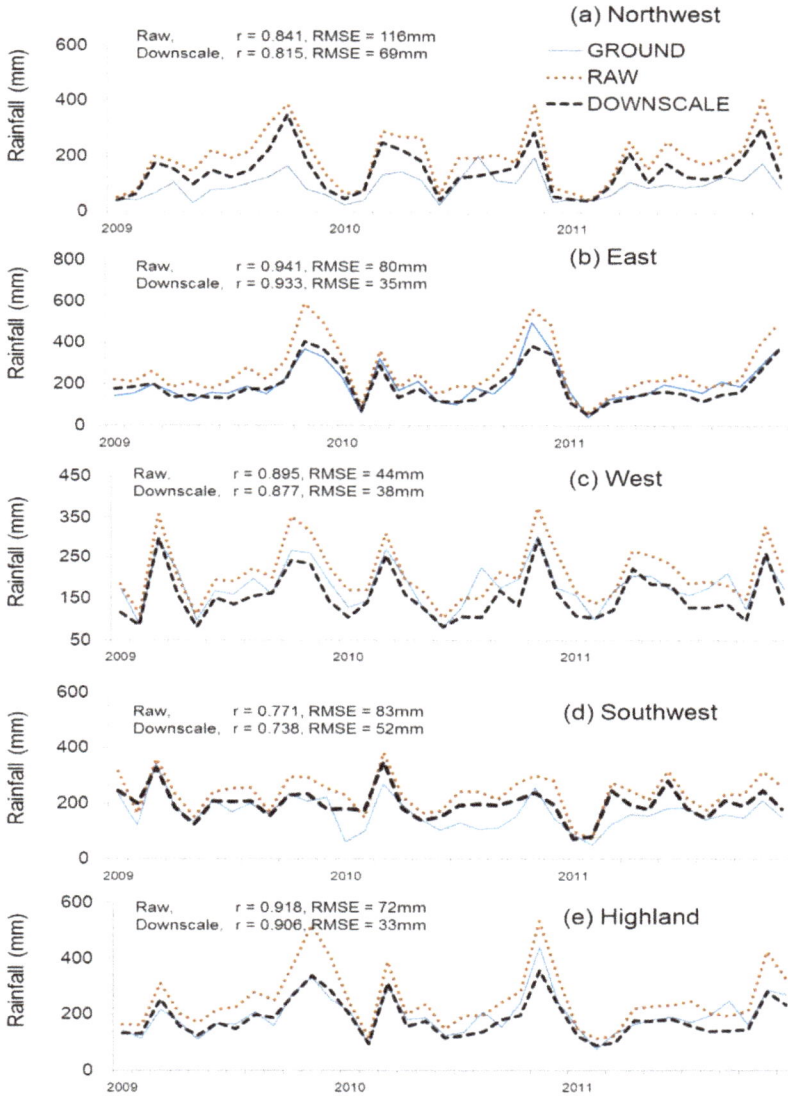

Figure 4. *Cont.*

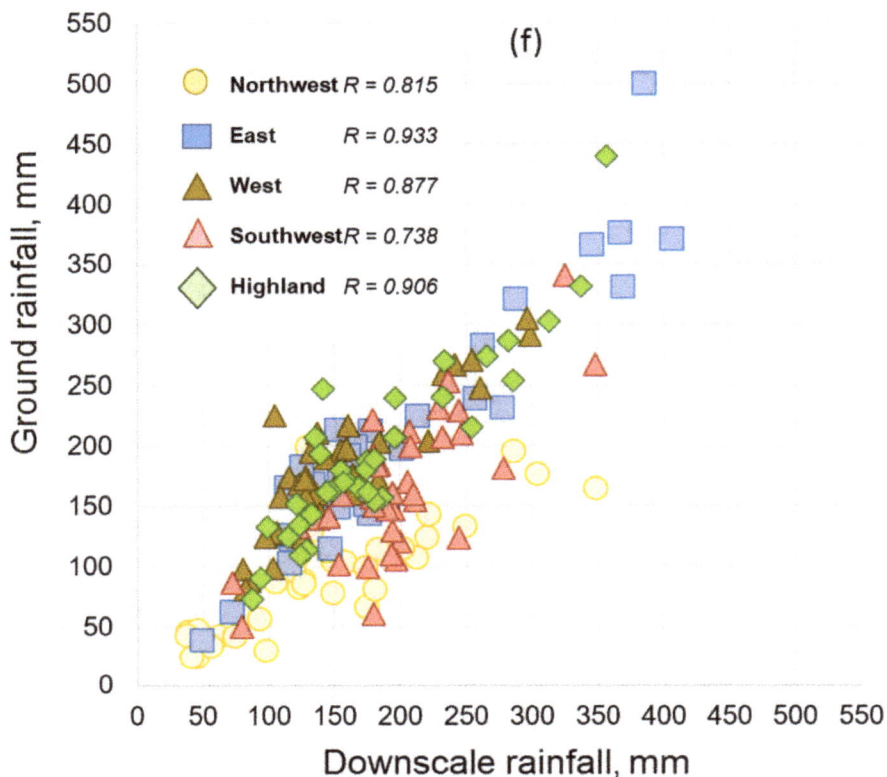

Figure 4. Time series between the ground areal rainfall, raw TRMM, and the SSC-downscale product from 2009–2011.

Table 2. Spatial autocorrelation of Moran's I value (transformed into a Z-score) of the interpolated rain gauge data (reference), raw TRMM, and the downscaled TRMM data.

Monsoon Season	Rain Gauge-Interpolated (a)	SSC-Downscale TRMM (b)	Raw TRMM (c)	Differences (%)	
				$(\|a - b\|/a) \times 100$	$(\|a - c\|/a) \times 100$
NEM	36.75	35.35	37.22	3.8	1.3
IM1	33.80	33.49	36.14	0.9	6.9
SWM	33.81	33.50	31.44	0.9	7.0
IM2	34.73	33.49	30.74	3.6	11.5

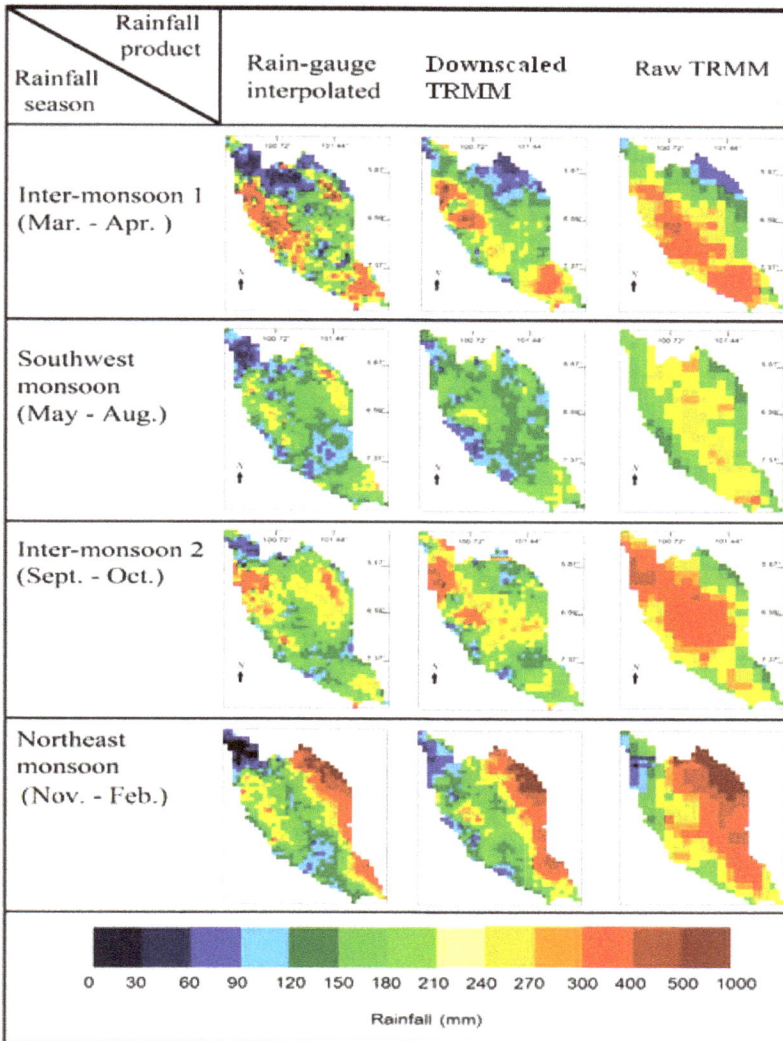

Figure 5. Seasonal rainfall maps of Peninsular Malaysia, from the interpolated rain gauge, site-specific coefficient-downscaled TRMM product, and raw TRMM product.

3.2. Coefficient of Variance of Historical Bias Ratio Records and Downscaling Performance

Hypothetically, the COV of the bias ratio records, which were the basis of the downscaling coefficient, should have a low variance (<35%). This 35% threshold as borrowed from [27], who used this value to classify the local hydro-climatic zone over the Malaysian peninsula. The regional average COV between the satellite and ground rainfall data was 37%, slightly higher than the preferred threshold. Nevertheless, this value was contributed by the large COV in the northwest (54%), while the other regions had relatively lower COV values (east: 34%; west: 34%; southwest: 31%; highland: 36%). One significant observation was that the COV was higher during the dry season, which takes place in February, in all regions (Figure 6). This finding showed that most of the downscaling coefficient was derived under a low-variance bias (<35%), except in February. Nevertheless, we found no evidence or trend that related the low-variance condition of the downscaling coefficient and the effectiveness of the result. Therefore, the results of the lower bias might be achieved due to the site-specific coefficient, which minimized the bias in a robust fashion.

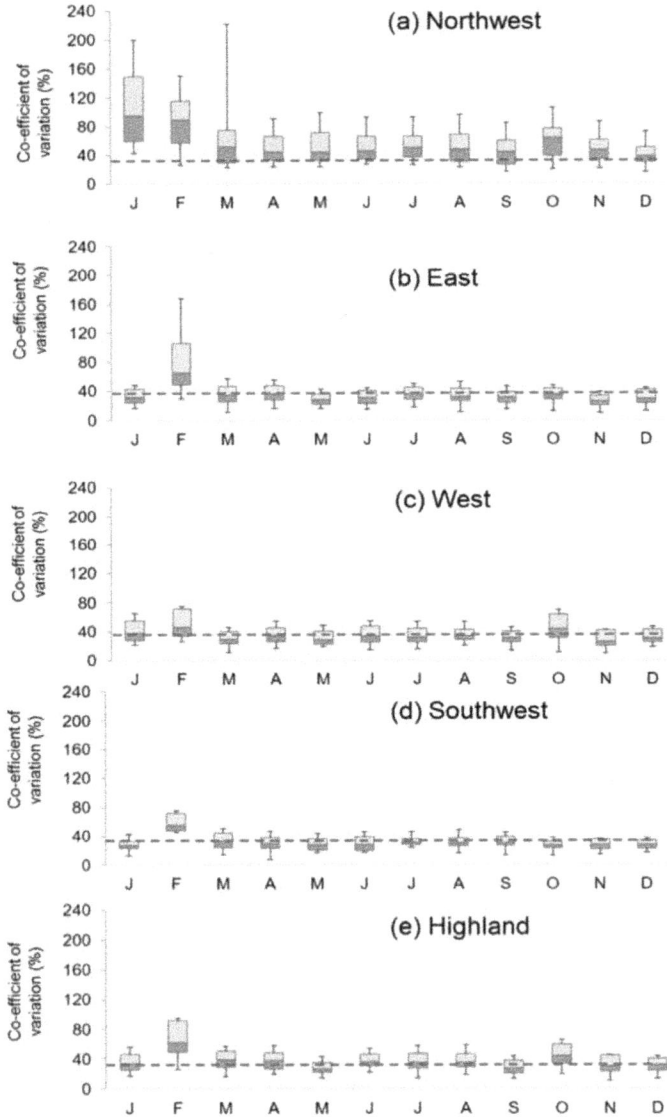

Figure 6. Coefficient of variance of the historical bias record, from the satellite and ground areal rainfall. The red line represents 35%, the threshold value for low variance.

3.3. Comparison of the Site-Specific Coefficient-Downscale and Other Satellite Precipitation Products

Discrete rain gauge comparison between the other satellite precipitation products were conducted, in order to determine the relative effectiveness of the SSC-downscaled precipitation. At the peninsular scale, the SSC-TRMM downscaled precipitation had a lower bias ratio than the other satellite precipitation products (Table 3). Only in the northwest and southwest did the GsMAP outperform the SSC-downscaled precipitation. However, the resolution of the GsMAP was relatively coarser (0.1°). We also found that the higher resolution of the SSC-TRMM-downscaled precipitation (<0.05°) was relatively better for depicting the local spatial rainfall in the western region, where many other precipitation products had failed to represent it.

Table 3. Comparison between the SSC-downscale product and other satellite precipitation products. N represents northwest, E is east, W is west, S is southwest, and H is highland.

Satellite Precipitation Products *	Grid Size (Deg.)	Ratio				
		N	E	W	S	H
TRMM V7—SSC	0.05	2.2	0.9	0.8	1.5	0.8
GsMAP	0.10	1.5	1.4	19.8	0.9	1.0
PERSIANN	0.25	15.0	1.8	43.5	1.9	1.4
CMORPH	0.25	10.6	1.8	20.5	2.0	1.3
CPC	0.50	21.0	1.9	23.7	1.4	0.8
GPCP	1.00	27.0	25.0	24.0	24.0	23.0
CMAP	2.50	15.0	1.8	43.5	1.9	1.4

* GsMAP: Global Satellite Mapping of Precipitation, PERSIANN: Precipitation Estimation from Remotely Sensed Information using Artificial Neural Networks, CMORPH: CPC Morphing Technique, CPC: Climate Prediction Centre Precipitation, GPCP: Global Precipitation Climatology Project, CMAP: CPC Merged Analysis of Precipitation.

3.4. Effects of the Interpolation Process

The cross-validation results showed that the interpolation scheme on both the SSC (Table 4) and ground areal rainfall were small (<10%) (Table 5). The very densely- and well-distributed samples could be the reason. A minor variation was found, where a higher error was indicated as rainfall intensity increased in specific monsoons. This minor effect was identified for interpolated ground rainfall. Nonetheless, we assumed that the interpolation process did not influence the results.

Table 4. Cross-validation analysis of the interpolated data. This evaluates the effect of interpolation to the derived site-specific coefficient. Zero percentage means that the interpolation has no effect to the value of the coefficient.

Monsoon Season	Cross-Validation Metrics
	MPE (%)
NEM (November–January)	5
IM1 (March–April)	3
SWM (May–August)	3
IM2 (September–October)	4
Average	4

Table 5. Ground areal rainfall (2009–2011). This result determines the effects of interpolation process to the ground areal rainfall, which was used in verifying the downscaled rainfall data from 2009–2011. Zero percentage of mean percentage error (MPE) means the interpolation process had no effect. RMSE is used in determining the quantitative effect of the interpolation in a standard unit (millimeters).

Monsoon Season	Cross-Validation Metrics		Average Ground Rainfall (mm)
	RMSE (mm)	MPE (%)	
NEM (November–January)	24	13	200
IM1 (March–April)	13	8	115
SWM (May–August)	11	9	118
IM2 (September–October)	21	10	176
Average	17	10	152

4. Discussion

The use of the SSC-downscaling method was able to produce a high-resolution precipitation map (0.05°) with improved quantitative accuracy. In addition, it was effective in spatially downscaling the future dataset without the input from rain gauges. Most of the present or previous merging, or other spatial resolution improvement methods, require multi-dataset or the ground in-situ preferences' surrogate information [5,6,24,32]. In the context of a tropical region, our results had a better performance compared to the downscaling based on multivariate regression done by [12] in mountainous, coastal, and forested environments. On the other hand, although our results' performance slightly underachieved in using the super ensemble method developed by Yatagai et al. (2014), we successfully produced higher-resolution precipitation data. Furthermore, our computation was less complex, and fewer input variables were required. Therefore, we can conclude that incorporating a dense rain gauge network [39], as well as monsoon rainfall seasonality and variability proved to be the effective in robustly downscaling satellite precipitation for various environmental contexts in the humid tropics.

Prior to positive results, this technique can be useful to the humid tropical regions, which have small land–sea ratio, many islands, and highly-variable seasonal rainfall patterns. Those characteristics are common and significant, especially for many areas in Southeast Asia [2,40,41]. It is also one of the regions in the world that receives large rainfall excess with high intensity [42], and is prone to extreme rainfall events [43]. The availability of high-resolution precipitation information would be significant for understanding the dynamics of tropical rainfall at a microscale.

In addition, smaller tropical catchment or sub-basin hydrology modelling from space will be possible. Current satellite precipitation data had limitations to representing the catchment scale rainfall, due to coarse resolution [2,44]. From a water resources perspective, with the availability of the global data, many humid tropical catchments for important reservoirs were categorized as smaller catchments (<10 km²) [45]. They were located in thick, remote, and mountainous tropical forests, which are difficult to access. Utilizing an operational infrastructure could be expensive and laborious. Literature had showed that a substantial number of them were inadequately monitored and require support mechanisms [46,47].

Another positive implication of the downscaling method is the opportunity to develop higher-resolution historical tropical precipitation data from satellite datasets. This was a critical parameter that was missing from precise regional climate modelling, which is the primary domain of future climate and environmental sustainability efforts [25]. There were substantial amounts of coarse-resolution satellite precipitation data before TRMM, especially from the early METEOSAT missions [48–50]. Performing our SSC-downscaling technique to those datasets is plausible, under the condition that the site-specific coefficient should be derived first.

Despite the promising outcomes of this study, there were a few limitations. First was the requirement of a large rain gauge dataset. Because the downscaling coefficient was eventually derived by correcting the bias factors at a smaller grid, it is necessary to have as large a rain gauge network as possible. This could be a limitation for hydrological data conflict areas (HDCS). An HDCS is an area which has experienced one or more of these conflicts: sparse rain gauges, missing rain gauge data, inefficient data sharing policies, or ineffective data management. The second limitation was whether the downscaling coefficient could be used for other satellite precipitation data besides TRMM. Hypothetically, it can be used, but a further investigation is needed.

The third limitation is that there was emerging evidence on the change in seasonal monsoon rainfall patterns, due namely to an external factor: El Nino Southern Oscillation (ENSO) [22,51,52]. This effect, however, was neglected in our study, due to lack of ENSO data at local scale. The final limitation was the effect of decreasing temporal correlation after the downscaling process. It was believed to be caused by the high-resolution output grid. Because the original TRMM gridded data was coarse, it tended to homogenize the local rainfall pattern. Therefore, as the grid was transformed to be smaller, the high heterogeneity of local rainfall patterns appeared. This effect, however, was minimal, and did not affect the output performance.

Anticipating the second and third limitations by testing the usability of the coefficient on other satellite precipitation data, and excluding samples that affected by ENSO, could be future work in to improve this study. In an effort to further localize the satellite precipitation data, utilizing the role of topographic control as a proxy variable is promising. This is especially true for high-altitude regions in the tropics. In addition, experimenting with the similar downscaling procedures at a higher temporal scale (i.e., weekly) could be worthwhile, because the rainfall in humid tropics is highly dynamic.

5. Conclusions

We tested the hypothesis that higher-resolution data on historical bias records for low-variance seasonal monsoon rainfall can be used to spatially downscale TRMM satellite precipitation data. The use of the site-specific coefficient successfully transformed the initial TRMM satellite precipitation data resolution from 0.25° to 0.05°, with smaller errors and increased similarity with the ground rainfall pattern. With the availability of the SSC, the downscaling of the future satellite precipitation data can be done without any ground reference or rain gauge data. However, it caused a small decline in the temporal correlation. The simplistic and effective procedure described in this study can be applied to spatially downscale satellite precipitation data in regions with low variability in seasonal rainfall in the humid tropics.

Acknowledgments: The authors would like to thanks all the stakeholders and respective agencies, particularly the Department of Irrigation and Drainage for their support regarding the in-situ data. Our utmost gratitude goes to Universiti Teknologi Malaysia and Ministry of Higher Education for supporting this study through the research grants (QJI30000.2727.02K83 and QJ130000.2427.04G12).

Author Contributions: Mazlan Hashim provides the datasets including the required supporting geoinformation software needed for the analyses; S. Numata, H. Matsuyama and T. Hosaka cooperated in designing and improving the concept of the research project and related processes; Mohd Rizaludin Mahmud conceived the research project, conducted the data processing and analysis. All the authors participated actively in preparing and reviewing the manuscript; moderated by Mohd Rizaludin Mahmud.

Conflicts of Interest: The authors declare no conflicts of interest

References

1.   Behrangi, A.; Khakbaz, B.; Jaw, T.C.; AghaKouchak, A.; Hsu, K.; Sorooshian, S. Hydrologic evaluation of satellite precipitation products over a mid-size basin. *J. Hydrol.* **2011**, *360*, 225–237. [CrossRef]
2.   Mahmud, M.R.; Numata, S.; Hosaka, T.; Matsuyama, H.; Hashim, M. Preliminary study for effective seasonal downscaling of TRMM precipitation data in Peninsular Malaysia: Local scale validation using high resolution areal precipitation. *Remote Sens.* **2015**, *7*, 4092–4111. [CrossRef]
3.   Bidin, K.; Chappell, N.A. First evidence of a structured and dynamic spatial pattern of rainfall within a small humid tropical catchment. *Hydrol. Earth Syst. Sci.* **2003**, *7*, 245–253. [CrossRef]

4.   Chadwick, R.; Boutle, I.; Martin, G. Spatial patterns of precipitation change in CMIP5: Why the rich do not get richer in the tropics. *J. Clim.* **2013**, *26*, 3803–3822. [CrossRef]

5.   Chen, F.; Liu, Y.; Liu, Q.; Li, X. Spatial downscaling of TRMM 3B43 precipitation considering spatial heterogeneity. *Int. J. Remote Sens.* **2014**, *35*, 3074–3093. [CrossRef]

6.   Cho, H.; Choi, M. Spatial downscaling of TRMM precipitation using MODIS product in the Korean Peninsula. In Proceedings of the AGU 2013 Fall Meeting, Francisco, CA, USA, 9–13 December 2013. Abstract, H43G-1537.

7.   Immerzeel, W.W.; Rutten, M.M.; Droogers, P. Spatial downscaling of TRMM precipitation using vegetative response on the Iberian Peninsula. *Remote Sens. Environ.* **2009**, *113*, 362–370. [CrossRef]

8.   Jian, F.; Du, J.; Xu, W.; Shi, P.; Li, M.; Ming, X. Spatial downscaling of TRMM data based on orographic effect and meteorological condition of mountainous region. *Adv. Water Resour.* **2013**, *61*, 42–50.

9.   Park, N.-W. Spatial downscaling of TRMM precipitation using geostatistics and fine scale environmental variables. *Adv. Meteorol.* **2013**, *2013*, 1–8. [CrossRef]

10. Shi, Y.; Song, L.; Xia, Z.; Lin, Y.; Myeni, R.B.; Choi, S.; Was, L.; Ni, X.; Lao, C.; Yang, F. Mapping annual precipitation across mainland China in the period of 2001–2010 from TRMM 3B43 product using spatial downscaling approach. *Remote Sens.* **2015**, *7*, 5849–5878. [CrossRef]

11. Park, N.-W.; Kyriakidis, P.C.; Hong, S. Geostatistical integration of coarse resolution satellite precipitation products and rain gauge data to map precipitation at fine spatial resolutions. *Remote Sens.* **2017**, *9*, 255. [CrossRef]

12. Ulloa, J.; Ballari, D.; Campozano, L.; Samaniego, E. Two-step downscaling of TRMM 3B43 V7 precipitation in contrasting climatic regions with sparse monitoring: The case of Ecuador in tropical South America. *Remote Sens.* **2017**, *9*, 758. [CrossRef]

13. Zhang, Y.; Li, Y.; Ji, X.; Luo, X.; Li, X. Fine-Resolution Precipitation Mapping in a Mountainous Watershed: Geostatistical Downscaling of TRMM Products Based on Environmental Variables. *Remote Sens.* **2018**, *10*, 119. [CrossRef]

14. Yatagai, A.; Krishnamurti, T.N.; Kumar, V.; Mishra, A.K.; Simon, A. Use of APHRODITE rain gauge–based precipitation and TRMM 3B43 products for improving Asian monsoon seasonal precipitation forecasts by the superensemble method. *J. Clim.* **2014**, *27*, 1062–1069. [CrossRef]

15. Tao, K.; Barros, P. Using fractal downscaling of satellite precipitation products for hydrometeorological applications. *J. Atmos. Ocean. Technol.* **2010**, *27*, 409–427. [CrossRef]

16. Anders, A.; Nesbitt, S.W. Altitudinal precipitation gradients in the tropics from Tropical Rainfall Measuring Mission (TRMM) Precipitation Radar. *J. Hydrometeor.* **2015**, *16*, 441–448. [CrossRef]

17. Muller, C.J.; Back, L.E.; O'Gorman, P.A.; Emanuel, K.A. A model for the relationship between tropical precipitation and column water vapor. *Geophys. Res. Lett.* **2009**, *36*, L16804. [CrossRef]

18. Chang, C.; Ding, Y.; Lau, N.; Johnson, R.H.; Wang, B.; Yasunari, T. *The Global Monsoon System*; World Scientific Series on Asia-Pacific Weather and Climate; World Scientific: Singapore, 2011; Volume 5.

19. Lee, H.S. General rainfall patterns in Indonesia and the potential impacts of local season rainfall intensity. *Water* **2015**, *7*, 1751–1768. [CrossRef]

20. Singhrattna, N.; Rajagopalan, B.; Kumar, K.K.; Clark, M. Interannual and interdecadal variability of Thailand Summer Monsoon Season. *J. Clim.* **2005**, *18*, 1697–1708. [CrossRef]

21. Varikoden, H.; Preethi, B.; Samah, A.A.; Babu, C.A. Seasonal variation of rainfall characteristics in different intensity classes over Peninsular Malaysia. *J. Hydrol.* **2011**, *404*, 99–108. [CrossRef]

22. Aldrian, E.; Dumenil, G.L.; Widodo, F.H. Seasonal variability of Indonesian rainfall in ECHAM4 simulations and in the reanalyses: The role of ENSO. *Theor. Appl. Climatol.* **2007**, *87*, 41–59. [CrossRef]

23. Fein, J.S.; Stephens, P.L. *Monsoons*; John Wiley and Sons: New York, NY, USA, 1988.

24. Huffman, G.J.; Adler, R.F.; Arkin, P.; Chang, A.; Ferraro, R.; Gruber, A.; Janowiak, J.; Rudolf, B.; McNab, A.; Schneider, U. The Global Precipitation Climatology Project (GPCP) combined precipitation dataset. *Bull. Am. Meteorol. Soc.* **1997**, *78*, 5–20. [CrossRef]

25. Manzanas, R. Statistical downscaling in the tropics can be sensitive to reanalysis choice: A Case Study for Precipitation in the Philippines. *J. Clim.* **2015**, *28*, 4171–4184. [CrossRef]

26. Thompson, R.D. *Atmospheric Processes and Systems*, 1st ed.; Routledge: River Thames, UK, 1998.

27. Wong, C.L.; Venneker, R.; Uhlenbrook, S.; Jamil, A.B.M.; Zhou, Y. Variability of rainfall in Peninsular Malaysia. *Hydrol. Earth Syst. Discuss.* **2009**, *6*, 5471–5503. [CrossRef]

28. MACRES & UTM. *Satellite Atlas Malaysia*; Malaysian Centre for Remote Sensing: Kuala Lumpur, Malaysia, 2000.
29. Okamoto, K.; Iguchi, T.; Takahashi, N.; Iwanami, K.; Ushio, T. The Global Satellite Mapping of Precipitation (GSMaP) project. In Proceedings of the 2005 IEEE International Geoscience and Remote Sensing Symposium, Seoul, Korea, 25–29 July 2005; pp. 3414–3416.
30. Hsu, K.; Gao, X.; Sorooshian, S.; Gupta, H.V. Precipitation Estimation from Remotely Sensed Information Using Artificial Neural Networks. *J. Appl. Meteorol.* **1997**, *36*, 1176–1190. [CrossRef]
31. Chen, M.; Xie, P.; CPC Precipitation Working Group CPC/NCEP/NOAA. CPC Unified Gauge-based Analysis of Global Daily Precipitation. In Proceedings of the Western Pacific Geophysics Meeting, Cairns, Australia, 29 July–1 August 2008.
32. Joyce, R.J.; Janowiak, J.E.; Arkin, P.A.; Xie, P. CMORPH: A method that produces global precipitation estimates from passive microwave and infrared data at high spatial and temporal resolution. *J. Hydrometeorol.* **2004**, *5*, 487–503. [CrossRef]
33. Xie, P.; Arkin, P.A. Global precipitation: A 17-year monthly analysis based on gauge observations, satellite estimates, and numerical model outputs. *Bull. Am. Meteorol. Soc.* **1996**, *78*, 2539–2558. [CrossRef]
34. Reynolds, R.W. A real-time global sea surface temperature analysis. *J. Clim.* **1988**, *1*, 75–86. [CrossRef]
35. Wagesho, N.; Goel, N.K.; Jain, M.K. Temporal and spatial variability of annual and seasonal rainfall over Ethiopia. *Hydrol. Sci. J.* **2013**, *58*, 354–373. [CrossRef]
36. Mitchell, A. *The ESRI Guide to GIS Analysis*; ESRI Press: Redlands, CA, USA, 2005; Volume 2.
37. Hastie, T.; Tibshirani, R.; Friedman, J. *The Elements of Statistical Learning, Data Mining, Inference, and Prediction*, 2nd ed.; Springer: New York, NY, USA, 2001.
38. Mair, A.; Fares, A. Comparison of rainfall interpolation methods in a mountainous region of a tropical island. *J. Hydrol. Eng.* **2011**, *16*, 371–383. [CrossRef]
39. Krishnamurti, T.N.; Kishtawal, C.M.; Simon, A.; Yatagai, A. Use of a dense gauge network over India for improving blended TRMM products and downscaled weather models. *J. Meteorol. Soc. Jpn.* **2009**, *87*, 395–416. [CrossRef]
40. Prasetia, R.; As-syakur, A.R.; Osawa, T. Validation of TRMM precipitation radar satellite data over Indonesian region. *Theor. Appl. Climatol.* **2013**, *112*, 575–587. [CrossRef]
41. Roongroj, C.; Long, S.C. Thailand daily rainfall and comparison with TRMM products. *J. Hydrometeorol.* **2008**, *9*, 256–266.
42. Endo, N.; Jun Matsumoto, J.; Lwin, T. Trends in Precipitation Extremes over Southeast Asia. *SOLA* **2009**, *5*, 168–171. [CrossRef]
43. Ono, K.; Kazama, S. Analysis of extreme daily rainfall in Southeast Asia with a gridded daily rainfall data set. Hydro-climatology: Variability and Change In Proceedings of Symposium J-H02 Held During IUGG2011 in Melbourne, Melbourne, Australia, 28 June–7 July 2011; p. 344.
44. Mahmud, M.R.; Hashim, M.; Mohd Reba, M.N. How effective is the new generation of GPM satellite precipitation in characterizing the rainfall variability over Malaysia? *Asia-Pac. J. Atmos. Sci.* **2017**, *53*, 375–384. [CrossRef]
45. Sidle, R.C.; Tani, M.; Zeigler, A.D. Catchment processes in Southeast Asia: Atmospheric, hydrologic, erosion, nutrient cycling, and management effects. *For. Ecol. Manag.* **2006**, *224*, 1–4. [CrossRef]
46. Liu, X.; Liu, F.M.; Wang, X.X.; Li, X.D.; Fan, Y.Y.; Cai, S.X.; Ao, T.Q. Combining rainfall data from rain gauges and TRMM in hydrological modelling of Laotian data-sparse basins. *Appl. Water Sci.* **2017**, *7*, 1487–1496. [CrossRef]
47. Musiake, K. Hydrology and water resources in monsoon Asia: A consideration of the necessity of establishing a standing research community of hydrology and water resources in the Asia Pacific region. *Hydrol. Process.* **2003**, *17*, 2701–2709. [CrossRef]
48. Jobard, I.; Desbois, M. Satellite estimation of the tropical precipitation using the METEOSTAT and SSM/I data. *Atmos. Res.* **1994**, *34*, 285–298. [CrossRef]
49. Levizzani, V. Precipitation estimates using METEOSAT second generation (MSG): New perspectives from geostationary orbit. In Proceedings of the 1999 EUMETSAT meteorological Satellite Data users' Conference, Copenhagen, Denmark, 6–10 September 1999; pp. 121–128.
50. Turpeinin, O.M. Monitoring of precipitation with METEOSAT. *Adv. Space Res.* **1989**, *9*, 347–353. [CrossRef]

51. Hendon, H.H. Indonesian rainfall variability: Impacts of ENSO and local air-sea interaction. *J. Clim.* **2003**, *16*, 1775–1790. [CrossRef]

52. Loo, Y.Y.; Billa, L.; Singh, A. Effect of climate change on seasonal monsoon in Asia and its impact on the variability of monsoon rainfall in Southeast Asia. *Geosci. Front.* **2014**, *6*, 817–823. [CrossRef]

MDPI

St. Alban-Anlage 66

4052 Basel

Switzerland

Tel. +41 61 683 77 34

Fax +41 61 302 89 18

www.mdpi.com

Water Editorial Office

E-mail: water@mdpi.com

www.mdpi.com/journal/water